GENETICS AND PHYSIOLOGY OF MICROBES

GENETICS AND PHYSIOLOGY OF MICROBES

By

DR. S. SUNDARA RAJAN

Principal

Reva Institute for Science and Technology Studies

R.T. Nagar

Bangalore – 560 024

ANMOL PUBLICATIONS PVT. LTD.

NEW DELHI - 110 002 (INDIA)

ANMOL PUBLICATIONS PVT. LTD.
4374/4B, Ansari Road, Daryaganj
New Delhi - 110 002
Ph.: 3261597, 3278000
Visit us at: www.anmolbooks.com

Genetics and Physiology of Mircobes

First Edition, 2003

ISBN 81-261-1367-7

PRINTED IN INDIA

Published by J.L. Kumar for Anmol Publications Pvt. Ltd., New Delhi - 110 002 and Printed at Tarun Offset

CONTENTS

PREFACE

Metabolic patterns of living organisms are based on the underlying genetic machinery. The variety of physiological processes in living organisms both micro and macro have been built on the plasticity and adaptability of their genome.

Genetics and Physiology of Microbes primarily deals with the varying mechanisms of metabolic processes and an equally varying array of genetic patterns.

The book consisting of 17 chapters gives a detailed outlay of genetic phenomena and metabolism that characterize the various microbes.

The book is meant to be a text book for students of life science, pharmacy, agriculture etc. at the graduate and post graduate level.

I am thankful to Mr. J.L. Kumar, Managing Director, Anmol Publications Pvt. Ltd., New Delhi for publishing this book.

—Dr. S. Sundara Rajan
Principal
Reva Institute for Science and
Technology Studies, R.T. Nagar
Bangalore – 560 024

1

INTRODUCTION

"Like begets like" and "Chip of the old block" are two of the adages quoted often whenever we come across a situation to compare parents an children. These popular adages are indeed so commonplace that we may not be able to appreciate the complete truth behind it. Indeed why should a human being give birth to a human being and not to any other organism. A mango seed sown in soil produces only a mango tree and not a guava tree. What is it that is hidden in the mango seed which makes it grow into another mango tree only. When children are born we often hear statements like "the nose is like the father, the eyes are like the mother, but the mouth and hair are different". This shows that the new born in addition to having parental characters has its own haracters too. again what is the mystery behind the fact, that the child has sufficient similarity with the parents and yet sufficiently different from the also?

Genetics is a branch of biology that aims to find answers to the above questions. In fact, genetics is still trying to unravel the total mystery behind the statement, like begets like, even though most of the facts both at the morphological and at the molecular level have been understood.

The factg that children resemble their parents is due to heredity and they also differe from their parents to some extent is because of Variation. Heredity may be defined as the potential of an individual to transmit its characters to the offspring while variations are the differences that the individual acquires due to the interaction with the environment. The main aim of Genetics is the study of these seemingly opposite phenomena namely heredity and variation. While heredity tries to maintain uniformity, variation brings in plasticity.

The term Genetics was first used by Bateson in 1906, even though the science of genetics made its firm beginnings in the form of Mendel's epoch making work with pea plants. The rediscovery of Mendel's work independently by

Devries. Tschermark and Correns indeed paved the way for founding the science of Genetics even though the name was coined only in 1906.

Organisms are capable of transforming their characters to the next progeny as a result of reproduction. Indeed ever since life first arose on earth (due to a chance combination of elements), it has maintained itself till to day in myriad forms only because of its capacity of reproduction. Untill 1868 when Mendel working with pea plants came out with coinvincing explanation regarding the mechanisms of heredity, it was largely speculative as to how organisms pass on their characters to the progeny.

Earlier concepts of heredity

Earlier concepts of heredity made their beginning first with the discovery of sexualityin organisms-both plants and animals. The invention of the microscope gave a great impetus to probe into the world of microcosm and get a visual knowledge about the link between one generation and the toher bringing about hereditary transmission of characters.

Harvey (1578-1657) speculated that all animals arise from eggs, with the semen only playing a vitalizing role. R.deGraff (1641-1673) opined that since the offspring has characters of both the parents, both of them should contribute to the heredity. Leeuwenhoek (1677) who studied the sperms in several animals suggested their association with the eggs.

Reproductive organs in plants were reported for the first time by Grew in 1682. It was Camerarius however who for the first time described sexual reproduction in plants. In 1717 Fairchild produced a hybrid having characters of both the parents. The hybrid was called *Fairchild's mule.*

Performation and epigenesis

The discovery of sexuality, revealed the physical link between one generation and another, thus providing a physical basis for the transmission of heredity from parents to progeny.

The homunculus theory of Wammerdam suggested that the development of an organism is a simple enlargement of a minute but completely preformed individual. According to Swammerdam, the preformed individual or homuncullus could be present either in the sperm or in the ovum. During 18th and early part of 19th centuries there were two groups among "Preformationists". Those who attached importance to sperm were called *animalculists.* Subsequent studies however clearly showed, that not only ovists and animalculists were wrong, but the entire preformation theory was nothing but a figament of imagination.

Epigenesis theory

Wolff (1738-1794) who opposed the preformation concept opined that, neither the egg nor the sperm had a preformed organism, but both of them have undifferentiated mass of living substance which developed into the organism after fertilization. According to this theory called epigenesis, new organs and tissues are formed *de novo* (from the mass of living substances in thegametes) due to vital forces.

Pangenesis theory

Hippocrates (400 BC) believed that reproductive material is produced from all parts of the body so that all the characters are represented in theprogeny. Aristole (350 BC) also was a believer of direct inheritence of characters.

Charles Darwin (1809-1882) was also a votary for this type of inheritance. He modified the views of Hippocrates slightly and called it the theory of pangenesis. According to this, all parts of the body productive invisible gemmules or pangenes that are carried into sex organs *via* the blood stream. There the pangenes are assembled into gametes. During fertilization, pangenes of mother and father combine to produce the characters of the offspring.

Lamarckian theory (J.B. Lamarck 1744-1829) of inheritance, often called inheritance of acquired characters opines that any new character acquired during the life time of an individual is inherited. It will be seen from the above the Charles Darwin's pangene theory is nothing but a version of Lamarckian theory.

Germplasm theory of Weismann

August Weismann (1834-1914) disproved the pangene theory with his experiments on rats. He cut the tails of rats and observed the progeny for 22 generations. In every generation the rat progeny continued to have tails in spite of the fact that the parents were tailless. If pangene theory were to be corect, the tailless parents could not have sent the tail pangene to the gametes and consequently the progeny should have been tailless.

In order to account for inheritance, Weismann proposed his germplasm theory, which is essentially correct even today. According to this theory, the body of an organism can be divided into two - **somatoplasm** and **germplasm.** Somatoplasm constitutes thevegetative body of the organism, while the germplasm constitutes the reproductive part. Somatoplasm develops newly in every individual and there is no continuity; while the germplasm is continuous and links the generations. With the information present in germplasm, every individual develops its own somatoplasm. Hence any change or variation occuring in somatoplasm cannot be transmitted to the next generation, while those of germplams can be inherited. The fermplasm theory provided a convincing physical basis for distinction of variations into heriditary and environmental (see later in the chapter).

Variations

Variations appearing in organisms are of two categories. *Hereditary variations* and *environmental variations*. Before the classical explanation of Wismann (germplasm and somatoplasm) it was baffling the biologists as to how variations of some kind are inherited while others are not. The explanation of Weissman that the body consists of germplasm and somatoplasm and only the former constitutes the link between one generation and the other provided for the first time a basis to classify the variations into two categories (named above).

Hereditary variations

These arise in the germplasm and consequently can be passed on to the next generation. These are also called heritable or permanent variations as they go from one generation to the other. It is primarily the hereditary variations that are responsible for the variations between one species and theother.

Environmental variations

These arise in thesomatoplasm due to the influence of environmental factors and consequently cannot be passed on to the next generation. They are called transient variations becausethey will last only for one generation.

Between the two types of variations, from the evolutionary point of view, heritable variations are of some significance as they tend to carry the variation for several generations.

Some interesting experiments have been made by Clausen *et al* in the Californian (USA) plant *Potentilla glandulosa,* a member of Rosaceae, concerning the difference between heritable and environmental variations. It is generally known that the individuals of a plant species growing at diferent altitudes ina mounainous region tend to show morphological difference. The individuals that grow at sea level have long stem and leaves and take nearly six months to produce flowers from the seedling stage. A much taller form is encountered at about 4000-45000 ft above sea level and it takes only 50-60 days to flower in its natural habitat. At about 10,000 ft above sea level (alpine region) there is a dwarf form of *Potentilla glandulosa* which takes about 50-55 days to produce flowers. To what extent these differences encountered are due to environment and to what extend due to heredity? In order to find out an answer, the plant from each locality was cut into there pieces and replanted in three experimental gardens at coasta level, at 4600 feet and at 10,000 feet. The alpine race (form) grown at sea level resembles the original race, but grows to a taller height at 4500 ft. The ming altitude race (4500 ft) becomes dwarf at the coastal level and extremely dwarf at the alpine region. The coastal race grown at 4500 ft altitude is dwarfer and at 10,000 ft the plant dies. The above experiment clearly illustrates that the members of the samestock (clone) react differently to different environments as shown in the case of height. Hence variations of this type are called environmental variations.

Genotype - Phenotype

An individual organism has tow sets of attributes-an endowment which it gets from the parents and an individuality which reflects a reaction of the hereditary potentiality with the surroundings. In order to clearly demarcate the two in genetic terms, Hohansen (1909) a Danish scientist proposed the *genotype - phenotype concept.* Genotype of an individual has been defined as thesum total of all the hereditary potentialities that the organism inherts from the parents; while the phenotype is the morphological and physiological attributes of the organism - like the size shape, colour, metabolic reaction rate etc. Phenotype reflects theinteraction between the genotype and the environment. Since the individuals of a species may grow in different environments their interactions are different. In other words the same genotype may interact variouslyin different environment and produce various phenotypes. In the case of *Potentilla glandulosa,* the same genotype behaved differently in different environments.

To sum up, we can say that as far as an individual is concerned its genotype is permanent (barring extreme circumstances) while its phenotype is transient. Individuals with different genotypes some time show similar phenotypes in a similar environment.

The diverse phenotypes possible for a given genotype under various environmental influences (without causing lethality to the individual) is called the *norm* or *reaction range.* The norm or reaction range extends from one extreme to the other (with reference to all factors or parameters of the environment) to theother (with reference to all factors orparameters of theenvironment) in which a given genotype can interact and survive. The full reaction range of a genotype perhaps can never be understood in its entireity, because in order to assess this, one has to know all parameters of the environment. This is easier said than done.

Hereditary diseases

These are inborn diseases that the individual has in its genetic mechanism. It reflects an inbuilt fault in the genetic machinery. Many of the human diseases are known to be hereditary - Taste blindness to PTC, colour blindness, Diabetes, albinism, Haemophilia, Xeroderma, phenyl Ketonuria, Alkaptanuria, Sickle cell anaemia are all due tto a fault in the genetic machinery and an individual suffers from these whether one likes or not. But there is a distinction in these diseases. Some of them like Haemophilia, colour blindness etc., will affect the individual irrespective of age, physiological condition etc of the individual. On the other hand, diseases like diabetes (not all types of diabetes), even though hereditary may not affect the individual until quite late in life. This means that hereditary diseases also interact with the environment and express themselves only in a suitable environment.

In two other human diseases

Albinism and xeroderma, the former is recognised by complete loss of pigments in skin, hair, iris of the eyes etc; the latter is also a skin ailment in which exposure to sun rays results in freckles on the skin. But both these can be withstood. Albinos can protect themselves by not exposing too much tgo light, the same is the case with xeroderma people. Till now there has been no total cure for a hereditary disease - the remedies are only pelliative. But recent techniques of genetic engineering have brought a ray of hope that the faulty genes can be identified and corrected very early. However this is still in the experimental stage.

Phenocopy

In order to cure a hereditary disease or rectify a genetic abnormality; it will be necessary for us to know the norm or range of reaction of the genotype in all the environmental conditions. Even though it is difficult to fully analyse the entire reaction range of a genotype it has been possible to do so to some extent atleast in some organisms.

In the fruit fly *Drosophila melanogaster,* the normal body colour is brown with black margins. A hereditary mutant for this fly was discovered by Morghan (1`910) where the body colour was yellow. This was a genotypic character in that the body colour was constant in both the flies in all environments. In 1939, Rapport discovered that, if larvae of these flies are fed on silver salts, they develop into yellow flies irrespective of the fact whether they were genetically brown or yellow. The yellow bodies fly which is genotypically brown is a variant of the original yellow bodies fly. This is now called a phenocopy. It is easy to distinguish between the original yellow body and its phenocopy. The offspring of a phenocopy, would have brown body, if they are not fed again silver salts. The temporary development fo body colour (yellow) in brown flies is an example of norm of its genotype.

MICROBIAL GENETICS

Gentetics studies during the early post mendelian period were concentrated mainly on higher organisms. The reasons for this are not far to seek. Manipulation of organisms for breeding and study of the chromosomes can be easily carried out in higher organisms. Further, phenotypic variations are discreet and offsprings after mating could be easily analyzed for genetic varaitions. Visual observation of segregation of characters also can be done easily. In microorganisms however the picture is totally different. Organisms themselves are too small to handle. Further, when the entire organism is made up of a single cell what sort of a phonotypic variation one can visually observe in microbes? Genetic studies in microbes indeed call for an entirely different approach than the conventional methods used for higher organisms.

In microbes where no morphologically variable traits are available within the individuals of a species for the purpose of hybridization and genetic analysis, the physiological traits are the only dependable characters. What is a physiological trait? When microbes such as bacteria are cultured on a medium continuously, they undergo mutation and this can be detected by their nutritional requirements. For instance if a bacterial colony does not require the addition of biotene to the medium even though it is required for its growth, the assumption is, it is able to synthesize biotene on its own. If such a colony sometimes does not grow on a medium which does not have specially added biotene the conclusion is, the colony has lost the ability to synthesize biotene due to mutation. Such a colony can we grown on a medium which has biotent in it. By this analysis we now have two genetic variations of a bacterial colony (I) colony that can synthesize biotene and (ii) colony that can not synthesize biotene. These two can now be used for genetic studies as they have two 'phenotypic traits'. The traits however are not morphological but nutritional. In other words the two colonies mentioned above are biochemical variants. In microbial genetics the phenotypic traits are nothing but biochemical traits and genetics of microbes is essentially biochemical genetics.

Beginnings of biochemical genetics

Earliest cases of biochemical mutants in metabolic pathways were studied by A.E. Garrod in 1909 which has been published in his classical book *'Inborn errors of metabolism'*. He studied three abnormalities associated with the metabolic breakdown of phenylalanine. The three abnormalities are *Phenyl ketonuria, Alcaptonuria* and *Albinism.*

The next significant step in biochemical genetics pertaining to microbes was that of G.W. Beadle and E.L. Tatum who were awarded Nobel prize for their outstanding work on the biochemical genetics of fungus *Neurospora.* Beadle and Tatum discovered several biochemical mutants in *Neurospora* and proposed their celebrated concept - **one gene, one enzyme hypothesis.** According to this hypothesis, if a biosynthetic pathway involves several steps, each step is controlled by a specific enzyme, which in turn in governed by a particular gene. Mutation in any one of the genes would alter the biochemical pathway. In addition, elucidation of biosynthetic pathways also will be made easy by the study of mutations. This is particularly true for microbes. For the study of biochemical mutations and mutations that bring about phage resistance or drug resistance, microbial systems are admirably suited. Large populations of microbes can be handled easily and mutations may be detected. Thus biochemical genetics of microbes came to be recognized as microbial genetics.

Genetic notations used for microbes. In order to discuss genetics one must be able to distinguish phenotypically one cell from the other. This is hardly possible in microbes as they do not manifest any morphological traits.

In microbes a phenotype constitutes a biochemical trait. Thus a cell which can synthesize methionine is called Met^+ and if it cannot synthesize methionine it is called Met^-. Obviously a Met^- cell must have a defective (mutated) gene which prevents the synthesis of methionine.

A typical microbe such as a bacterium has only one set of genes; it is haploid. Enkaryotic organisms on the other have two sets of genes i.e., they are diploid. A diploid cell might have normal copy of the gene and one mutated copy. In this case the heterozygote will have both copies and the gene symbols are separated by a diagonal line Met + Met^-. Alternatively a cell may have both defective copies or both normal copies. To summarize, the following conventions are used (Friefelder, 1987).

1. Abbreviations of phenotypes contain three Roman letters (the first capitalized) with a superscript + or - to denote the presence or absence of a designated character and a non superscript + or - for drug resistance.

2. The genotypic designation is always lower case, except for superscripts, all components of the symbol are itacilized.

The above conventions, however has not been adopted for bacteriophages, for which one and two letter symbols are widely used.

Some milestones in the History of Genetics

Name	Year	Contribution
Empedocles-		Semen theory
Aristotle	200 BC	Semin is highly purified blood
Harvey	1651	Living beings originate from egg.
Hartsoikes and	1694	Supported preformation theory
Datempotius	1692	"Observed" miniature human being in
a sperm's head		
Charles Darwin	1809-1882	Pangenesis theory
August Weismann	1889	Germplasm theory
Gregor Mendel	1822-1884	Laws of Heredity; regarded as father
of Genetics		
Hertwig	1875	Discovered fertilization
Flemming	1879	Described longitudinal splitting of
chromosomes		
Benden E.W.	1883	Gametes of round worm have half the
number of		chromosomes as the
body cells		
Waldeyer	1888	Coined the term chrosome
Tschermark		
Devries and		
Correns	1900	Rediscovering of Mendelian laws
Landsteiner	1900	Blood groups
Devries	1901	Mutation in Oenothera
McClung	1902	Sex chromosomes
Bateson. W	1906	Coined the term genetics
Bateson and		
Punnet	1906	Linkage in peas
Johanssen	1909	Coined the word gene
T.H.Morghan	1910	Sex linkage in Drosophila
Bridges and Tatum	1941	Biochemical genetics in Neurospora
Stern	1931	Cytological proff of crossing over
Avery, Mcclead and		
McCarty	1944	DNA - genetic material
Muller	1946	Radiation genetics
Chargaff	1950	Base pairs are equal in proportion in
DNA		
Watson and Crick	1953	Molecular model of DNA
Otchoa and Kornberg	1959	DNA synthesis
Jacob and Monad	1961	Operon concept
Nirenberg, Holley		

and Khorana	1968	Genetic code
Calude and Palade	1974	Ultra structure of cells.

2

CHEMICAL BASIS OF HEREDITY

The cells have diverse chemical components of which most were known to scientists by the first quarter of the nineteeth century. These are classified into three major groups namely–carbohydrates, proteins and lipids. Of these threc– proteins are the most complex and diverse and many biologists thought that they may constitute the basis of biological diversity.

Eventhough another group of macromolecular compounds – nuclei acids were discovered in 1869 by Sweedish biochemist Fredrich Miescher, it was only in 1952, their (nuclei acids) role in inheritance was proved.

Meischer collected pus filled hospital bandages and from them he isolated and purified nuclei. On chemical fractionation he found that a cellular chemical content had a very high component of phosphorous compound than what was known before. As the new compound was obtained from nucleus, Miescher named it *nuclein*. Subsequently, nuclein was isolated from many sources and due to its acidic properties, it was named nucleic acid. Upto 1952, however, interest in nuclei acids was limited to only its physiochemical studies. When its role in inheritance was proved nucleic acid became the focus of intensive research. Comparative studies of nuclei acids and proteins revealed that while proteins are of diverse types, there are only two types of nuclei acids. It was hard to believe that nuclei acids which apparently had no diversity could be the basis of heredity and not the proteins.

Presently, however, there is not even an iota of doubt as to the fact that nucleic acids alone constitute the hereditary factor. Three major experiments helped provide an evidence for this.

Bacterial Transformation

Fred Griffith, an English bacteriologist, worked with *Diplococcus pneumococci* and found out that there are two strains in this – *Virulent and Avirulent.* When injected to experimental mice the virulent strains cause the death of the mice, while avirulent strains do not casue any harm. Griffith showed that heat killed virulent strains injected into the mice along with avirulent strains cause the death of the mice and from the bodyof the mice live virulent strains can be obtained. Dawson and Sia (1931), showed that the heat killed virulent strains can transform avirulent strains on a nutrient broth. Alloway (1933), showed that even a lysed cell of virulent strain retains the ability of transformation thereby indicating that some component of the cell rather than the entire cell was involved in transformation.

Nature of the transforming agent

Three scientists – O.T. Avery, Colin Macleod and Maclyn McCarty (1944), went into serious search of the transforming agent from the lysed cells of bacteria. Even after digestion of polysaccharide coat, the cell retained the ability of transformation. This shows that the capsule has no role in transformation. Similarly, protein degradation also had no adverse effect on transformation. With the two important components out of the way, it thus became clear that nucleic acids perhaps might play a role in transformation. It was shown that the DNA fraction of nucleic acids was responsible for transformation. It was also shown that if the cells are treated with DNAse (a DNA degrading enzyme), the transforming ability will be lost. Enzymes which digest RNA do not disturb transformation.

Bacteriophages and chemical basis of heredity

Bacteriophages are viruses that attack a bacterial cell and multiplty inside the (bacterial) cell. The phage ataches itself to the bacterial cell and insects its components into the cell, where replica of phages are produced. Hershey and Chase (1952) conducted an experiment to find out which component of the bacteriophage carries the genetic information necessary for the replication of viruses. They cultured the host bacterium *Escherichia coli* in a medium containing radio isotopes of Sulphur (S^{35}) and phosphorus (P^{32}). When T_2 phages attacked these cells subsequently, the labelled compounds get incorporated in phage DNA which always has radioactive phosphorus but not any sulphur. In the next set of experiment only the protein coat of T_2 phage was labelled with S^{35} and the DNA core of another was labelled with P^{32}. Hershey and Chase separated the radioactive phages by centrifugation and allowed them to infect a population of non radioactive bacteria. A few minutes after initial infection, they separated the phages from bacteria by agitating the medium. After centrifugation, the host and the parasite were separated by centrifugation and their radioactive measured. About 95% of P^{32} was found in bacteria while about 95% S^{35} was seen in phages. This experiment conclusively proved that during infection, only DNA is injected into the medium. This is the reason why the non radioactive bacteria showed labelled P^{32} (as they obtained it from labelled phage DNA). For this work, Hershęy was awarded Nobel prize in 1969.

Thus, the experiments of Hersehy and Chase, Avery, Mcleod and McCarty. Griffith etc., clearly proved that the DNA constitutes the chemical basis of heredity.

TMV and Chemical Basis of Heredity

The discovery that DNA is the hereditary material posed some problems to explain the inheritance in instances such as TMV (Tobacco mosaic virus), where there is no DNA (TMV is a RNA virus). But, the absence of DNA has not prevented TMV from transmitting its characters from one generation to the other. TMV is made up of a RNA core and a helical protein coat. The protein RNA component of TMV can be easily separated by shaking the viral particles in a mixture of water and phenol. Frankek Conrad and Wichams (1955), separated the protein and RNA from TMV and reassembled the particles from the individual components thus indicating the purification process (of the components) does not affect their function. Subsequently, Gierr and Shramm (1956), showed that on contact with the host only the RNA can bring about the disease and not the protein coat. In TMV therefore, it is RNA as the hereditary material. It can therefore be concluded that nucleic acids in general, and DNA in particular is the hereditary material in most of the organisms; and in the absence of DNA (as in some viruses) RNA can constitute the genetic material.

Nucleic Acids

These are the most important constituents of the nucleus playing a vital role in the inheritence of characters from one generation to the other. There are two types of nucleic acids – Deoxyribose nucleic acid (DNA), and Ribose nucleic acid (RNA). While both constitute the genetic material in different organisms, it is predominantly the DNA which is the genetic material in most of the organisms. The following are some of the important milestones in the discovery of Nucleic acids.

1. Meischer (1869), isolated nucleic acids and called them nuclein.
2. Hertwig (1884), suggested a heriditary role to the nuclein
3. Kossel (1884), recognised purines and pyrimidines in the nucleic acids.
4. Feulgen (1912), devised basic fushsin dye positive only to DNA.
5. Astbury and Bell (1938) showed that the bases in DNA molecule are arranged one above the other.
6. Avery, Macleod and Macarty (1944), presented evidence for DNA to be the genetic material.
7. Chargaff (1950), showed that DNA contains equal amount of purines and

pyrimidines.

8. Wilkins *et al* (1950), identified the molecular distance between the bases.
9. Furberg (1952); DNA is a coil of single chain.
10. Pauling and Corey (1952); DNA is a helix of three chains.
11. Watson and Crick (1953) demonstrated the double helix model for DNA molecule.
12. Meselson and Stahl (1958), demonstrated the semi conservative replication in DNA.
13. Jacob and Monod (1961), postulated the presence of RNA in protein synthesis.

DEOXY RIBOSE NUCLEIC ACID (DNA)

DNA is the main genetic material in almost all organisms except for certain viruses. In prokaryotes like *Escherichia coli,* the genetic is a single giant molecule of *DNA*. It is mainly concentrated in the nucleus, it has also been reported in cytoplasmic organelles like chloroplast, mitochondria etc.

Chemical Composition of DNA

DNA extract from many living organisms have revealed it to be a complex macromolecule. It is a long chain polymer cmposed of monomers (units), called nucleotides. Each nucleotide inturn is composed of three sub units. These are – (a) nitrogen base, (b) sugar and (c) phosphate.

Nitrogen bases

The nitrogen bases are of two kinds viz., pyrimidines and purines. Both pyrimidines and purines are heterocyclic compounds, with a basic six membered ring in which one or the other carbon position is replaced by nitrogen.

The pyrimidines have a basic six membered ring of benzene in which, position 1' and 3' are replaced by nitrogen. There are two kinds of pyrimidines – *Cytosine* and *thymine.* Structurally, cytosine has a single six membered ring with a hydroxyl group at the 2' position and an amino group at the 6' position (*2 hydroxy, 6 amino pyrimidine).* Thymine also has a single six membered ring with hydroxyl groups at the 2' and 6' position, and a methyl group attached to the 5' position *(2, 6 hydroxy 5 methyl pyrimidine).*

The purines have a double ringed structure. Attached to the six membered benzene ring is a 5 membered imidazole ring. The imidazole ring is joined to the hexagonal ring at 4' and 5' positions. The benzene ring has nitrogen in the 1' and 3' position (replacing Carbon).

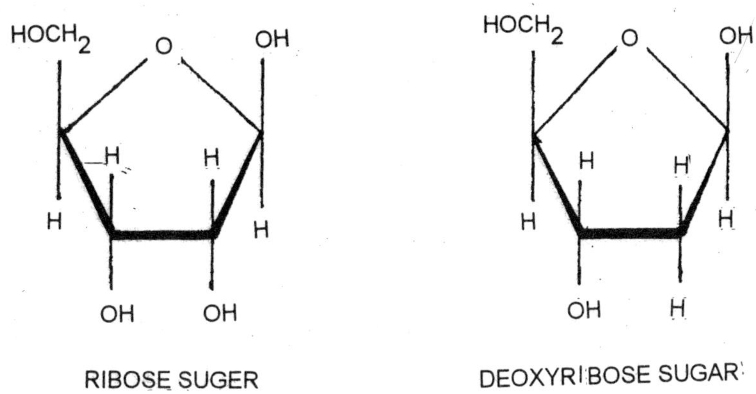

Fig. 2.1 Chemical basis of heredity
Pentose sugars of nucleic acids (left, robose; right deoxyribose)

There are two types of purines viz. *Adenine* and *guanine.* Adenine has an amino group at the 6' position (*6 aminopurine),* while guanine has an aminogroup at the 2' position and a hydroxyl group at the 6' position (2 *amino, 6 hydroxy purine).*

Sugar

The sugar molecule in DNA is basically a ribose sugar. The ribose sugar is a *Pentose Sugar.* It is a 5 membered ring. The ring consists of four carbon atoms while the 5' C atom is outside the ring. Positions 1', 3' and 5' in the ring have OH groups, while the 2' carbon position has only H. There is no O; hence the sugar is said to be Deoxyribose (no oxygen in the second carbon position).

Phosphate

It has one central phosphorus with four oxygen atoms.

Nucleosides

A nitrogen base coupled with sugars forms a nucleoside. DNA has four nucleosides –

(a) Deoxycytidine
(b) Deoxythymidine
(c) Deoxyadenosine and
(d) Deoxyguanosine

The sugar molecule is attached to the nitrogen base to the 3' position in the benzene ring in the case of pyrimidines and to the 9' position of the imidazole ring in the case of purines.

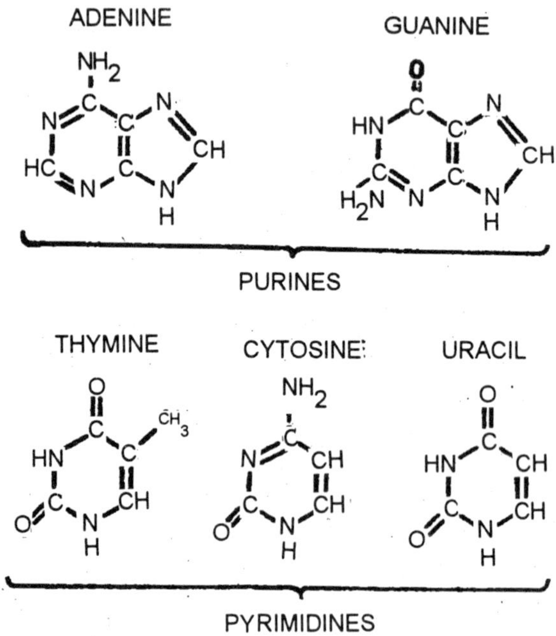

Fig. 2.2 Chemical basis of heredity
Nitrogen bases of nucleic acids

Nucleotides

A nucleotide is derived from the nucleoside by the addition of phosphoric acid molecule to the latter. The phosphate group is usually linked to the two sugar molecules above and below. The linkage is with the 3'C atom of one deoxyribose sugar and with the 5' C atom of another. DNA has four nucleotides – *Deoxycytidilic acid, Deoxy thymidilic acid, deoxy adenylic acid* and *Deoxyguanilic acid.*

Polynucleotide

DNA is said to be a polynucleotide chain with a number of nucleotides joining together with the help of DNA polymerase.

Molecular model of DNA structure

A molecular model for DNA structure first proposed by J.D.Watson and F.H.C

Crick (1953) has been universally accepted. They were awarded Nobel prize for their discovery. The work of Crick and Watson has been supported by X-ray diffraction picture obtained by Wilkins and his associates.

According to this model, a DNA molecule is composed of two chains that are spirally coiled and form a double helix, much like a spiral stairway. The distance between the two chains is uniform and it is maintained by base pairing. The back bone of the chain is made up of alternating sugar and phosphate molecules. At specific intervals, the chains are joined by 'steps'. Each step is made up of two nitrogen bases. Each nitrogen base is attached to the sugar molecule of the opposite chain. Between themselves, the nitrogen bases of a set are joined by hydrogen bonding. The pairing of the nitrogen bases is always specific. It is made up of a purine and a pyrimidine i.e., a double ring and a single ring. Two purines would occupy too much space and can not be accomodated between two chains; similarly two pyrimidines would form a short step and leave a gap between two chains. Hence the pairing is always between a purine and pyrimidine. Because of this pairing,in a DNA molecule, the total purine content is always equivalent to total pyrimidine content.

The purine, pyrimidine base pairing is further specific. It is always Adenine with Thymine and Guanine with Cytosine. Adenine and Thymine are joined by two hydrogen bonds through atoms attached to position 6' and 1'. Cytosine and Guanine are joined by three hydrogen bonds at positions 6', 1' and 2', the bonding to hydrogen is weak; but this facilitates easy separation of two strands during replication. According to Crick and Watson, the helix has a diametre of 20 A^0 and there is one complete turn to the helix at every 34 A^0 and has a stack of 10 nucleotides at every turn.

One unique character of DNA is its unity and yet infinitismal variety. The DNA structure mentioned above is uniform in all organisms, whether it is man or virus. At the same time it is also true that DNA is the genetic material that decides the characters of individual organisms. Then how does DNA decide the individuality? The answer lies in the base pair sequence. The combination of A/T, G/C is uniform, but their sequence is not; for instance in one individual it may be AT, G/C, GC, AT, GC< AT etc. There is no limit to the variety in which base pair sequences are arranged.

This is the secret of individual differences. In no two individuals, the base pair sequence is same.

Unusual bases in DNA

The normal bases in DNA are A, T, G and C. In some viruses like PBS1 and PBS2, uracil replaces thymine. In some bacteriophages, C is replaced by 5 hydroxy methyl cytosine.

Fig. 2.3 Chemical basis of heredity

F.H.C. Crick (left and J.D. Watson (right)

Single strand DNA

Although DNA in most organisms is double stranded, in some phage viruses (ϕ x 174), single stranded DNA is found. The following table summaries the differences between double stranded DNA and single stranded DNA.

Character	Double Strand DNA	Single Strand DNA
1. U.V. Absorption	Constant from 10-80^{0}C and then rises rapidly	Steady increase from 20^0 to 90^{0}C
2. Action of formaldehyde	Resistant	Not Resistant

3.	Ratio of	equal, A=T and 1.33:0.98:0.75	A:T:G:C is
4.	Shape	Linear	Circular
5.	Respiration	Two strands separate and replication begins	Becomes double (replicativeform) temporarily.

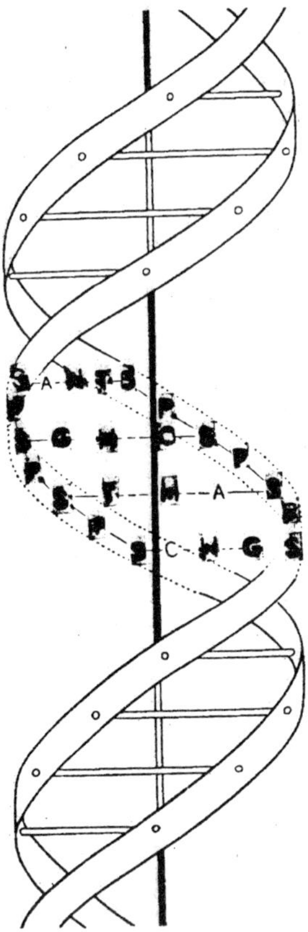

Alternative forms of DNA double helices

The structure of DNA molecule described above as visualized by crick and Watson assumes that the helical coiling of the molecule is right handed and this form of DNA has been called the B-form. Subsequent researches, however, have indicated that other types of coiling are also possible for the DNA helex. These alternative forms of DNA differ from one another in the following respects.

1. Number of residues per turn (n), and
2. Spacing of residues along the helix (h)

The existence of DNA as double helix has also been confirmed by experiments used for measuring the number of base pairs per turn of the helix and this works out to be 10.4 instead of 10 in the B-DNA. This leads to the modification of angle of rotation from 36 to 34.6. Thus, it is possible that there could be structural variants of DNA with varying values for n and h. Three structural variants of DNA A,B and C are known since a long time and transitions between them are also possible. The following table gives a brief summary of the characters of these forms of DNA.

Fig. 2.4 Chemical basis of heredity
The double helix model od DNA

Helix type bryophytes	Condition	Basepair	Rotation per
A	75%	11	+32.7 (right handed)
B	92%	10	+36.0 (right handed)
C	66%	9.33	+38.6 (right handed)
Z	very high salt	12	-30.0 (left handed)

The forms of DNA are assumed to be found in all molecules depending on the surrounding conditions. In addition to the above, D and E forms of DNA are also found very rarely with only 8 and 7.5 base pair per turn. These are found only in some molecules of DNA lacking in Guanine. In contrast A,B and C forms are found in all molecules irrespective of DNA sequence.

Z-DNA

This type of DNA was first synthesized artificially, but subsequently has been reported from living cells also. As this molecule follows a zig-zag course it is called Z-DNA. The resemblances and differences between B-DNA and Z-DNA are as follows –

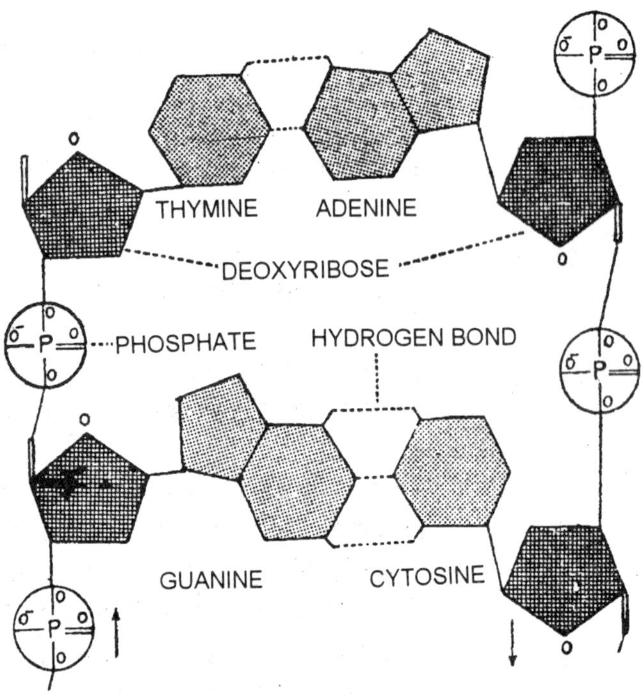

Fig. 2.5 Chemical basis of heredity
Arrangement of nucleotides in a segment of the DNA model

Resemblances between B-form and Z-form

1. Both are double helical.
2. The two strands are antiparallel in both forms.
3. Both forms exhibit A=T and G=C pairing.

Differences between Z-DNA and B-DNA

1. Z-DNA is left handed, while B-DNA is right handed.
2. The phosphate back bone is regular in B-DNA, while it is zig in Z-DNA.
3. In Z-DNA, sugar residues have alternating orientations as a result of which repeating, units are dinucleotide, whereas in B-DNA they are monocleotides.
4. Z-DNA has 12 bp for every twist of 360^0 while B-DNA has only 10bp.
5. As more base pairs are accommodated in one helix in Z-DNA, the angle of twist is 60^0 as against 36^0 in B-DNA.
6. One complete helix is $45A^0$ in Z-DNA, while it is $34A^0$ in B-DNA.
7. The diameter of the helix is $18A^0$ in Z-DNA, while it is $20A^0$ n B-DNA.

The role of Z-DNA is not certain, even though it is thought to play a role in regulation of gene activity, recombination or rearrangement of DNA.

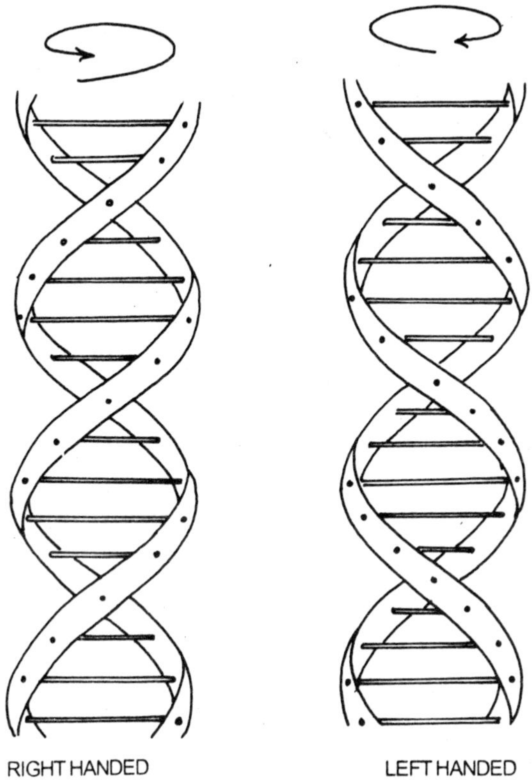

Fig. 2.6 Chemical basis of heredity
Two helices, one showing right handed coil as found in B-DNA and the other showing left handed sense as found in Z-DNA.

RL Model

A new model for the molecular structure of DNA has been proposed by Sasisekharan and associates (1976-89), working at the Indian Institute of Science, Bangalore. According to them, the right handed (R) helix of B-DNA cannot easily explain the unwinding of the coil due to topoligical differences. Hence they suggested a combination of right handed (R) and left handed (L) helices, which they called the RL model. In this model, a B-DNA molecule will have alternating left handed and right handed segments of approximately 5 base pairs in a repeat of 10 base pairs. This model provides conformational flexibility. Many molecular biologists opine that there is a strong experimental evidence for the RL model although one is not sure of the continuous length of right and left handed segments in a molecule of DNA.

Super Coiling of DNA

The double helix structure of DNA represents in linear molecule, but in living conditions the molecule is a closed structure without free ends. In bacteria, the DNA is circular while in eukaryotes also large loops appear to represent part of a circle. As a result of this, the DNA becomes supercoiled. One supercoil is introduced everytime when the duplex is twisted about its axis. The super coil may be negative (winding opposite to the helix coiling), or positive (winding in the same direction as the helix coiling).

RIBONUCLEIC ACID (RNA)

The RNA is also a polymer of nucleotides. These are called ribonucleotides. Each ribonucleotide has three parts viz., a pentose sugar, a nitrogen base and a phosphate group.

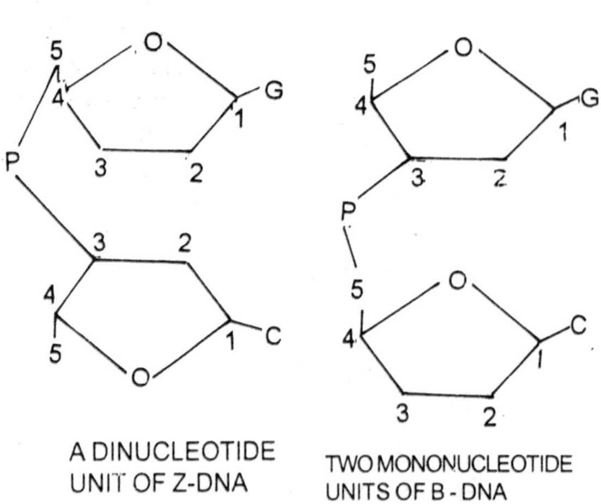

Fig. 2.7 Chemical basis of heredity
Orientation of adjacent sugar residues in Z-DNA and B-DNA showing opposite orientations in Z-DNA and same orientation in B-DNA. This results in dinucleotide units in Z-DNA as against mononucleotide units in B-DNA

The pentose sugar is a ribose sugar much like that of DNA, but with oxygen (present) attached to the 2' carbon position of the ring.

The nitrogen bases are – Adenine and guanine (purines), and cytosine and uracil (pyrimidines).

Molecular structure of RNA

Here also, the back bone is made up of alternating sugar and phosphate. The single chain folds to enable the base pairs to meet here and there to form the steps. The pairing is between A-U and C-G. Occasionally, the RNA may be double stranded but does not form a helix like DNA.

Single stranded RNA is the genetic material in plant viruses, poliomyletis viruses etc. Double stranded (non helical) RNA occurs as the genetic material in reoviruses.

Types of RNA

The following are the types of RNA found :

(a) Genetic RNA
(b) Ribosomal RNA
(c) Messenger RNA
(d) Soluble or Transfer RNA.

Genetic RNA

In certain viruses, RNA is the only genetic material. DNA being absent. This RNA is self- replicating; it is called RNA dependent RNA synthesis. The viral RNA directly functions as a messenger and produces RNA polymerase enzyme as well as proteins.

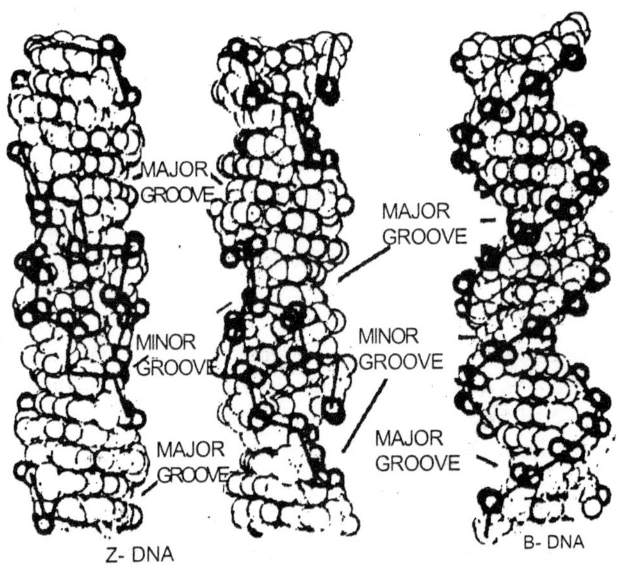

Fig. 2.8 Chemical basis of heredity

Side views of A-DNA and B-DNA. Two views of Z-DNA are 300 apart. Irregularity of backbone in Z-DNA is shown by heavy lines showing the path of phosphate residues, which is quite regular and smooth in B-DNA

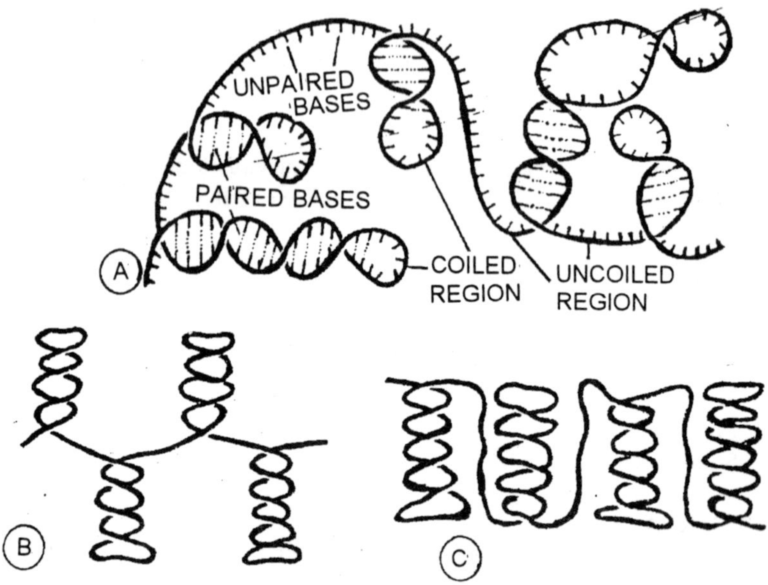

Fig. 2.9 Chemical basis of heredity

Secondary structure of ribosomal RNA (diagrammatic) **A.** Extended strand, **B.** Negative interaction showing repulsion, **C.** Positive interaction showing association

Ribosomal RNA (rRNA)

This is found in the ribosomes. It constitutes about 80% of total RNA of the cell. The base pairs of rRNA are complementary to that part of the nuclear DNA from where they are produced. rRNA consists of a single strand folding upon itself at some regions. The folded regions have paired bases, while the other regions have unpaired bases. Due to this in rRNA there is no purine pyrimidine equality.

There are three types of rRNA bases on sedimentation and molecular weight. These are (a) rRNA with molecular weight over a million, e.g. 21s-29s rRNA, (b) rRNA with molecular weight less than a million, e.g. 12s-18s rRNA, (c) rRNA with very low molecular weight (40,000) e.g. 5srRNA.

Messenger RNA (mRNA)

mRNA carries the message from DNA to ribosomes during protein synthesis. It is about 3-5% of total RNA content of the cell.

mRNA has a molecular wieght of about 5,00,000 with a sedimentation cefficient of 8s. This, however, is highly variable mRNA and is always single stranded and there is no base pairing, even though random coiling may be observed. The mRNA molecule is produced from the coding strand and its base sequence is complementary to the coding strand. The following are some of the general features of mRNA :

1. *Cap :* This is a blocked methylated structure found at the 5' end of RNA.
2. *Non coding region :* This follows the cap and consists of 10–100 nucleotides. There are no codes for proteins in this region.

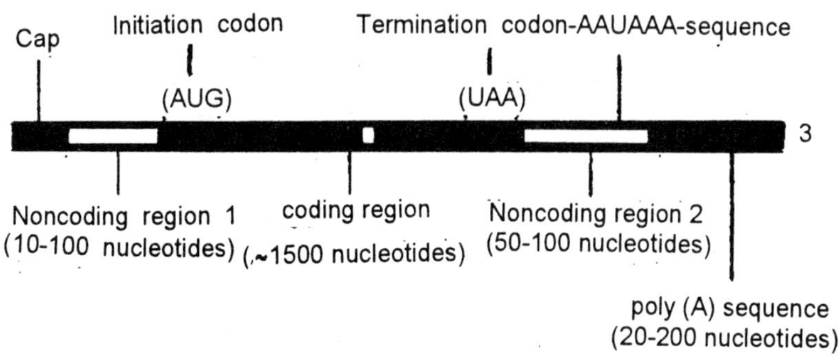

Fig. 2.10 Chemical basis of heredity
General features of eukaryotic mRNA

3. *Initiating codon :* This region begins the coding.
4. *Coding region :* Codes for proteins; and has about 1500 nucleotides.
5. *Termination codon :* The coding region is followed by a terminating codon, which stops translation.

Transfer RNA (tRNA or sRNA)

The transfer RNA also called soluble RNA, is next in quantity to rRNA. It has a low molecular weight of about 25,000–30,000 and the sedimentation co-efficient is 3.8s. It is made up of about 70–90 nucleotides.

tRNA is synthesised in the nucleus on a DNA template. Only a minute fraction of DNA (0.025%) codes for tRNA. The tRNA molecule consists of a single strand looped around itself with the 3' end always terminating in a C-C-A sequence. Some bases are paired.

The function of tRNA is to carry aminoacids to mRNA during protein synthesis. Each aminoacid couples with a specific tRNA. Since the protein molecule consists of about 20 types of amino acids, there must at least be 20 types of tRNAs. But there could be many more, sinced there are two tRNAs for every amino acid in several cases.

Secondary structure of RNA

This refers to the arrangement of the chain of the molecule to enable it to function in a proper way. Many models have been proposed for the secondary structure of tRNA. Of these, clover leaf model proposed by Holley (1966), is most widely accepted.

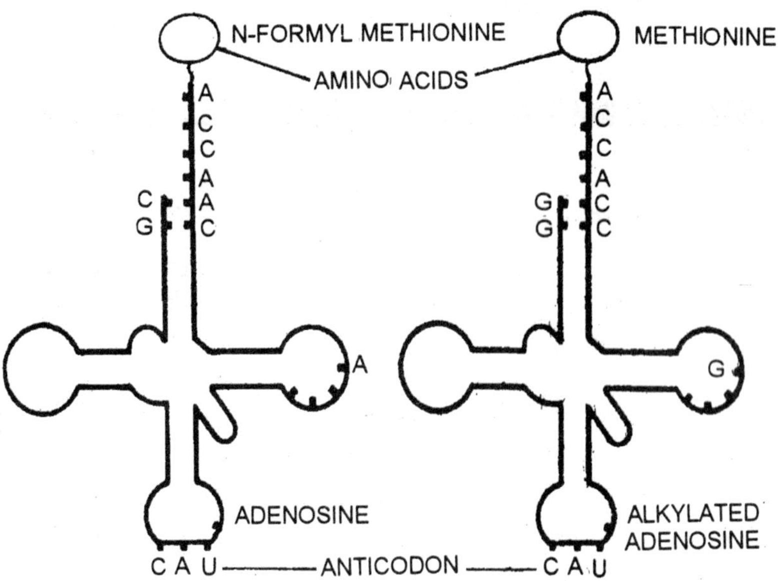

Fig. 2.11 Chemical basis of heredity
Secondary structure of tRNA (Clover leaf model)

According to the clover leaf model, the single chain of tRNA is folded on itself to form five arms. As a result of this folding, the 3' end and 5' end lie adjacent to each other. Each arm consists of a basal stalk, and a loop. The double stranded stalks show the typical base pairing. There could be an exceptional mismatch. The loops do not have base pairing. One of their arms which does not have a loop is called the acceptor as this is the binding site for amino acids. The second arm is called the DNA arrm. This has a synthetase site which recognises the amino acid activating enzyme. The fourth is a variable arm; it has a stem, but the loop may or may not be present. The Tψ arm has a ribosome recognition site.

There is also a tertiary structure to the rRNA. This is brought about by H bonding between bases. and between bases and sugar phosphate backbone.

3

GENES AND CHROMOSOMES

The coloured thread like structure that appear during cell division have hidden within them all the secrets of gėnetic characters that are passed on from one generation to another generation. Housing the particles of inheritance, the genes, these vehicles of heredity are so important for all organisms; because if it these that delimit one individual from another. These are the chromosomes (Gr. *Chromo* = coloured, *soma* = body) which appear like coloured threads on staining. In reality however these are not coloured. In this chapter we will learn about the structure and functions of these unique threads of life.

History

It was Waldeyer (1888) who first coined the term chromosome to the darkly staining bodies of the nucleus. Fleming (1879) actually had described the splitting of the chromosomes, but he had coined ther term chromatin to refer to them. Sutton and Boveri (1902) usggested that the chromosomes are the carriers of hereditary particles. Morgan (1933) discovered the role of chromosomes in the transmission of hereditary characters.

Chromosomes occur as a network of fibrils called chromonema in the interphase nucleus. At the time of cell division the chromonema breaks and condensation takes place forming specific number of chromosomes.

Number of chromosomes

The number of chromosomes is constant for a given species. This number has been evolved during the course of evolution and characterises the species. Chromosome number and morphology is one of the parameters in deciding the phylogeny and interrelationship of the group as well as the individual. The number of chromosome in various species is as varied as the organism themselves. They range fro 4 to 500.

Polidy of chromosomes

The total number of chromosomes present in the cell of an organism is called *ploidy.* In all sexuallyreproducing higher organisms the ploidy is of two types - *haploid* and *diploid.* The somatic cells contain diploid or two sets of chromosomes (2x). In these organism during reproduction, the chromosome number would be halved in meiosis. As a result, the gametes have haploid or one set (x) of chromosomes. When two gametes fuse, the resultant zygote has two sets of chromosomes (diploid). It is this zygote which develops into the adult individual.

In lower organisms like many algae and fungi, the somatic cells have haploid chromosome numbers. In these organisms only the zygote will have diploid chromosomes. The meiotic division which halves the number of chromosomes takes place in zygote rather than during gamete formation. In some unusual cases the cells have many sets (polyploid) of chromosomes, in which case it is a multiplication of the base number called x.

TABLE

Group	Name of the Organism	Chromosome number
Protozoa	*Paramecium*	30-40
Coelenterata	*Hydra vulgaris*	32
Nematoda	*Ascaris Lumbricoides*	24
Chordata	*Rana esculenta*	24
	Homo Sapiens	26
Fungi	*Mucor heimalis*	46
Pteridophyte	*Ophioglossum spp*	2
Gymnosperm	*Pinus Ponderosa*	500
Angiosperm	*Crepis Capillaris*	6
	Allium cepa	14

Size of the chromosomes

There is a great variety in chromosome size and hspae. In size they range from 0.1μ to about 30μ in length. In exceptional cases however (polytene chromosomes) the chromosomes may be as long as 2mm. All the chromosomes in a species may be of similar size or they may vary. In *Yucca arakansana,* there are two distinct size groups. In some human beings there is a gradation in size. In some instances (*Mediola*) the cells at different regions (roots and shoot tip) show different sizes of chromosomes. In general, the chromosomes of plants are larger than those of animals. Monocotyledons have larger chromosomes than dicotyledons.

Structure of the Chromosome

The following are the parts usually associated with the chromosome.

Pellicle and Matrix

Light microscopic studies have revealed that each chromosome is surrounded by a membranous is surrounded by a membranous structure called *Pellicle.* The membrane encloses a jelly like substance called *matrix.* The presence of matrix has been shown in *Luzula campestris* (Juncaceae). Electron microscopic studies however have revealed, that there is nothing like a pellicle enveloping the chromosomes.

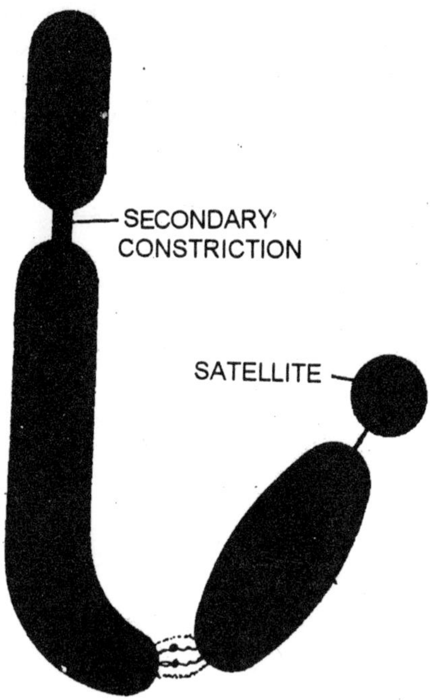

Fig. 3.1 Chromosomes
External view of a metacentric chromosome

Chromonema

When the nucleus is resting, the chromosomal material will be in the form of a network called chromonema. At the time of cell division these break up and condense to form chromosomes. The metaphase chromosomes have two thread like structures caled chromatids. The chromatids are held together at the region of primary constriction (for details see later in the same chapter). Treatment of chromosomes with a protein hydrolysing enzyme (trypsin) will reveal such chromatid to have two subchromatids; but there is no unanimity on this. Various chromosome models have proposed chromonema coils ranging from 2 to 32. The chromonema coils (subchromatids) were first observed by Baranetzky

(1880) in the pollen mother cells of *tradescantia.* It was Vejdovsky (1912) who first gave the name chromonema to these coils.The number of chromonema fibrils not only varies with different species, but at different growth stages in the same cell.

The chromonemal fibrils are coiled around each other in two different ways. These are *Paranemic coils* and *Plectonemic coils.*

The paranemic coils are loosely coiled and the threads are easily separable from one another. The plectonemic coils however are interwined very closely so that they are not easily separable.

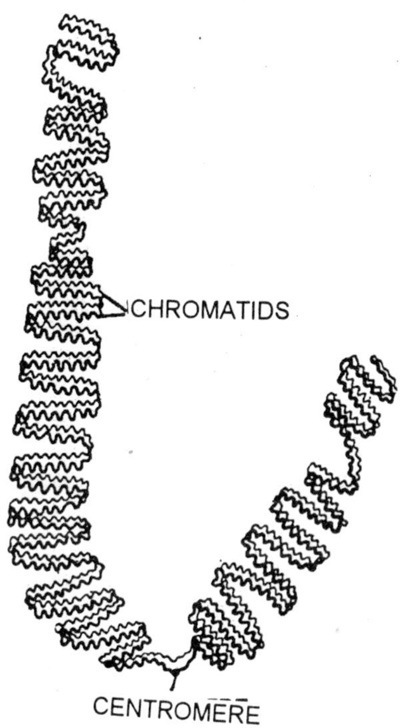

Fig. 3.2 Chromosomes
Internal structure of a metacentric chromosome

The extent of coiling in the chromonema fibrils during cell division depends on the length of the chromosomes. These coiling are of three categories **Major Coils** Each coil consists 10-30 gyres **Minor Coils** These are at right angles to the major coils and have many more gyres than major coils.

Standard or somatic coils : These are coils found in the chromonema of cells during mitosis. They are similar to major coils.

Chromomers

The chromonema in the chromosomes particularly during mitotic prophase appears to contain alternating thick and thin regions giving a bead like appearance in a chain. The thick bead like regions are called *chromomers* and the region in between the two chromomers is called the *interchromomere.* The position of chromomers in the chromonema appears to be specific to a given chromosome in an individual.

What are Chromomers

Cytologists have gien various interpretations; while some regard it as the region of condensed nucleoplasmic material, others interpret it as region of superimposed chromonemal coils. This view is supported by electron microscopic observations. Previously the chromomeres were mistaken for genes.

The chromomers were first described by Balbiani in 1876 and later by Pfitzner in 1881. Belling (1928) had equated chromomeres with the genes in liliaceous plants. But McClintock (1944) has shown that genes are located in the interchromomere region of the chromonema. However in case of polytene chromosomes the chromomeric bands are associated with certain regions.

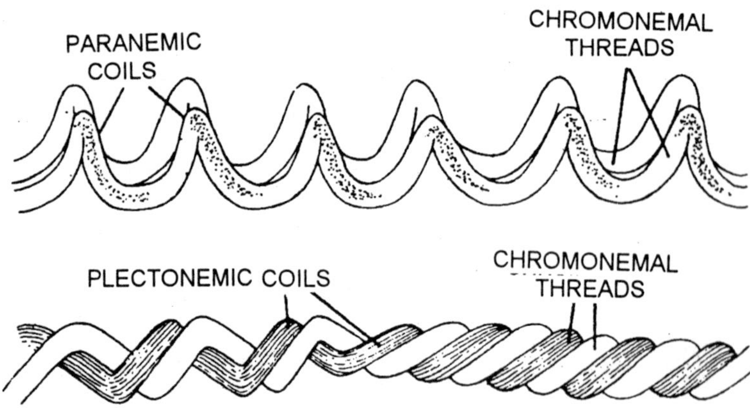

Fig. 3.3 Chromosomes
Diagrammatic representation of a paranemic and piectanomic coiling

Centromere (Primary Constriction)

The stained chromosomes reveal a certain region or constriction which stains lightly. This is called *centromere* or *kinetochore* or primary constriction. The centromeres occupy a constant position relative tothe ends of chromosomes and are responsible for the various shapes of chromosomes.

The centromere has a diameter of 0.2μ and may occupy various position on the chromosome. It (centromere) consists of spherules or granules (*kinosome*) and interchromomeral fibrillae. the numbeer of fibrillae per centromere varies. The chromonema are less coiled in the centromeric region and hence stain lightly. DNA however is present in the centromeric egion (Lime-de-Faria 1955)

Most of the organisma have their chromosomes with a single centromere (*monocentric)*. Certain other chromosomes have two centromeres (*dicentric)*.

In hemipteran insects however the chromosomes.have many centromeres (*polycentric)*. Such a centromere is said to be a *diffused centromere.* The proof for centromere being diffuse comes from the fact that when such a chromosome is fragmented by X-Rays, each bit functions like a typical chromosome, replicates itself and participates in the division cycle. Diffused centromere has also been reported in *Luzulla purpurea* of the family Junacaceae (Brown 1954). According to some cytologists, chromosomes with a diffused centromere are more primitive than the ones with a localized centromere. The polycentric chromosomes found in the nematode *Ascaris megalocephala* are unique in the sense they are restricted to only those cells which constitute the germ line (Swanson 1957)

The position of the centromere in the chromosome (which is constant to a given chromosome) varies i.e., it may occupy different position. Based on this, four morphological shapes have been identified in chromosomes. These are :-

1. Metacentric

The Centromere occupies a middle position with reference to the length of the chromosome. The two arms thus resulted are almost equal in length. During anaphase movement in cell division metacentric chromosomes appear 'V' shaped. e.g. : *Trillium, Tradescantia* etc.

2. Sub Metacentric

When the centromere is located some distance away from the middle region of the chromosome, the position is said to be sub median and the chromosome sub metacentric. As a result, one arm of the chromosome will be shorter than the other. During anaphase movement, sub metacentric chromosomes appear 'L' shaped e.g : *Human beings.*

3. Acrocentric

In this case, the centromere is situated is situated exactly at one end, the chromosome will be having only one arm. Telocentric chromosomes are very rare. Truly telocentric chromosomes have been identified by Marks (1957) in certain plants, protozoa and certain mammals. Cleveland (1949) has also reported in certain protozoa inhabiting the digestive tract of wood termites, the occurrence of telocentric chromosomes. The earlier report regarding chromosome IV of *Drosophila Melanogaster,* being telocentric has been disproved as a second arm has been seen.

Acentric Chromosome

In certain instances, chromosomes may undergo fragmentation resulting in one fragment without a centromere. Such a chromosome fragment is said to be *Acentric*.

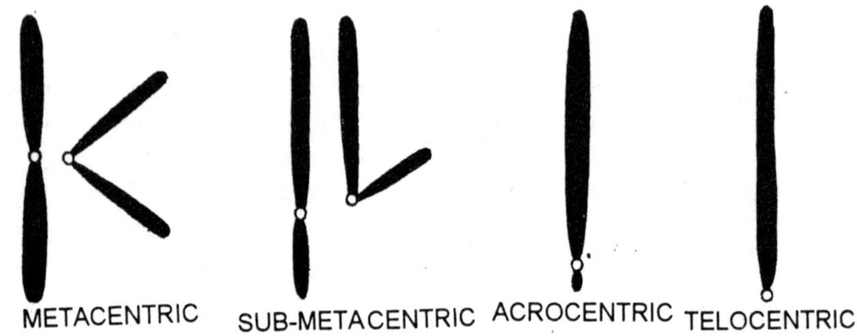

Fig. 3.4 Chromosomes
Types based on centromere position

Arm Ratio

The ratio of length between the long arm and the short arm is said to be the *arm ratio,* which is constant for a given chromosome. The arm ratio is less in metacentric chromosomes and high in acrocentric chromosomes.

Centromeric Chromomeres

The centromere should not be mistaken or equated with a chromomere as the formere stains lightly and is a constriction while the latter is a swelling on the chromosome found throughout its length. Metaphase chromosomes which have two chromatids exhibit in their centromere four granular structures. The granules called centromeric chromomeres are about 0.5μ in size and are arranged in a square.

Functions of Centromere

1. Centromere provides mobility to the Chromosome.
   Chromosomes lacking in centromere fail to orient themselves properly on the anaphase plate, lag behin and finally get eliminated.
2. Centromere decides the shape of the chromosome.
3. Centromeres attach themselves to the spindle fibres during anaphasic movement.
4. The centromere divides along the longitudinal axis of the chromosome at the beginning of anaphase. Very rarely the centromere may divide at right angles to the chromosome giving rise to two pieces of centromere, each bit holdng the arms parallel to each other and genetically alike (because of

duplication) as they represent the replicated parts of a single chromatid. Such chromosomes with genetically identical arms (having same genes on both arms) are called **isochromosomes** (Darlington 1939).

Secondary Constrictions

These are lightly straining areas on the chromosome other than centromere. One or more secondary constrictions may be found in the chromosomes. The part of the chromosome (arm) present beyond the secondary constriction is called a *Satellite.* There may be one or two satellites depending on the number of secondary constrictions. Chromosomes with satellites are called *SAT-Chromosomes (*sine acid thymonucleinico - without DNA). Generally in a genome atleast one chromosome will have a satellite.

Secondary constrictions associated with the nucleolus are called *nucleolar organisers.* Usually in a diploid organism two homologous chromosomes have additional constrictions where nucleolus will be located during the interphase stage of the nucleus. During prophase, the nucleolus disappears to appear again in the identical area. The nucleolar organising region is best seen during the pachytene stage and it is heteropycnotic in maize (McClintock). In polyploid organisms, nucleolar organisers will be more than two in number. Even though it ws believed earlier that the nucleolar organisers lack DNA, subsequent studies have shown that it has about 0.3% of nuclear DNA. The gent loci at the region are responsible for the formation of 28s ribosomal RNA.

Telomeres

The terminal part of chromosomes are called telomères which have polarity, preventing the tips of other chromosomes from fusing with it. Telomeres however attach themselves to the nuclear envelope during interphase.

CHEMICAL COMPOSITION OF CHROMOSOMES

The two main chemical components of the chromosome are nucleic acids and proteins.

Proteins

Besides nucleic acid, there are three types of proteins in the chromosomes. These are *protamines, Histones* and *non histone proteins.*

Protamines are basic proteins with a molecular weight less than 4,000. They are rich in amino acid arginine and are found in the spermatozoa of salmon, herring and other fish, snails and squids. The protamines are helically bound to be DNA. Each polypeptide of protamine consists of about 29 amino acids of which 19 residues are arginine.

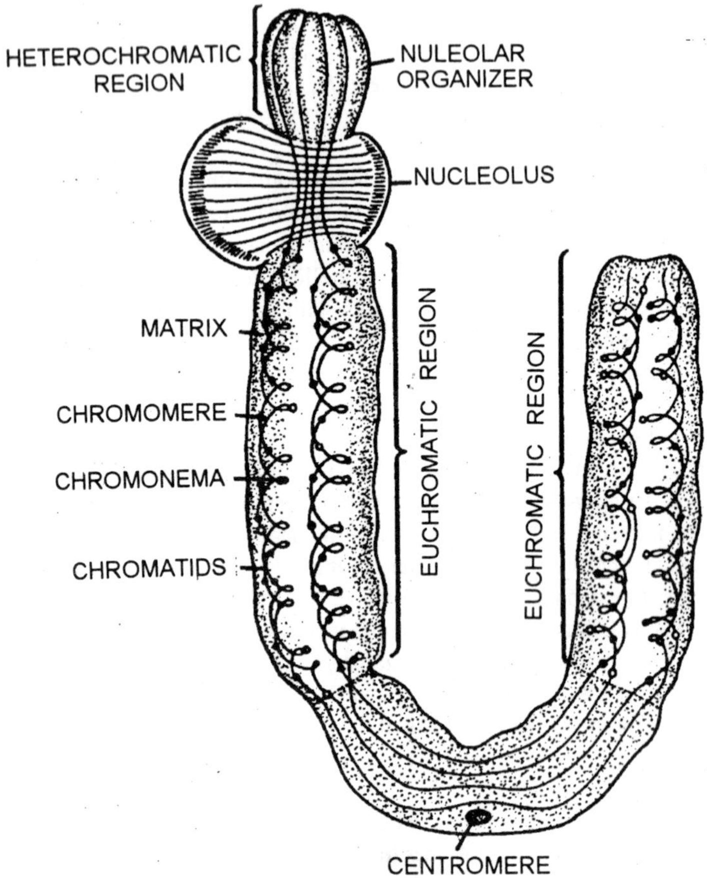

Fig. 3.5 Chromosomes
Structure of a typical chromosome showing a nucleolar organiser

Histones are basic proteins associated with eukaryotic nuclear DNA in the ratio of 1:1 in quite a number of plants and animals. They are the main structural proteins of chromosomes.

Based on the amino acid sequence they are categorized into five principal classes H1(H5), H2a, H2b, H3 and H4.

Histones are characterized by the total lack of *tryptophan,* but have an abundance of *arginine* and *lysine*. Other modifications of histone proteins areacetylation, methylation and phosphorylation.

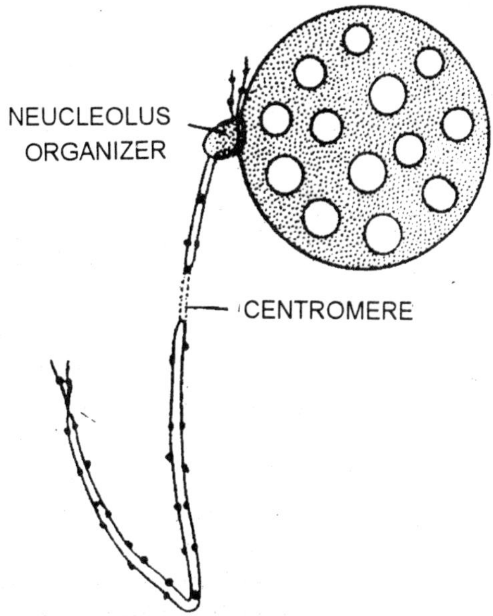

Fig. 3.6 Chromosomes
Nucleolar organiser (enlarged)

Functions of Histones

They act as gene repressors; but this may be a general activity rather than being specific as most of the histones have identical composition. Histones perhaps provide a structural framework for encompassing the DNA molecule.

Synthesis of Histones

In the interphase nucleus, histones are synthesised during the 'S' phase. Some histones may also be synthesized at other stages. The eukaryotic DNA has certain genes called repetitive genes coding for histones.

Non histone proteins

These are acidic proteins with a very high molecular weight when compared with that of histones. NH proteins are known to initiate genetic activity, while histones are known to depress genetic activity.

DNA, Protein Association in the Chromosome

There are different views regarding the association of DNA with the proteins. It is possible according to one interpretation that the protamines occupy only the small grooves in the DNA molecule, while histones occupy both the grooves (small and large). Another view is that thehistones follow the two helices of the DNA molecule coiling in the same way. According to Dupraw, the histones hold together the DNA molecule in the supercoiled position.

Euchromatin and Heterochromatin

Based on the staining reaction two regions may be identified in the chromatin material. These are - *Euchromatin* and *Heterochromatin.*

The euchromatin stains positively with the DNA specific stains (basic fuchsin) indicating a concentration of DNA. This region is gentically active and stains lightly. The euchromatic regions are supposed to represent areas of less condensation.

Heterochromatin stains more deeply than euchromatin and represent highly condensed regions on the chromosome. In the interphase nucleus, the heterochromatic regions form condensed structures called *chromocentres* or false nucleoli. The following table summaries the differences between the euchromatin and heterochromatin.

		Heterochromatin	**Euchromatin**
1.	Staining	Stains deeply	Stains less deeply
2.	Coiling	Found in the region of more coiling and condensation.	Found in the region of less coiling and condensation
3.	Replication	Late replication replication in late 'S' phase.	Early replication in early 'S' phase.
4.	Acetylation	Does not occur	Occurs.
5.	Effect of temperature	Pronounced	Not affected easily.
6.	Genetic effect	Inert genetically but few genes are present	Genetically active. More number of genes present.
7.	Cross over frequency	Less	Frequent.

Types of Heterochromatin

There are two main types viz., *constitutive heterochromatin* and *adaptive heterochromatin.* The first type shows dark staining reaction (Heteropynosis) always and in all cells, while the second type shows heteropycnosis only in certain cells or only at certain stages in the cycle. Constitutive heterochromatin was originally named satellite DNA (S-DNA). It is mostly inactive during protein synthesis and is repetetive from thepoint of view of base sequence. Adaptive heterochromatin is metabolically inert and comprises of about 2.5% of genome.

Heteropycnosis

This is the staining phenomenon seen in the chromosomes with reference to heterochromatic and euchromatic regions. The heterochromatic regions which stain deeply are said to exhibit heteropycnosis as against euchromatin which stain lightly. However it should be noted that the heterochromatin always does not stain deeply with reference to euchromatin. While the euchromatic region is always constant in its behaviour (non heterochromatin) towards staining (has constant coils), the heterochromatic region may stain deeply or lightly and this is known as *differential heteropycnosis.* Differential heteropycnosis is of two types - positive and negative.

The areas of heterochromatin showing positive heteropycnosis are more condensed and deeper staining than euchromatic region. In areas of positive heteropcnosis there is more of nucleic acids and proteins. Chromosomes with positive heteropycnosis in prophase appear like metaphase chromosomes.

The areas of heterochromatin showing negative heteropycnosis are possibly less condensed and stain lighter than euchromatic region. Chromosomes with negative heteropycnosis have a hazy outline and metaphase chromosomes appear like prophase chromosomes.

CHROMOSOME MODELS

The chromosome consists of DNA and proteins, but the way in which the two are arranged in the chromosome is a matter of debate. The main aspects of the chromosome structure are (a) the number of strand or strands of DNA (b) whether the DNA molecule is continuous throughout the length of the chromosome or whether it is interrupted and (c) what is the mode of association of protein and DNA.

Various chromosome models have been proposed and these fall into two basic categories viz., *Multiple strand models* and *single strand models.*

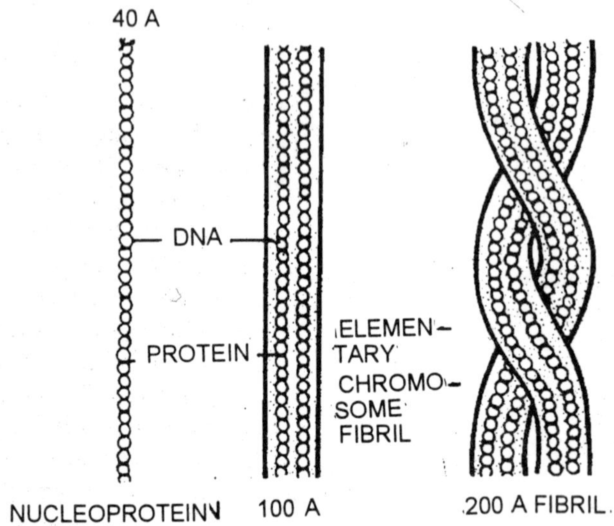

Fig. 3.7 Chromosomes
Ris' multistrand model

Multiple strand models

According to this, the chromosome consists of several DNA and protein strands. The work of Nebel (1938) on radiation effect on chromosomes have shown that there are at least four sub units. Similarly some models suggest the existence of 4,8,16,32 or even 64 fibrils with each fibril having about 20-40 DNA molecules. The following are some of the multiple strand models :-

Simple multistrand model

According to this model proposed by Steffensen (1961), the chromosome has 64 molecules of DNA arranged parallely and twisted around each other. If one agrees with this model, the replication of DNA would be a difficult process.

Ris' Multistrand Model

In this model, the DNA and histone form a complex 40 Å wide fibrlil Two such fubrils according to Ris (1961) combine to form an *elementary chromosome fibril,* two of these again combine to form a fibril which is 200 Å wide. These fibrils form the chromatids. Each chromatid has 16n to 32 elementary chromosome fibrils. But there isno experimental proof for this work. In fact Ris himselv rejected this model later.

Single Strand Models

Most of the current evidence available now indicates that the chromosomes (barring the abnormal chromosomes) are made up of a single strand of DNA. Experimental evidence by X-ray radiation have suggested the whole of the chromosome is composed of one single coiled molecule of DNA. The following are some of the single strand models.

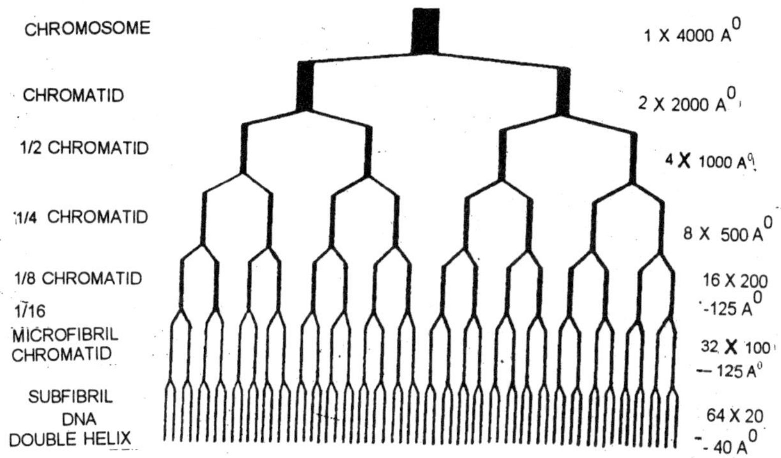

Fig. 3.8 Chromosomes
Arrangement within a chromosome based on multistrand model

The Freese - Taylore Model

According to this model originally proposed by Freese (1958) and later modified by Taylor (1959), there are two chains of proteins with the DNA molecule attaching itself to these like the rungs of a ladder. The proteins linkers maintain the shape of the DNA molecule. The two protein chains get closely appressedin the chromosome resulting in the coiling of DNA.

The original model suggested by Taylor (1957) is called the *side chain model* or *centipede model.* According to this model, the chromosome consists of a central backbone of protein from which the loops of DNA spring up like the legs of a centipede. According to him, replication is possible when the protein backbone made up of two chains, separates longitudinally into two and in theprocess separates also the DNA coils, with each strand having a single strand of DNA. But this is not in conformity with the linear arrangement of the genes in the chromosome.

Nebel's coiled coil model

According to this, the chromosome is made of a single fibril (see his earlier model) is folded to form a 100 Åwide nucleo - histone fibril. The nucleohistone

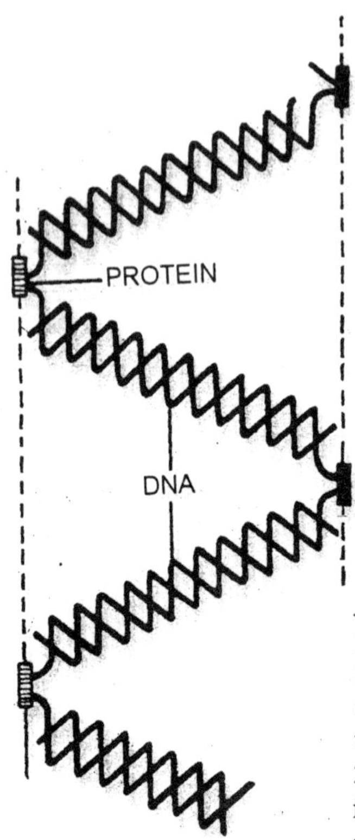

Fig. 3.9 Chromosomes
The Freese-Taylor model

fibril folds because of calcium bridges to form a 250 Å wide *basic fibril.* Further folding of the basic fibril forms the chromosome.

Folded fibre model of Dupraw (1966)
While studying leucocyte (human) chromosomes, Dupraw observed that the chromatids have folded fibres of 200-500 Å width. As he did not find many free ends it was concluded that there must be a single strand of DNA which is highly folded.

The essentials of Dupraw's model are basically like that of Ris' model with some variations. According to this, the 20 Å DNA is packed in protein to form a fibril. The fibril gets coiled and forms a 100 Å wide *type A fibre.* The DNA is packed in this fibre in the packaging ratio of 6:1 (packaging ratio is the ratio between the condensed length and the extended length). The type a fibre is coiled in the ratio of 10 : 1 to form type B fibre which is 200-250 Å wide. The type B fibre is extensively folded to form the chromatid. This has been supported by many workers (McDermot 1968, Fedric 1969, Lampert 1969 etc). Though many scientists have supported Dupraw's view, White, M.J.D. (1973) strikes a discordant note when he does not agree with the folded fibre model and he considers it extraordinary that the DNA of all the chromosomes are linked in the form of a ring. Dupraw's model is now regarded as untenable in view of the fact that the DNA itself forms a wrap around the histones to form *nucleosomes.*

Nucleosome Model
This model proposed by Woodcock (1973) suggests that the chromatid is a string of beads made up of repeating units. These beads called nucleosomes are oblate particles with a diametere of 100 Å. They have a core of histones 40 Å high and 80 Å wide, around which is found DNA in a super helical structure. Finch and Clug (1976) have proposed an arrangement of chromatin in the chromosome model of nucleosomes. The coiled DNA according to them is in the wtring of nucleosomes. The nucleosomes string is then coiled into 300 Å wide solenoid, which in turn coils to form a *supersolenoid* with a diameter of

4000 Å. This super solenoid structure is the unit fibre (chromonema seen in light microscope) whose coils constitute the chromosomes.

The eact nature of the chromosome arrangement of proteins and nucleic acids may be said to be still an open question.

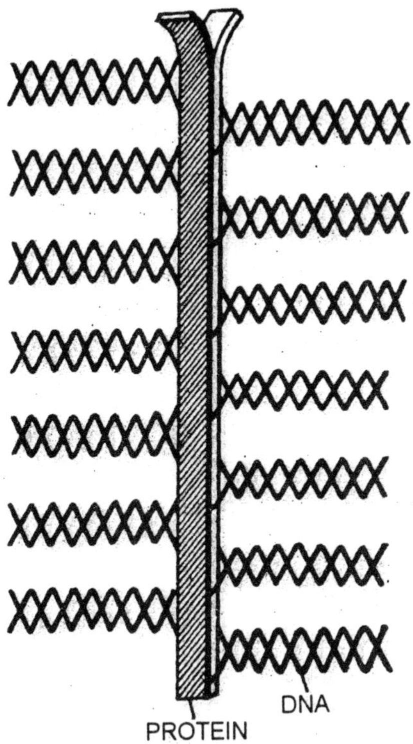

Fig. 3.10 Chromosomes
The centipede model of Taylor

UNUSUAL CHROMOSOMES

The above discussion on chromosomes pertains to normal ones. There are however abnormal or giant chromosomes encountered in many instances. These are the *polytene chromosomes* in the salivary glands of fruit flies, the *lamp brush chromosomes* in the oocytes ofamphibians, and the *supernumerary* or *B chromosomes*.

Polytene Chromosomes

These multithreaded chromosomes were first reported by Balbiani in 1881 while observing the suqash of salivary cells of *Chironomous.* Subsequently they have been seen in the larvae of dipterans (*Drosophila* etc). The name polytene was suggested by Kollar due to the apperance of many threads. Besides salivary glands the polytene chromosomes also occur in rectal epithelium and malphigian tubules.

The nuclei in the salivary gland cells are much large than the normal ones and are about 25μ in diametre with the giant chromosomes larger by 50-200 times than the normal chromosomes. The polytene chromosome in the salivary cells of *Drosophila melanogaster* (Painter 1933) are about 200μ in length. The enormous size is due to the duplication of chromonema which do not separate. According to an estimate, the polytene chromosomes have 1000 times more DNA than the normal somatic chromosomes.

Structure

Each polytene chromosome is tranversely striated by alternating bands and *inter band regions.* In *Drosophila,* over 5000 bands have been found in the four chromosomes of the salivary gland cells. The bands take an intense stain (Feulgen) while the inter band regions stain less intensely.

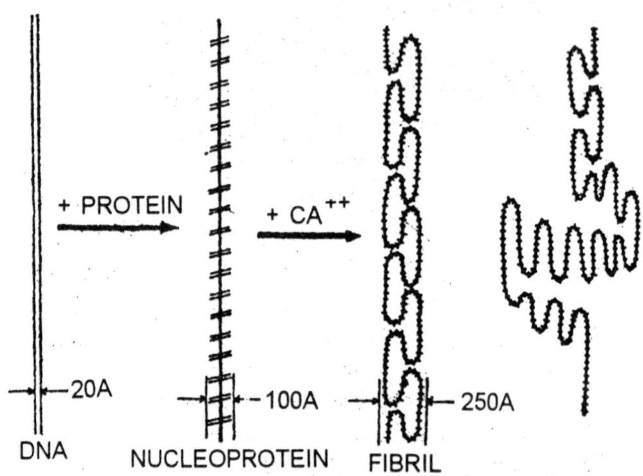

Fig. 3.11 Chromosomes
Ris' modified model

The bands have chromomeres arranged at right angles to the long axis of the chromosome. The bands actually represent regions of tight coils of chromonema. Van Herwerden (1911) interpreted these bands as gyres of a continuous spirl. This view was later supported by Kostoff (1930). The bands have ahigh DNA content and absorb U.V. light at 2600 Å.

The interbands stain lightly and they are negative to feulgen stain indicating a low concentration of DNA. The chromonema in the interband region are less coiled and absorb very little U.V. light.

Painter (1933) and Bridges (1936) have shown that in *Drosophila* the bands are clearly associated with the genes. Association of certain defects due to mutated genes have also been associated with the bands. Certain recent studies however have shown that together with the bands, the interband regions also have genes.

Chromosome puffs

These are certain interesting structures associated with the bands in the giant chromosomes called *chromosome puffs* or *Balbiani rings.* The swelling in bands at certain stages are given the name chromosome puffs. It is quite possible that these puffs are associated with the metabolic activities and represent areas of active RNA synthesis. According to Beerman and Bahr (1954), the coils in the puff region open out to form many loops. The larger chromosome puffs are called *Balbiani rings* and they show a high DNA and mRNA content.

Lamp Brush Chromosomes

The giant chromosomes found in the oocytes of fish, amphibians, reptiles and aves look like the brush used to clean the chimney of a kerosene lamp-hence the name *lamp brush chromosomes.* These are very large and are characterised by loops arising out of the main axis laterally. Discovered first by Flemming in 1882, in amphibian oocytes, they were studied in great detail by J.Rucket (1892) in the oocytes of shark. Rarely lamp brush chromosomes have also been found in some invertebrates (Sepia, Echinaster, insects etc) but they are not as large as those of vertebrates. It seems they are present in certain plants also (Grun 1958).

In somee organisms, lamp brush chromosomes as long as 1000 microns and 20 microns wide have been reported during early prophase I. Towards the end of prophase, they get reduced in size. There are reports of lamp brush chromosomes as long as 5900 microns in some salamander oocytes. These chromosomes are elastic and can be stretched.

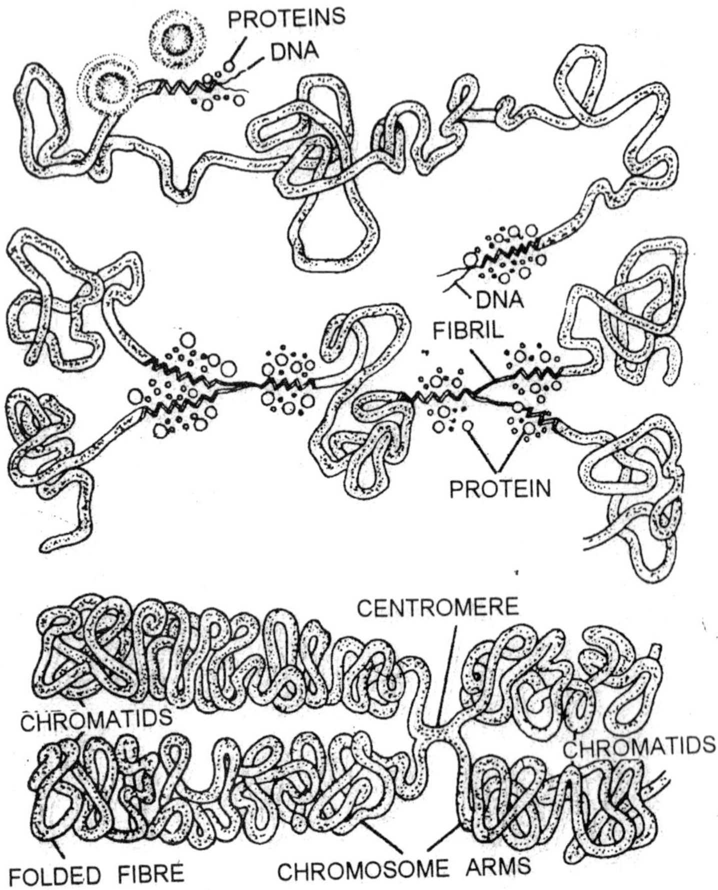

Fig. 3.12 Chromosomes
Folded fibre model of Dupraw

Structure

There are two main parts viz., the main axis and lateral loops. The main axis has two chromatids (the homologues have four) and is rich in DNA. The axis is in continuity with the loop axis. The tips of theaxis are without loops and are the regions of telomeres. The centromeric region is also without loops. At certain regions along their length the chromonema of the axis get coiled and form the chromomeres. The chromomeres are found in pairs. Electron microscopic studies of the lamp brush chromosomes of the salamander *Triturus* (Miller and Beatty 1969) have shown the axial fibre containing dense granules which are probably the molecules of the enzyme RNA *polymerase.*

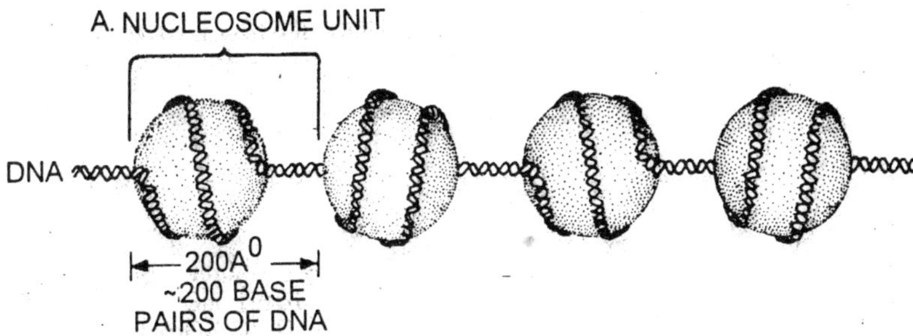

Fig. 3.13 Chromosomes
Nucleosome model

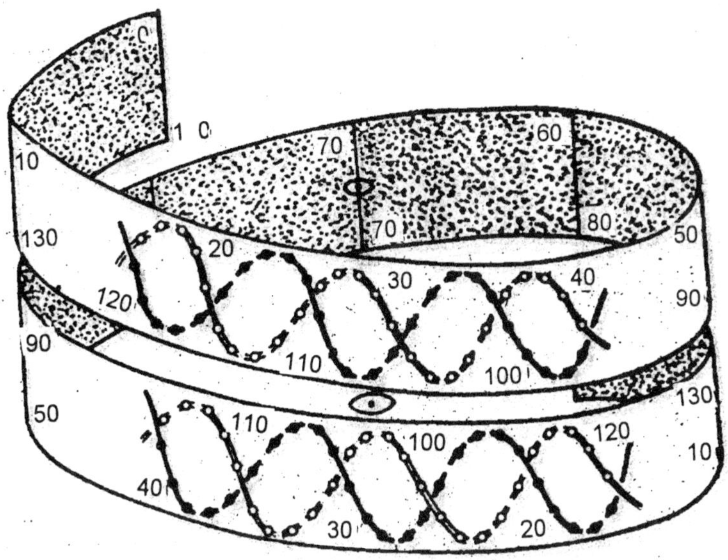

Fig. 3.14 Chromosomes
Schematic representation of DNA super helix in the nucleosome

The loops are of two main types viz., *typical* and *special*. Most of the loops which are typical consist of a bent central axis from which are given out RNA fibrils. These RNA fibrils are of progressively increasing length as a result one side of theloop is pronouncedly thick. The special loops are symmetrical and have granules at the end of RNA fibrils. Each loop axis has an axial fibre covered with a matrix. The diametre of the loop axis varies from 30-50 Å.
The axiaL fibre is made up of DNA< while the matrix has RNA and proteins.

Lamp brush chromosomes are involved in the synthesis of RNA and proteins. Each loop is believed to represent one long operon consisting of repetetive cistrons (structural genes). Each gene locus codes for RNA. The loop is supposed to synthesize at a A high rate because of repetetive gene sequence. There arealso reports that the lamp brush chromosomes help in the formation of yolk material in the egg.

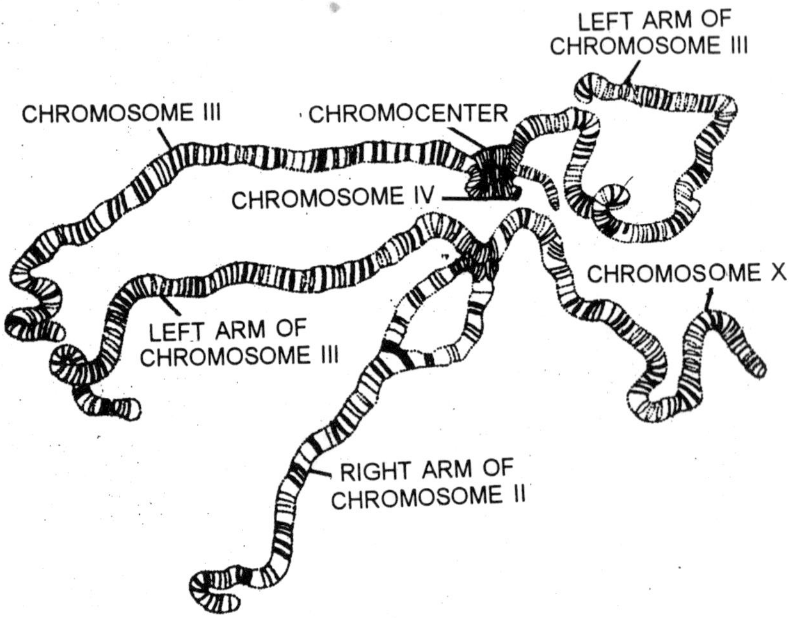

Fig. 3.15 Chromosomes
Polytene chromosomes of drosophila

Super numerary or B-chromosomes

In many plants and animals, the cells contain in addition to the normal complement, one or more extra chromosomes called supernumerary or accessory or B-chromosomes. These were first discovered in Hemipteran insects (Wilson 1905); since then they have been found in a number of insects and plants.

The supernumerary chromosomes are generally smaller than the others and are genetically inert hence produce little or no phenotypic expression. This suggest that these chromosomes may largely be heterochromatic which is supported by their differential staining activity. But this is not universal, for in *Tradescantia* they are entirely euchromatic, while in maize they are part euchromatic and part heterochromatic.

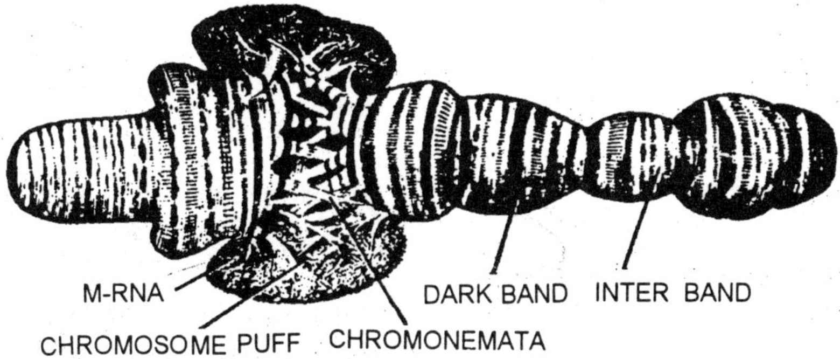

Fig. 3.16 Chromosomes
Polytene chromosomes of a Dipteran insect

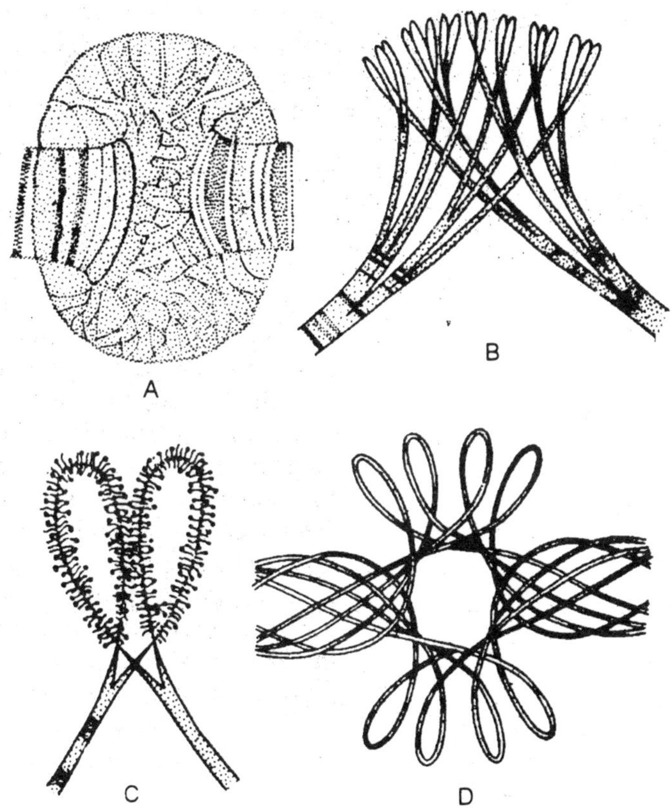

Fig. 3.17 Chromosomes
Balbiani ring in a polytene chromosome

While no function can be attributed to supernumeraries, their continued presencein the gene pool suggests that they may perform some as yet unidentified functions. The supernumeraries split up but undergo non disjunction thus both the halves being included in the same cell as in

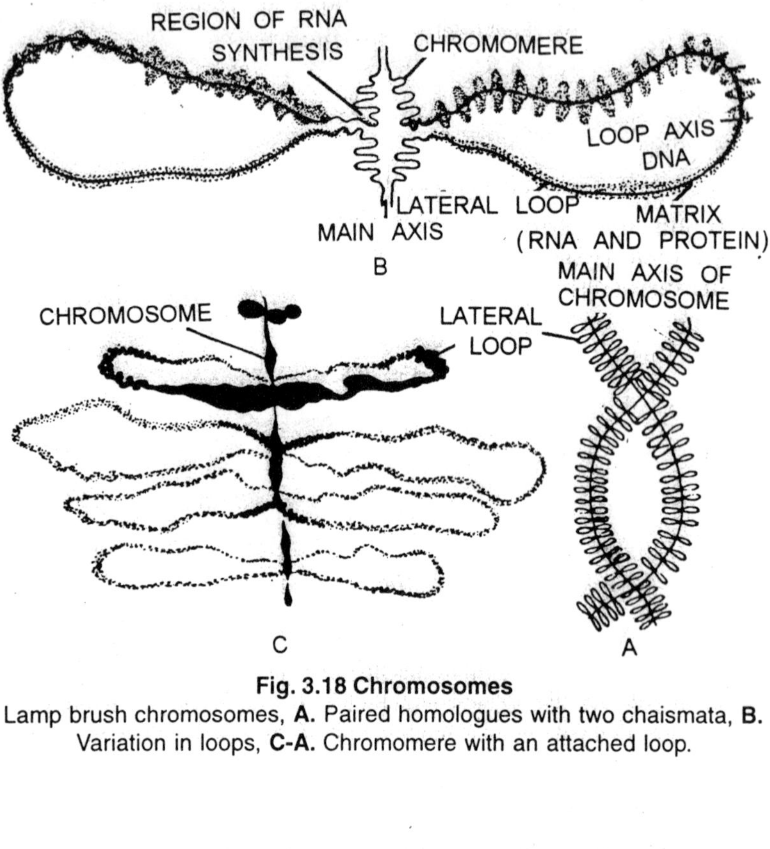

Fig. 3.18 Chromosomes

Lamp brush chromosomes, **A.** Paired homologues with two chaismata, **B.** Variation in loops, **C-A.** Chromomere with an attached loop.

Anthoxanthum aristatum and *Secale cerale.* A similar behavious has been seen in maize also. The non disjunction is attributed to improper division of the centromere in *Sorghum purpureosericeum.* Sperms with B-Chromosomes preferentially fertilize the egg. According to Randolph (1941) accumulation of too many supernumeraries in an individual will result in loss of vigour as seen in maize. The increase in the number of supernumeraries is due to irregular fragmentation. The origin of supernumeraries in plants is unknown, while in some animals they can be traced to the Y chromosome (Wilson 1905). This is the case in *Metapodius*, where the chromosomes have diffused centromere, hence any fragment can survive.

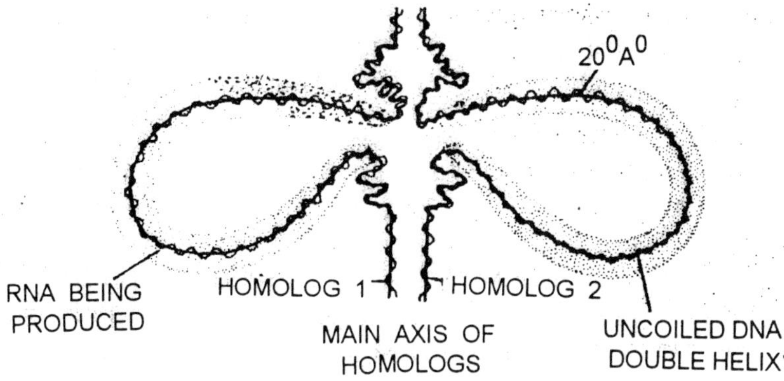

Fig. 3.19 Chromosomes
Lateral loop of a lampbrush chromosome enlarged

GENES

The possibility of the existence of discrete hereditary particles came to be known due to the path breaking work of John Gregor Mendel. It was he who provided some order into the concept of heredity which was otherwise an assemblage of weird theories proposed by many biologists others. Mendel, by his simple but ingenious experiments proved that the heredity is controlled by discrete particles called 'factors' which seggregate, recombine and seggregate again producing a variety of characters. Thus Mendel provided the proof of existence of a physical basis for heredity. The factors of Mendel later came to be known as genes. The concept of gene has undergone a sea change eversince Mendel introduced it, but its essentials remain the same.

The term *gene* was coined by Johannsen in the year 1909 for Mendelian factors. He however knew very little regarding the nature and structure of the gene.

Gene concept

The gene concept was first introduced by Sutton and works of Morghan, Bridges, Boveri etc proved this hypothesis. The essential features of gene concept are summarised below -

1. Genes decide the physical as well as physiological characters. These are transmitted from parents tooffspring, thus serving as a genetic link between one generation and the other.

2. Genes are located on the chromosomes. Many genes are located on a single chromosome. This is because, there are a large number of traits and consequently large number of genes, while the number of chromosomes are far less.

3. Each gene occupies a specific position or *locus* on the chromosome. Any alteration in the position and number of genes on the chromosome will bring in large scale genetic changes. (Eg., bar locus in *Drosophila)*.
4. Genes are arranged in a linear order on the chromosome much in the same way as beads on a string.

5. A gene may exist in only one state or may exist in or many states. The alternative state of existence of the gene is known as *allele.* Most of the genes are *diallelic* i.e., they exist in two states - *dominant* and *recessive;* of these one is the original and the other is mutant. In some genes, due to series of mutations in the same locus instead of two alleles, there will be a number of alleles. This is known as *multiple alleles* (eg., blood groups in man). They represent a series of less and less dominant genes starting with the wild one.

   In all cases however, even in dialletic series a gene need not be completely dominant or recessive.

6. Genes undergo change, resulting in their altered structure and function. This is known as mutation.
7. Genes duplicate themselves precisely andproduce copies. This is known as replication
8. Genes control the production of proteins (enzymes) and thereby regulate metabolic activity.
9. Chemically genes are made up of segments of DNA. This has been proved by the genetic transformation experiments of Avery Macleod and McCarty (1944). In some plant viruses however genes are composed of only RNA (eg., TMV).

Modern Definitions of Gene

Benzer clearly established the relation between various genetic phenomena and the DNA molecule. He coined different terms to denote the various aspects of gene structure and function. These are - *Recon, Muton* and *Cistron.*

Recon

Genetic research has indicated that on the DNA molecule, short segments exist which can be separated from the rest during crossing over. These segments are

refered to as genes. In other words, gene represents a unit of recombination of the DNA molecule. Within a gene crossing over or recombination is not possible. But studies have indicated that within a gene also, recombination is possible. Hence a sub unit of gene which is capable of recombination is called a *recon*. Recombination studies on microbes have indicated that structurally, recon consists of one or two pairs of nucleotides.
Crossing over within a gene was demonstrated by Benzer in T4 bacteriophages.

Cistron
Garrod (1908), thepioneer who established the relationship between gene and its biochemical function argued that the gene directly controls enzyme synthesis. This was later elaborated by Beadle and tatum (1941) as '*one gene one enzyme hypothesis*'. The gene was regarded as a single unit to code for a single enzyme protein.

According to Benzee, the cistron represents a segment of the DNA molecule and consists of a linear sequence of nucleotides to control the production of a protein. In most cases thecistron is equivalent to the gene structurally. In *Escherichia coli,* eachcistron has 1500 base pairs. Some cistrons may be as long as 30,000 base pairs. As the genetic code is a triplet, each gene has 3 times the base pairs as compared to thenumber of aminoacids in the polypeptide chain (it codes).

Researches have however indicated that not all genes code for proteins, the hypothesis one gene one enzyme is no longer tenable.

Complon
This is a term used to replace cistron. A complon is said to be a unit of complimentation. Complimentation is a phenomenon, where in the active groups of enzymes (when they are made up of more than one polypeptide chain) are complimentary to one another.

A few other aspects of gene structure and function are given below.

Discontinuous genes
Researches carried out in 70s indicated that biological informationmay nor be carried by the gene continuously all along its length. A gene is split into several sections separated by non coding segments of DNA. Such genes are called *discontinuous genes* or *split genes* or *mosaic genes*.

In split genes there are two sections - *introns* and *exons*.

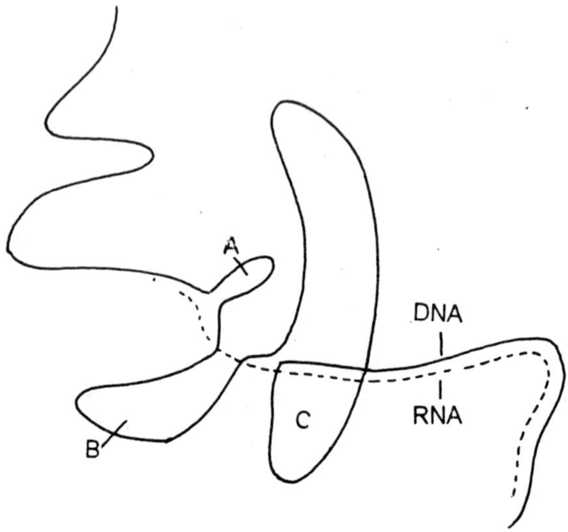

Fig. 3.20 Chromosomes
The hexon gene of Adenovirus. The loops of DNA represent regions of intron

Introns are non coding sections of DNA segment. These are present in genes of eukaryotes and their viruses, but absent in viruses, but absent in prokaryots and their viruses. In eukaryotes mitochondrial genome does not have introns. The mammalian α collagen gene has as many as 52 introns.

Gilbert, has proposed a hypothesis for the origin hypothesis for the origin of new genes by the shuffling of exons between different discontinuous genes.

Transcription from a split gene which has several introns would initially produce a precursoe mRNA which will have several non coding regions. Before translation, the introns have to be removed and the exons reatached to produce a final copy of mRNA with continuous codons. This is called *gene splicing.* Examples of split genes include *Hexons gene* of adenovirus, *ovalubumine gene* of chicken, β globin gene of mice and rabbits. In the ovalbumin gene the intron proteins of mRNA loopout and then these loops are excised.

Overalpping genes
According to the one gene one protein hypothesis of Beadle and Tatum (1940), each gene will separately code for one protein and that genes do not overlap.

But the work of Barrel *et al* (1976) gave the first evidence of overlapping genes in the bacteriophage ϕ X174. They found that two genes code for a particular protein. Usually in organisms the length of DNA should be more than the length of the polypeptide as three nucleotides code for one aminoacid. In phage ϕ X174, however the length of the polypeptide chain is more than that of DNA. This indicates that the proteins coded by two genes is specified by the same of DNA.

Jumping genes

Generally, the gene loci on the chromosome are fixed. But the work of Nobel laureate Barbara Mclintock has shown that genes keep changing their location on the chromosome. The occurrence of jumping genes has also been discovered in bacteria. Streches of DNA in bacteria can move from one location to another, therby altering the expression. The jumping segments of DNA have been called **insertion elements.**

Barbara Mclintock, working on maize has shown the movement of certain streches of DNA known as controlling elements from one location to another.

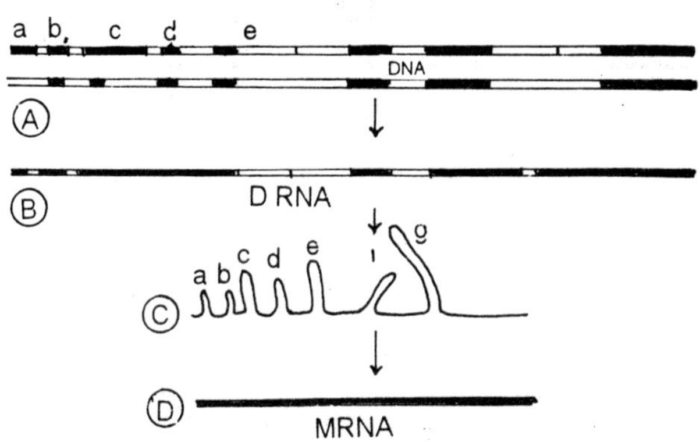

Fig. 3. 21 Chromosomes

Transcription and mRNA in ovalbumin. gene-Differentstages.(In A and B blank representsnon coding regions while black represents coding regions)

Pseudogenes

In multicellular organisms, a wide variety of DNA sequences are found which apparently have no function. Some of these sequences are defective copies of functional genes and are therefore called pseudogenes. Pseudogenes have been reported from mouse, *Drosophila* and human beings. A typical example of human pseudogene is α globin and β globin pseudogene.

4

GENOME IN PROKARYOTES

The hereditary material in all organisms is remarkably similar. Having been evolved in the remote past, life has been able to maintain itself because of the ability of the hereditary materials to maintain a basic uniformity in structure and function. The hereditary material in all living beings (whether it is a border line entity like a virus, or an evolved human being) is Nucleic acids without exception, even though, there may be some stray instances of proteins (as prions) believed to carry out inheritance. The structure of the nucleic acids is unique; it has a basic uniformity, yet sufficient plasticity to allow for its manifestation from most primitive to the advanced organism.

In eukaryotic organisms, the genetic material is organised into large entities called chromosomes which allow for proper replication and distribution of genetic material during cell division and reproduction.

In prokaryotic organisms also,the nuclei acids constitute the chromosome, but the chromosome of prokaryotes is a naked strand of DNA (not associated with proteins) lying free in cytoplasm. The entire genetic system of each species is encoded in a single chromosome. Even in viruses, the genome is either DNA or RNA.

The prokaryotic genome is refered to as the chromosome, centrosome, nucleoid, chromatin body or the gneophore. Some traditional cytogeneticists object to the word chromosome and prefer to use the term genophore or nucleoid. However, as Swanson (1981), points out "this is more a matter of semantics than of cytogenetics, since the prokaryotic chromosome not only has the same fundamental structure based on nucleic acid sequences, but it also carries out, the same three basic functions of its eukaryotic counter part. That is, it is the organelle which stores, replicates and transmits the coded information of biological inheritance; which is so regulated in its action that its primary products appear at appropriate times and in appropriate amounts during the life-cycle of the organism; and which is capable of recombining its genetic content with other similar individuals in such a manner that the seggregating progeny of a hybrid union can have different genetic properties". Swanson (1981), therefore, opines that the term "chromosome" can be applied to both eukaryotic and prokaryotic species with distnctions in structure being made at appropriate times.

The chromosome of *Escherichia coli,* and *Salmonella typhimurium,* are probably the most thoroughly mapped genomes of any cellular organism. Hundreds of identifiable genes are located in a single circular linkage group. The entire genome of a bacterium is made up of a single chromosome. Occasionally, a single bacterial cell may have more than one chromosome at a given time. This should not be taken as the cell having two chromosomes. This represents a stage wherein the chromosome has replicated and formed two units, but the cell is yet to divide.

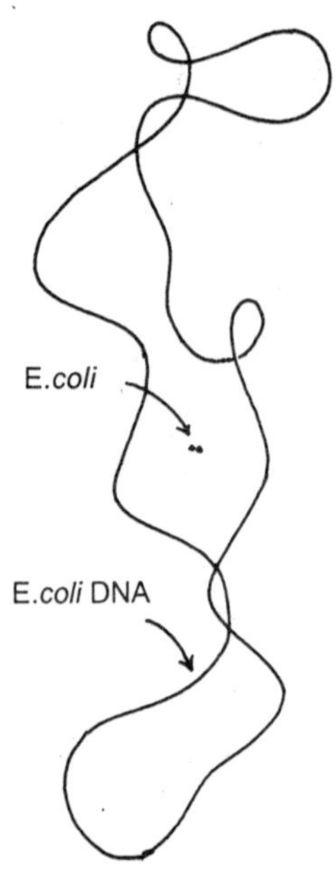

Fig. 4.1 Genome in Prokaryotes

Schematic diagram showing the relative sizes of E.coli and its DNA molecule

The chromosome of *E.coli* is a single, large double helix which is circular in outline. The DNA molecule is supercolied. It (DNA) has no association of histones as in eykaryotic chromosomes. However, small amounts of basic proteins have been found to be associated with the isolated DNA. It is quite likely that these might prevent some sections of the chromosome from transcription thus functioning like repressor proteins. Thus, it can be argued that the absence of DNA-histone association may not be such a fundamental point of difference between prokaryotic and eukaryotic DNA.

In *E.coli,* the total length of the circular DNA is 1300μm, whereas cylindrical bacteria have a maximum diameter of 1μm only. This clearly indicates that the chromosome must be highly folded in the cell.

When the DNA of *E.coli* is isolated by techniques that avoid both DNA breakage and protein denaturation, a highly compact structure called *nucleoid* can be seen. This structure contains a DNA molecule, a fixed amount of protein and variable amounts of RNA. As has been pointed out already, the proteins might play a part in repression of transcription.

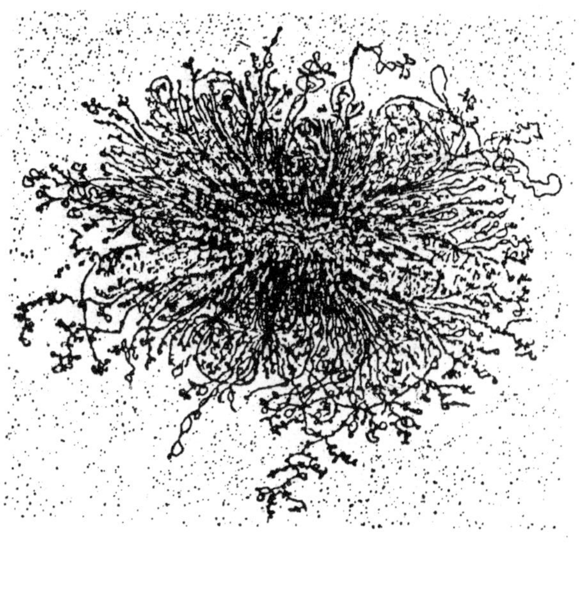

Fig. 4.2 Genome in Prokaryotes
An electron micrograph of an E.coli chromosome showing the multiple loops merging from a central region

The DNA molecule is about 1000 to 1100µm in length with a diameter of 20 Å to 40 Å. Such a molecule has more than 3.235 x 10^6 nucleotide pairs (each pairs is 3.4 Å long and there are 2941 base pairs per micrometer). The molecule has a molecular weight of 2.5 x 10^9 daltons (one dalton is equivalent to the weight of one hydrogen atom). This entire molecule is limited in the nuclear area of the cell which is no more than one micron and has no limiting boundary (such as a nuclear envelop), hence the molecule must be enormously folded.

It has been pointed above, that a small amount of RNA and basic proteins are associated with the DNA molecule. But the fact that the action of RNAse or protease does not alter the basic structure of DNA, indicates that neither RNA nor proteins are necessary for the basic circular natrue of the chromosome.

Detailed Structure of the Chromosome

The chromosome (DNA) of *E.coli* on isolation appears to be a highly folded compact structure. These folds appear to be highly regular and not random. If they were to be random, replication and proper seggregation would not be possible. The chromosome obviously has a high level of organization. Two features of the chromosome are noteworthy. These are – (1) The DNA is arranged in a series of loops and

(2) Each loop is supercoiled. A dense region containing membranous material is seen in the central part. Some people think it to be an artefact resulting out of the isolation procedure. But it is quite likely that it represents fragments (measosomes) of the plasma membrane to which the chromosome is attached in the intact cell. It is possible that enzymes required for the replication are located at this point of contact (with the membrane). Mesosomes do play a role in the separation of the replicated molecules of DNA during cell division. See chapter 2 reproduction in bacteria for more details).

The DNA has four usual bases – adenine, guanine, thymine and cytosine. In addition, the genome of bacteria may also have some amounts of methylated bases such as *6-methylaminopurine* and *5-methylcytosine.*

Introduction of a single strand break into the supercoil by a DNAse enzyme results in an abrupt change from supercoiled to non-supercoiled condition. This is because of free rotation about the opposing sugar phosphate bond. If the nucleoid, however, is nicked (cut), it does not become an open circle. But with each nick it passes through 40–50 intermediate stages in which the supercoiled loops turn into open loops one by one. This is a clear pointer to the fact that a single strand break does not cause free rotation of the entire molecule. The DNA is assumed to be fixed to the proteins in such a way that does not allow overlapping of the individual loops.

The enzyme (DNA gyrase) that brings about DNA replication is also responsible for supercoiling. This can be proved by the addition of *coumermycin,* an inhibitor of DNA gyrase, to the culture medium containing *E.coli* cells. The chromosomes of *E.coli* quickly lose supercoiling (on addition of coumerycin). In addition to DNAgyrase, another enzyme *topoisomerase* I, has the ability of unwinding the supercoils. Hence, it is believed that the two enzymes –*DNA gyrase* and *topoisomerase.* I act in a coordinated fashion and regulate the degree coiling. *In vitro* experiments have shown that purified topoisomerase I enzyme relases supercoiling to a non-supercoiled convalent circle. Mutants of *E.coli* lacking *topiosomerase* I have been seen to possess increased supercoiling to the extent of 32 percent more than the normal.

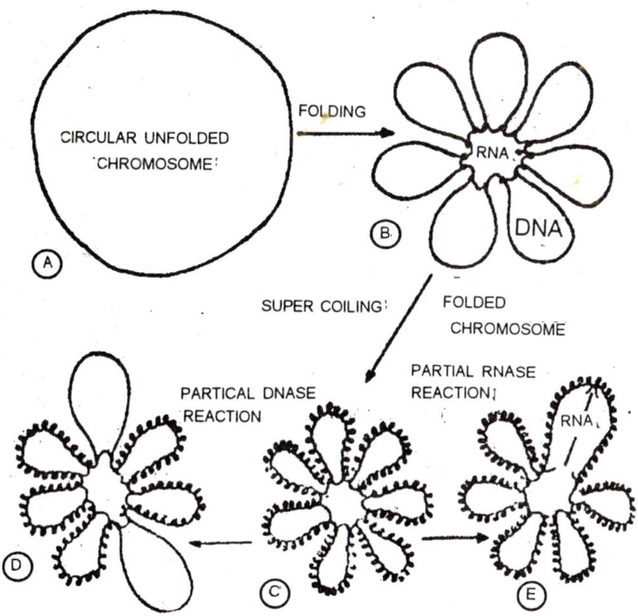

Fig. 4.3 Genome in Prokaryotes
Model of bacterial chromosome

A. Circular unfolded chromosome
B. Folded chromosome
C. Folded and supercoiled chromosome
D. Supercoiling relaxed into loops by DNAse nicks
E. RNAse cuts one RNA molecule

Some interesting observations have been made on these mutants lacking in topoisomerase. I. In *in vitro* conditions, the mutants undergo a further secondary mutation in or near the gyrase gene which seems to reduce the gyrase activity and return the supercoiling to near normal state.

Other Associated Structures of Chromosomes

The isolated genome of *E.coli* has about 30 percent by weight RNA and 10 percent protein. The RNA is largely a newly transcribed mRNA, tRNA or ribosomal RNA which is yet to be released into the cell machinery. A small amount of additional RNA in this is responsible for the folded character of the genome. RNAse activity can partially unfold the genome.

Most of the proteins have been identified as RNA polymerase engaged in transcription. This is quite understandable as a metabolically active bacterial cell would be virtually transcribing throughout its life cycle. A small portion of the protein represents the mesosome.

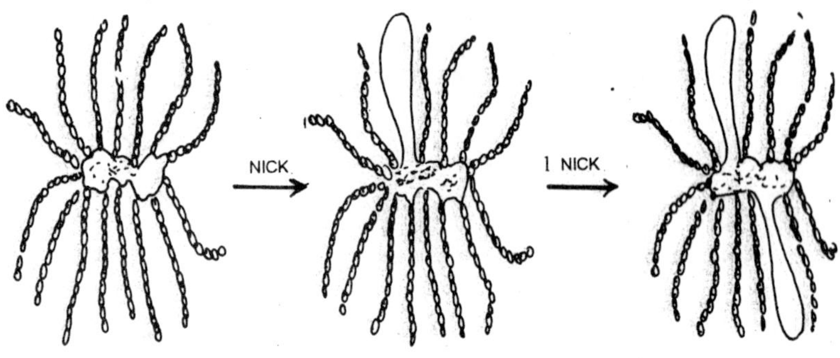

Fig. 4.4 Genome in Prokaryotes

A schematic drawing of the highly folded supercoiled *E.coli* chromosome, showing only 15 of the 40-50 loops attached to proteins (stippled region) of unknown organization, and the opening of a loop by a single-strand break (nick)

Chromosome of Other Bacteria

The thermophilic bacterium *Thermolplasma acidophilum,* possesses a chromosome that is quite unusual for a prokaryote. The genome is associated with a histone like protein which can complex with about 20% of the chromosome. The purpose of such an association is to maintain the stability of DNA during unfavourable (denaturing) environmental conditions.

The chromosome of *Bacillus subtilis,* is circular and has approximately the same dimensions and molecular weight like that of *E.coli.* There are suggestions that the genome of *B.subtils* replicates as a linear molecule rather than as a circular molecule. *Streptomyces coelicolor* and *Salmonella typhimurium* are also known to possess circular genome with the latter acomodating 323 identified genes. Species of *M.mycoplasma* have a small circular chromosome of about 265μ in length. *Haemophilus infleuzae,* the whooping cought causing bacterium, has a chromosome as long as 865μm which often appears linear with two free ends. Whether these ends really represent the points of a linear molecule or represent the ruptures of a circle during extraction, it is difficult to say.

PLASMIDS

The bacterial cell has an additional fragment of nucleic acids in addition to its normal chromosome. This additional genetic material was given the name episome (Jacob, Sccaeffer and Woolmann, 1960). Episome can exist either independently or it can integrate itself with the bacterial chromosome. Plasmids, however, are always independent and never integrate with the bacterial chromosome. Plasmids control a number of traits in bacteria such as –toxin production, antibiotic resistance, pilus production etc. Some people regard that episomes and plasmids are synonymus.

The word plasmid was first used by J.Lederberg (1952). However, the term is restricted only to bacterial cells where circular DNA molecules exist independent of the main genome.

Plasmids are circular double stranded molecules of DNA, and exist independently of chromosomal DNA in bacterial cells. They undergo replication during cell division and are carried to both the daughter cells. The plasmid has a right handed superhelical coil during the resting stage. There is one super helical turn for every 400–600 base pairs. The twisted conformation is known as **covalently closed DNA (CCD).** Cleavage results in the conversion of CCD to open circular form.

The following are some of the properties of the plasmids.

1. They are made up of DNA (double stranded).
2. They are capable of independent replication.
3. Some plasmids control bacterial conjugation.
4. Capable of reversible insertion into bacterial chromosome (only some plasmids).
5. They have the ability to get transfered from one bacterial cell to the other

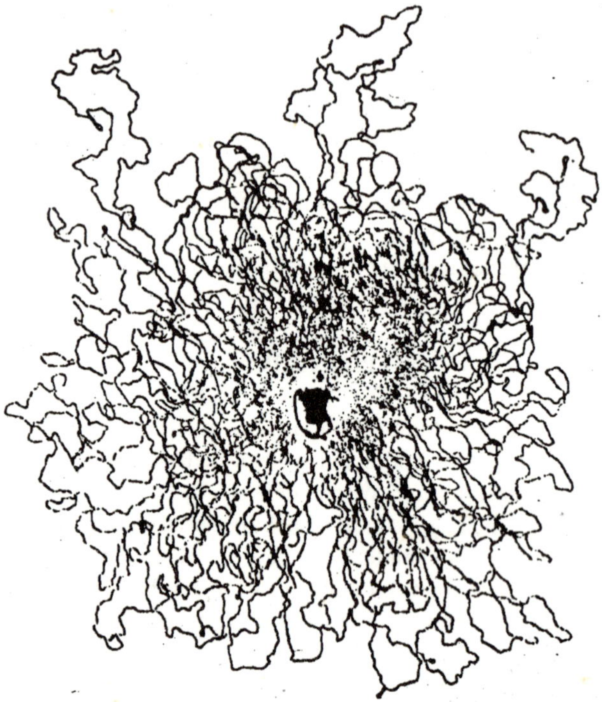

Fig. 4.5 Genome in Prokaryotes
An electron micrograph of the chromosome of the bacterium *Hemophilus infleunzae*

Plasmids are refered to as *conjugative* and *non-conjugative* based on the ability/ inability to promote their transfer during conjugation.

Classification of Plasmids

Basically plasmids are classified into three categories –

1. The F factor or sex factor.
2. The R Plasmid or resistance transfer factor (RTF), and
3. Col factor. There are some plasmids which may share the properties of more than one type of plasmid. Some bacteriologists recognise more than the three types based on some other properties.

F factor

The F factor plasmid confers the ability to the bacterial chromosome to get transferred to another cell (recepient). Hence, the donor is said to be F^+. F factor integrates with the bacterial genome facilitating its transfer. Because of this integration, it is often referred to as an episome.

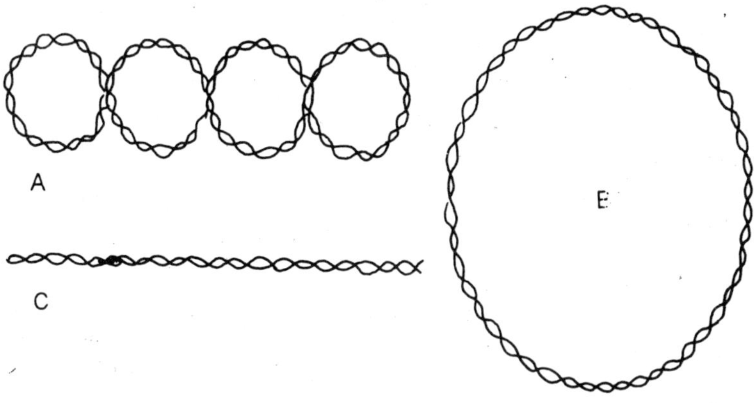

Fig. 4.6 Genome in Prokaryotes
Plasmid DNA, **A.** Supercoiled form, **B.** Open nicked circle, **C.** Linear duplex

R Plasmid
These plasmids, when present, confer on the bacterial cell resistance to several antibiotics such as streptomycin, tetracycline, chloramphenicol, sulphonamid etc. R plasmid mediated antibiotic resistance was first discovered in Japan in 1956, when *Shigella* became resistant to several antibiotics at one step. This was due to the transfer of R plasmid from one strain to the other resulting in not only antibiotic resistance but also the ability of a donor (it can donate R factor to other sensitive strains making them resistant).

R plasmid is a small extrachromosomal DNA ring. Sometimes, the R plasmid has two parts – (a) Part havin drug resistant genes and (b) Part having plasmid transfer genes.

The col factor (Plasmid)
Certain bacteria like *Escherichia coli, Shigella, Salmonella* etc., produce toxins called *colicins.* These are toxic proteins and can kill other bacteria. The ability to produce *colicin* rests in the plasmid called col plasmid or *colicinogenic factor.* Col factors resemble phages since they cause the death of bacteria, but they differ from phages as they are not released out during the lysis of the bacterial cell. Like the F factors, colplasmids also can be transferred from one bacterial cell to the other.

The number of plasmids per bacterial cell varies. It may be *a single copy* or *multiple copies.* Multiple copy plasmids which replicate independently have been used in gene cloning experiments. A plasmid is cleaved and a desired gene is inserted into the plasmid DNA. The plasmid DNA during its replication produces multiple copies of foreign DNA. Such plasmids are called vectors.

GENOME OF CYANOBACTERIA

In cyanobacteria, the nucleoplasmic region can be identified by its low electron density when observed under electron microscope. The DNA fibres are organized into a complex helical and folded structure and are distributed uniformly throughout the centroplasm. It is quite possible that DNA is associated with histione like proteins and RNA (Fay, 1983).

The size of the genome varies widely in cyanobacteria having a molecular weight ranging between 1.6 x 10^9 and 8.6 x 10^9 daltons. Most of the unicellular cyanobacteria possess genomes of about 1.6 x 10^9 to 2.7 x 10^9 daltons. This compares favourably with similar genomes found in other prokaryotes. Some filamentous cyanobacteria, however, have larger genomes. The distribution of genome sizes in cyanobacteria falls into four categories – 2.2, 3.6, 5.0 and 7.4 x 10^9 daltons. Members with a more complex morphology obviously possess larger genomes than those with a simple structure. There is a theory that the modern day prokaryotes evolved from a primitive bacterium with a small genome size of about 1.2 x 10^9 daltons. Replication of this genome during the course of evolution is supposed to have given rise to simple integer multiples seen in modern bacterial genome.

THE VIRAL GENOME

The core of the virion is made up of nucleic acids. A virus has only DNA or RNA, never both together. Within these parameters, however, considerable diversity exists. The nucleic acids may be double stranded, single stranded, linear or circular. Some have plus polarity; others may have minus polarity. Four types of nucleic acids are found in viruses with reference to the number of strands. These are –

1. Single stranded DNA — (ss DNA) e.g., Colliphage virus
2. Double stranded DNA (ds DNA) e.g., Herpes virus
3. Single stranded RNA (ssRNA) e.g., TMV
4. Double stranded RNA (dsRNA) e.g., Reovirus

Single stranded DNA is found in ϕX174 virus. It was first discovered by Sinscheimer *et al,* in 1950's. ssDNA may be linear (Paroviruses), or circular (ϕX174 virus). The ssDNA becomes double stranded during replication and at that time it is known as the replicative form.

Double stranded DNA is found in a number of animal viruses and bacteriophages. It has a variety of forms – linear (bacteriophages), cross linked (vaccinia virus) or closed circular duplex as in papova Viruses.

Single stranded RNA is found in a variety of animal viruses and icosahedral plant viruses. The strand may be plus (infectious), as in RNA bacteriophages, togaviruses etc. or minus (non infectious), as in rhabdoviruses and paramyxoviruses. Plus ssRNA acts directly as mRNA and translates proteins on the ribosomes of bacteria, whereas minus ssRNA first transcribes an mRNA and then it (mRNA) translates proteins (hence called non-infectious).

Double stranded RNA is found in animal viruses like reovirus, blue tongue virus etc. The dsRNA has 10 or more segments.

The nuclei acid core of Viruses may be summarised as follows –

1. Plant Viruses have only RNA (ss or ds), with the exception of cauliflower mosaic Viruses (DNA virus).
2. Animal Viruses have RNA (ss or ds), and DNA (only ds no ss).
3. Bacteriophages have DNA (ss or ds) or RNA (ss or ds). However, most phages have DNA Viruses.

5

REPLICATION OF DNA
(*In Prokaryotes*)

The DNA molecule regulates all the activities in the cell which includes both metabolism as well as reproduction. Because of its unique structure, it is able to control the synthesis of other macromolecules of the cell. At the same time when the cell reproduces, the DNA also should duplicate itself to be able to get distributed equally to the daughter cells. These two unique properties of the DNA molecule can be called *Heterocatalytic* and *Auto catalytic* functions. In the heterocatalytic function, DNA regulates the synthesis of other molecules like proteins. In the autocatalytic function which is perhaps very unique to DNA, it is able to make copies of itself in a precise and accurate manner. It is this ability of DNA that is responsible for the maintenance of life and inheritance of characters from one generation to another generation.

In this chapter, the various aspects of DNA replication in prokaryotes is discussed together with a brief account of the enzyme responsible for replication.

Types of Replication

Three different types of replication have been proposed at the molecular level for DNA. These are –
Semiconservative, Conservative and Dispersive.

1. Semiconservative

In this model of replication, the DNA helix unwinds but there is no rupture of a separated polynucleotide strands. Thus, the primary helices of DNA are not completely changed. According to this mechanism, the unwinding of the original strand would be followed by the synthesis of complementary strands for each of the two separated original polynucleotide strands. This mechanism was proposed by Crick and Watson, the discoverers of the molecular structure of DNA. In this method of replication, after replication of the original molecule, two molecules of DNA are formed. Each one will have one parental strand and one newly synthesised strand. Thus, in the semiconservative process, after replication the molecules of the next generation can be regarded as hybrids. The mechanism is called semiconservative because of the two strands of the parental molecule, only one strand remains in the molecule whereas the other goes to another molecule.

2. Conservative

In this mechanism, the original molecule is completely conserved and thus forms a part of the second generation molecule. Both the strands of the parental helix will synthesise an entirely new molecule, the new molecule synthesized would have both of its strands produced newly. After one generation of application, 50% of the molecules will be parental and 50% newly produced molecules. This is called conservative, because the original combination of both the parental strands will remain together and will not be separated.

3. Dispersive

In this method of replication, the parental DNA helix breaks up at several points forming many fragments. Each fragment of the helix would replicate and recombine at random to produce two daughter molecules of DNA. This is called dispersive because the original combination of the strands is completely disturbed as they get fragmented into bits.

Evidence for the Semiconservative Mode of Replication

Of the three types of replication mentioned above, there are overwhelming evidence to support the semiconservative mode of replication. Taylor *et al* (1957), demonstrated by autoradiography that the two chromatids during prohpase have one strand old and one strand new material. This clearly indicates that the replication is semiconservative. The following three experiments have clearly given experimental evidence for the semiconservative mode of replication.

1. Meselson and Sthal's Experiment

Meselson and Stahl (1958), conducted experiments on *E.coli* and demonstrated that the semiconservative mode of replication is correct. In their experiments, they used the following two basic principles.

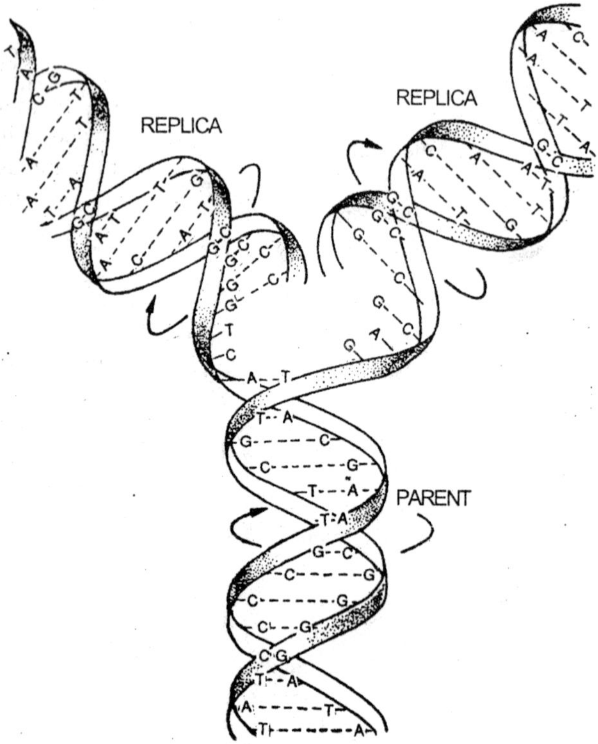

Fig. 5.1 Replication of DNA
Model of replication of DNA

(i) They cultured *E.coli* on a culture medium containing N^{15} (N^{15} is an isotope while normal nitrogen is N^{14}). The bacteria replicated in this medium for a number of generations and as a result, both the strands of the DNA molecule now contain N^{15}. In other words, the DNA of the bacterial populaton was now labelled with heavy nitrogen (N^{15}). The labelled bacteria were then transferred to a culture medium containing N^{14} and are allowed to replicate. After the first generation of replication, of the molecules of DNA, each one possesed one strand with N^{15} and the other strand with N^{14} (this should be the composition of the molecules if the mode of replication is semiconservative). In these molecules, the N^{15} strand represents a parental strand and the lighter N^{14} strand presents a newly synthesised strand. Thus, each molecule can be regarded as a hybrid DNA (having both N^{15} and N^{14}).

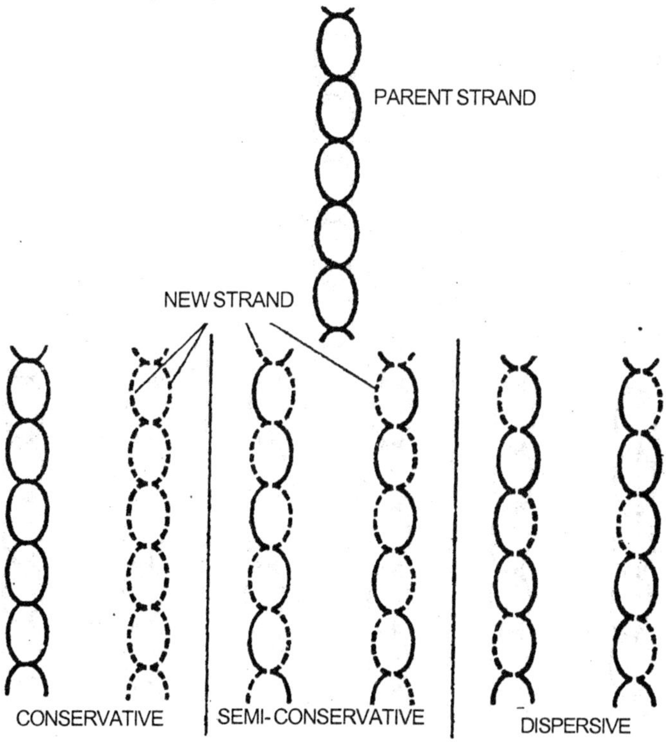

Fig. 5.2 Replication of DNA

Models of DNA replication - Conservative, semiconservative and dispersive

(ii) Meselson and Stahl used the principal of density gradient centrifugation and prepared gradients of Caesium chloride. This density gradient is based on the fact that when heavy salt solutions are subjected to ultra centrifugation, a continuous density gradient is set up (heavy would be at the bottom and lighter ones will be at the top with a range in between in the centrifuge tube). When a substance having a density within the range of this gradient is dissolved in salt solution, the substance will find place at its own level of density. Thus, with this technique even very slight differences in density can be detected.

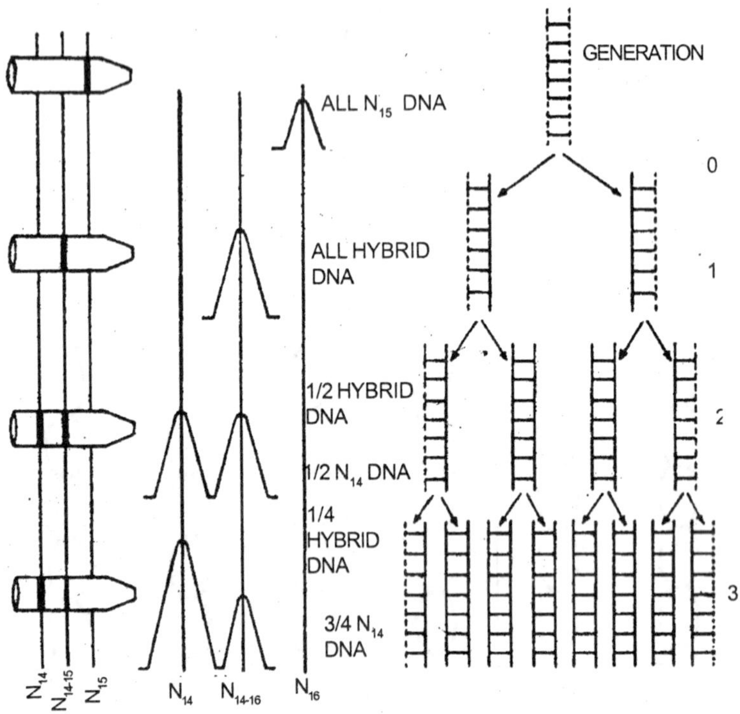

Fig. 5.3 Replication of DNA

Meselson and Stahl experiment in support of the semiconservative model of DNA replication

Based on the principles menthioned above, Messelson and Stahl subjected the bacterial cultures for density gradient centrifugation. At the first instance, N^{15} labelled bacteria were allowed to replicate on a medium containing N^{14}. After the first division, the centifugation showed the hybrid DNA had an intermediate density and formed a band inbetween N^{14} and N^{15} indicating clearly that each molecule contains one N^{14} and N^{15}. In the second generation two bands were found, one having normal N^{14} and the other having a hybrid (N^{14} and N^{15}). This is expected because the labelled nitrogen - N^{15} is getting reduced as the medium has only N^{14} and the original labelled N^{15} gradually gets decreased in the molecules. In the third generation, 3/4ths of the DNA and N^{14} and 1/4th of the DNA was hybrid. If the replication continues for some more generations N^{15} would disappear.

In the above experiment, if we assume that the replication is conservative there

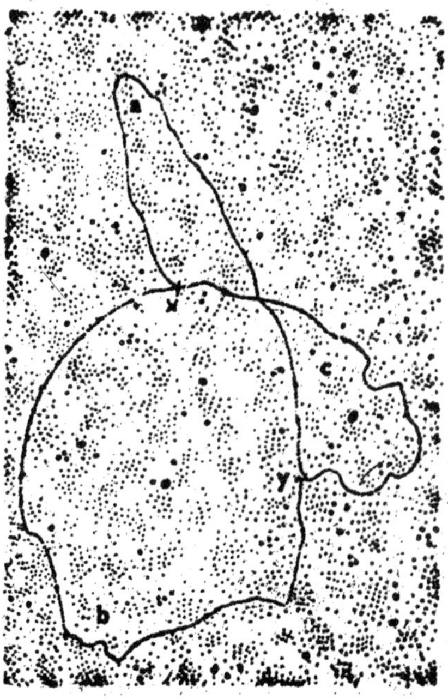

Fig. 5.4 Replication of DNA
Autoradiography of the single circular chromosome of *E. coli* labelled with tritiated thymidine-Cairans experiment

should be two bands in the centrifuge one with N^{15} and the other with N^{14} (newly formed molecule). On the other hand, if the mechanism is dispersive the molecule should spread all over the medium and should form only one band as both N^{14} and N^{15} in different combinations would be there in all the molecules. If the mechanism is semiconservative, the first generation would show only one layer having a mixture of N^{14} and N^{15}, and in the second generation there would, be two layers one N^{14} and the other N^{15} and N^{14}. Gradually, N^{14} would show greater band width and N^{15} would decrease. (Note. In the dispersive mode also only one band is formed in the first generation and it continues to be so for the second generation also. Where as in semiconservative mode, two bands are formed in the second generation).

The above explanation clearly demonstrates that the semiconservative mode of replication proposed by Watson and Crick is the actual method of replication.

2. Cairn's autoradiographic experiment

The semiconservative mode of DNA replication has been proved in the chromosomes of *E.coli* by autoradiographic experiments conducted by Cairns.

Autoradiography is based on the fact that decaying radioactive elements emit electrons which collide with the silver grains of a photographic plate and turn it black. Thus in a autoradiogram, black dots represent positions of decaying radioactive elements. DNA of an organism can be labelled selectively by growing cells in a medium containing radioactive thymidine. Radioactive thymidine is obtained by using H_3– a heavy isotope of hydrogen known as tritium. Cairns used tritiated thymidine as it selectively labels only DNA and not RNA, because thymidine base is found only in DNA and not in RNA. By growing *E.coli* in the culture medium containing tritiated thymidine, radioactivity was incorporated in the daughter DNA molecules. The material (DNA) was extracted and mounted on a photographic emulsion for autoradiography. The labelled DNA exposes the film producing a diffused profile of DNA molecule. The duplicating molecule showed a replication fork. After the replication, the radioactivity was found incorporated in only one strands of DNA and in the second generation, both the strands become labelled and this will be clearly seen on the photographic emulsion because of the presence of black dots.

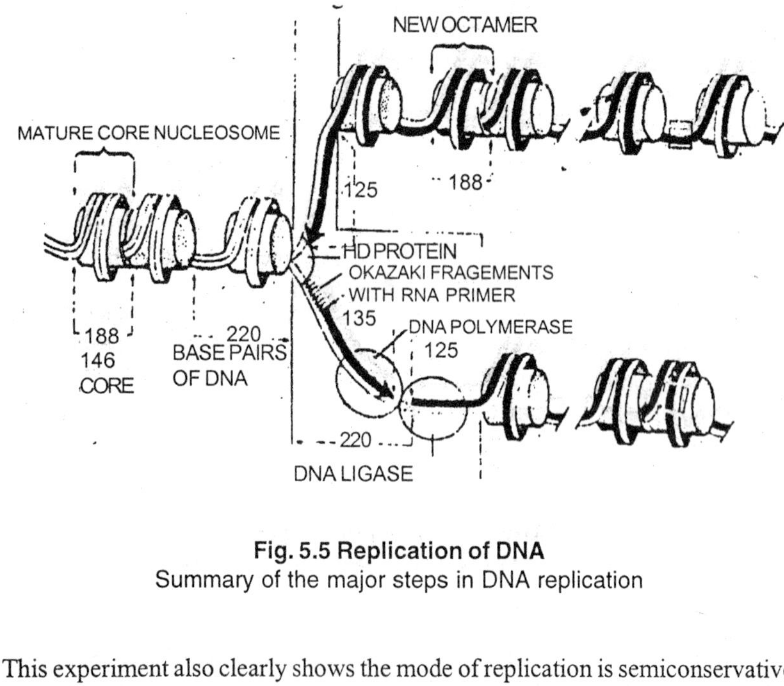

Fig. 5.5 Replication of DNA
Summary of the major steps in DNA replication

This experiment also clearly shows the mode of replication is semiconservative because after the first replication, each molecule will have only one labelledand the other non-tritiated strand. In the second generation, both the strands are tritiated in some molecules and in the other hybrid – one tritiated other non tritiated strands occur. and the other non-tritiated strand. In the second generation, both the strands are tritiated in some molecules and in the other hybrid – one tritiated other non tritiated strands occur.

Mechanism of replication

The mechanism of DNA replication in prokaryotes is as follows –

1. *Stage of replication during cell division* – It has been observed that

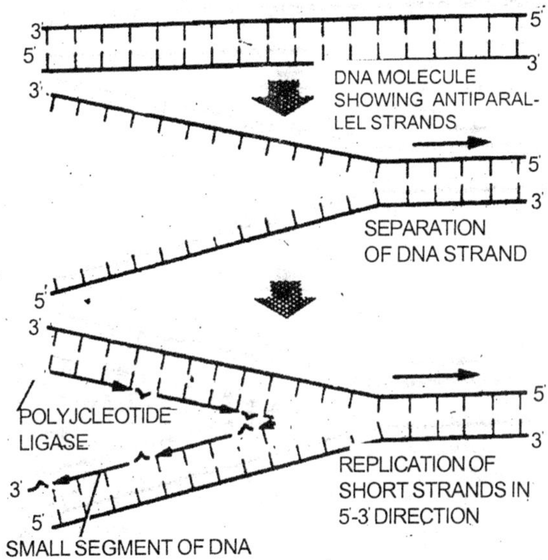

Fig. 5.6 Replication of DNA
Diagrammatic representation of discontinuous synthesis of DNA

replication takes place during interphase between two mitotic cycles. During interphase, the amount of DNA doubles due to synthesis. However, DNA synthesis does not take place during the entire interphase. It is confined to the S phase.

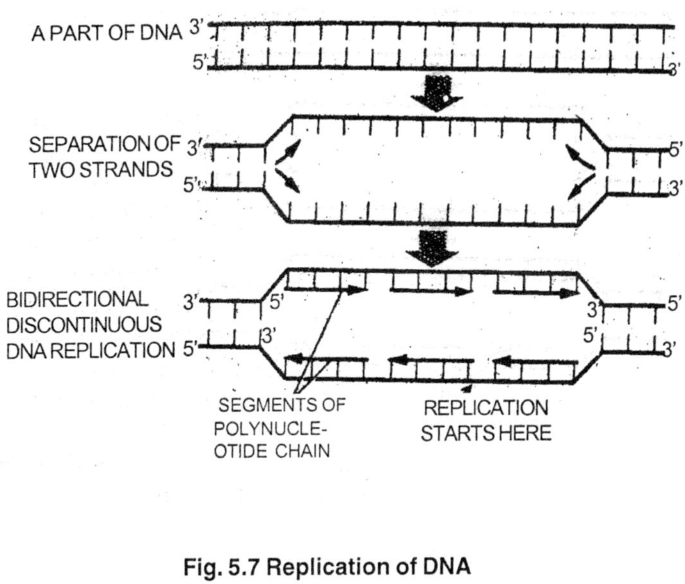

Fig. 5.7 Replication of DNA
Diagrammatic showing bidirectional DNA replication

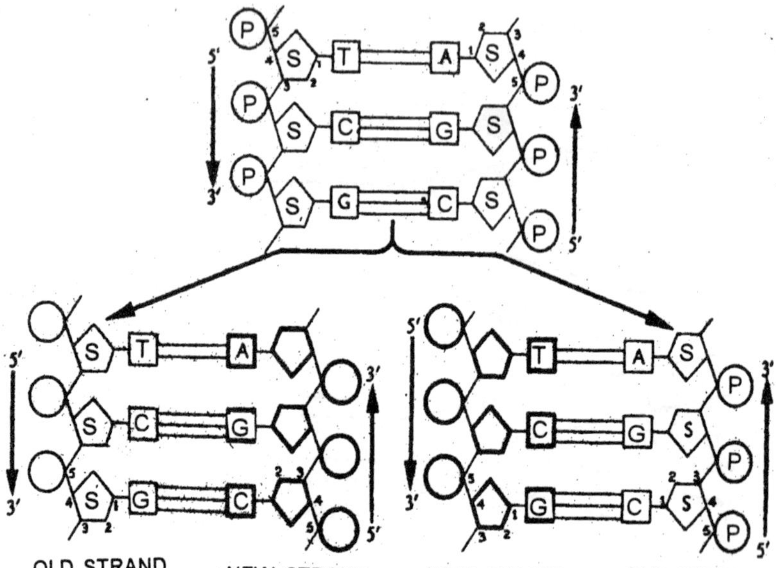

Fig. 5.8 Replication of DNA

Replication of DNA molecule. The two strands of DNA separate and each of them synthesizes its counterpart, thereby forming two similar molecules of DNA

2. *DNA Synthesising Genes* – In *E.coli* the overall process of replication requires the products of the genes *dnaA, dnaB, dnaC-D, dnaE,* and *dnaG.* The gene products of AC and D are required for the initiation of fresh replication and the products of E and G are required for the fork movement for replication. The product of G appears to be necessary for initiating the synthesis of DNA fragments. The *E.coli* also possesses a gene for the origin of DNA synthesis which is about 74 minutes from the beginning. There is also a gene for the termination of the replication. There are additional genes to control the synthesis of the three DNA polymerase enzymes. PolA and polB regulate the synthesis of polymerase I and polymerase II enzymes while G and E regulates polymerase III enzyme.

3. *Initiation of replication with a nick* – Unlike in the case of eukaryotes

when the molecules are linear in prokaryotes the DNA molecules are circular. As such, there has to be break in the circle to begin replication. The DNA consists of many replicating units or replicons. According to one hypothesis, replication is initiated only after a break or nick in one of the two parental strands. The cut is presumed to take place in the helix at a specific initiation point. This cut or incision is made by a specific endonuclease enzyme.

4. *Origin of replication* – Replication does not begin randomly, it always begins with a specific site called ori site and is regulated by a gene and its product. It has been discovered that in *E.coli*, the origin of replication point as about 74 minutes on the map.

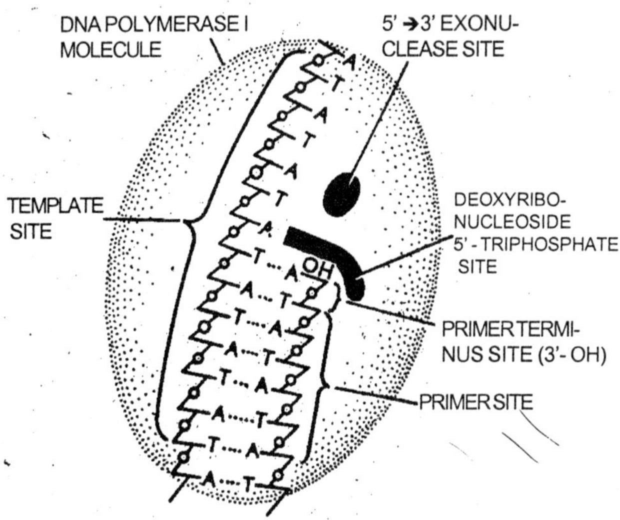

Fig. 5.9 Replication of DNA

A molecule of DNA polymerase-I showing specific binding sites

5. *Role of plasma membrane* – There are evidences to indicate that *E.coli* consists of an enzyme complex at the point of attachment of DNA to the membrane. This point of attachment is known as the replicating point. The DNA thread moves through the replicating point and duplicates itself in the process. In this process, the replication is exactly opposite of what is seen in eukaryotes where it is the enzyme that moves along the strand. The exact relationship between the plasma membrane and the DNA is not known. Quite probably, the membrane lipids may have some role to play in the process of initiation. Polymerase II and polymerase III enzymes are seen at the point of attachment.

6. *Strand unwinding* – The DNA helix has to first unwind in order to separate the two strands for beginning the replication. A study of the unwinding process in T_2 bacteriophages shows that each DNA molecule has to undergo 20,000 rotations in unwinding during replication. As the molecule itself consists of 2 lakh nucleotide pairs, in these phages, replication must go on simultaneously along with unwinding.

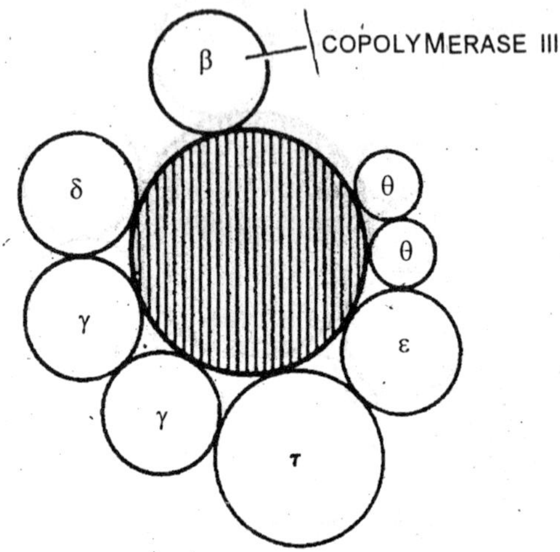

Fig. 5.10 Replication of DNA
DNA polymerase-III holoenzyme, it consists of subunits α, β, γ, ε, θ, δ and ε

The unwinding of the helix is controlled by enzyme proteins commonly referred to as DNA unwinding proteins. These proteins selectively bind to the single strands of DNA and promote unwinding of the coil. The region of the double helixwhere the proteins bind forms a Y shaped structure called the replication fork. It hasbeen foundout that nearly 200 molecules of the proteins are found at each fork. Chemical analysis of the DNA unwinding protein shows that each molecule is a etramer having 4 sub units. Experimental evidence in *E.coli* has shown that the addition of DNA unwinding proteins to the medium increases the phase of the unwinding of the DNA molecule.

7. *Role of Swiveling Protein* – As the replication fork moves downwards along the parental strands, unwinding begins. For every ten nucleotides, there will be one rotation of unwinding. As the original configuration of the molecules is in the form of a wound, helix in circular DNA molecules present in prokaryotes untwisting or unwinding imposes a strain. To some extent, this strain can be received by additional twisting called super twisting in the unreplicated part of the DNA. However, this super twisting has a limit and ultimately even the supertwisting should be unwound for the replication fork to continue. The relieving of the strain imposed by unwinding in brought about by proteins called superhelix relaxing proteins. These proteins introduce small cuts in the non-replicating region of DNA thereby making one strand to rotate upon the other. Thus, relieving the strain. The breaks can then be healed by ligation.

8. *DNA Template* – To synthesise a fresh strand of DNAS apart from the enzymes, the most important requirement are the nucleotides and a DNA template. The 4 nucleotides are adenine, guanine, cytosine and thymine. The previously existing DNA can serve as a template on which can be assembled fresh nucleotides.
The template DNA can be either single stranded or double stranded. Single stranded DNA is found naturally in some bacteriophages. Template DNA acts as a mould on which new strand of DNA is synthesised by complementary base pairing.

9. *RNA Primer* – DNA synthesis cannot be initiated on a DNA primer. DNA polymerase enzymes can initiate DNA synthesis only if there is a RNA primer. In other words, synthesis of a fresh strand can begin only if there is a small piece of a strand already. This small piece which is RNA is called RNA primer. The RNA primer is a short polynucleotide chain synthesised by the DNA template close to the point of replication. In all probability, this is catalysed by a modified RNA polymerase enzyme. It has been noticed that in *E.coli*, RNA polymerase is necessary to begin the replication.

10. *Elongation of chain* – Synthesis of new DNA strand takes place by the addition of fresh nucleotides to the 3'-OH group of the last nucleotide in the RNA primer. This synthesis takes place in 5'→3' direction and the enzyme that catalyses this is DNA polymerase III.

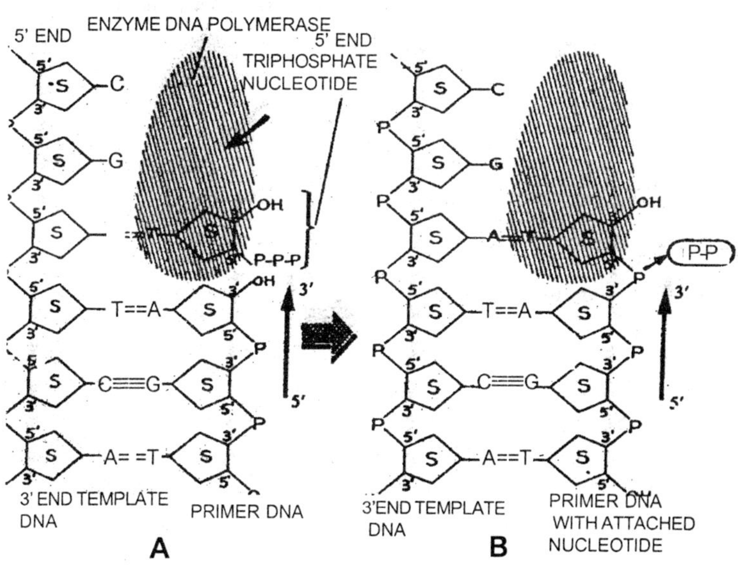

Fig. 5.11 Replication of DNA
Diagram showing action of DNA polymerase in DNA replication

The newly synthesised DNA chain has an RNA primer attached to its 5' end. After the synthesis of the chain, the RNA primer is hydrolysed by the exonuclease activity of DNA polymerase I. The resulting gap left by the RNA primer is filled by DNA nucleotides by the catalytic activity of DNA polymerase I. The freshly formed nucleotide chain is joined to the existing chain by the lygase enzyme.

Several opinions have been put forward to account for the presence of an RNA primer to initiate the synthesis of DNA chain. According to one opinion, if any errors occur during the initial synthesis, it can be eliminated when the RNA primer is removed.

11. *Direction of Replication* - It has been seen that in prokaryotes the replication may proceed in one or both directions from the point of origin. Thus, replication may be unidirectional or bidirectional. Bidirectional replication has been seen in *E.coli, Bacillus subtilis, Salmolella, Typhimirum, E.coli* phages etc. Unidirectional replication has been reported is bacteriophages P2 and 186, and also in the mit DNA of mouse LD cells. Bidirectional replication has an advantage in that, the origin is completely duplicated before the complication of replication.

12. *Formation of Replicating Forks* – At the point where the two strands get separated to initiate replication, a replicative fork is found. The shape of the replicating fork is in the form of the English alphabet Y. When bidirectional replication takes place, the separated strands between the two forks appear like the bubble or a convex lens under the electronic microscope. The speed of the movement of the two forks need not be the same.

    In organisms like *E.coli,* where the chromosome is circular, there will be only two replicating forks, where as in eukaryotes there could be several thousand replicating forks. These large number of forks is due to the fact that replication begins simultaneously at many points along the length of DNA. Occasionally, however, even in bacteria when replication is going on at one point another replication may start at another point. This will facilitate in completing the replication in shorter period.

13. *Discontinuous Replication* (Okazaki fragments) - The duplication of the strand of DNA takes place by the movement of replication forks. Replication of the two parental strands will take place at the same time as the fork moves along. However, all the DNA polymerase enzymes can enhance the length of the strands only in the 5'→3' direction. The strands of DNA on the other hand, polarise in opposite directions. This naturally poses a question as to how can DNA polymerase bring about simultaneous

replication. While the enzyme moves forward on one DNA strand, simultaneously it cannot move backward on the other.

Okazaki et al (1968), have provided an answer to the above problem. According to them only one strand 3'→5' is continuously replicated, while the other 5'→3' replicates in a discontinuous manner and synthesis, only short fragments of DNA. These short fragments have been called Okazaki fragments. The direction of synthesis of Okazaki fragments in opposite to the movement of the replicating fork. Eventually, Okazaki fragments are joined together by the enzyme polynucleotidfe ligase. In prokaryotes, the Okazaki fragments are as long as 1,000–7,000 nucleotides.

According to another suggestion (Okazaki et al), both the strands may replicate discontinuously and the fragments formed may join up later.
Evidence obtained from *in vitro* culture of *E.coli,* have shown that both the strands are synthesisd discontinuously.

Enzymes involved in DNA Replication

1. **DNA Polymerases** – Kornberg (1957), isolated an enzyme from *E.coli* which had the ability to polymerise the nucleotides dATP, dTTP, dCTP and dGTP to form a polynucleotide in the presence of a primer and DNA template. This enzyme was capable of synthesizing new DNA strands by proceeding in 5'→3' direction. Subsequent works of DeLucia and Carirns (1969), Knippers (1970), and Gefter et al (1969), showed that in prokaryotic cells there are at least three DNA polymerases – *poly I, pol II,* and *poly III.* Kornberg's DNA polymerase is equivalent to *Pol I.*

Polymerase I – This enzyme does not seem to be necessary as bacterial mutants lacking this are able to synthesize DNA. *Poly I* seems to be taking part in DNA repair. Structurally, poll is a single polypeptide chain with a molecular weight of 1,09,000. Electron microscopic studies of the enzyme have shown that the enzyme has a number of functional sites –

(i) A template site for binding template DNA.
(ii) Primer site for binding primer RNA segment.
(iii) A primer terminus site for 3' - hydroxyl terminus of primer.
(iv) A triphosphate site - locus for incoming deoxy ribonucleotide 5' group.
(v) Exonuclease site for 5'→3' exonuclease activity.

Polymerase II - This is a single polypeptide chain with a molecular weight of 90,000. *E.coli* cell has about 40 molecules of *pol III.* It helps in 5'→3' polymerization.

Polymerase III - Discovered by Kornberg and Gefter (1972), it is the most active among all the three enzymes. The enzyme is chiefly responsible for DNA chain elongation. It has a molecular weight of 5,5000. The enzyme also acts as 5'→3' and 3'→5' exonuclease..

DNA ligase– Also known as polynucleotide ligase, this enzyme joins the two fragments by catalysing the synthesis of a phosphodiester bond between a 3'-OH group at the end of one chain and 5'→3' - OH group at the other end of the other chain. The enzyme is necessary to join Okazaki fragments which are produced after discontinuous replication.

The enzyme has been isolated from *E.coli*. It is made up of a single polypeptide chain and has a molecular weight of 70,000. Each cell of *E.coli* is known to posses 2000–4000 molecules of this enzyme. DNA ligase has the following functions.

1. Joins Okazaki fragments during DNA replication.
2. Helps in sealing single strand nicks in DNA duplex.
3. Joins free ends of DNA (linear molecule) to form circular chromosome.
4. Joins the segments of DNA during recombination during meiosis.

MODELS FOR DNA REPLICATION (IN PROKARYOTES)

Jaboc and Brenner's Model

Studies with phase contrast microscope have shown that in the dividing cell, the sister chromosomes get oriented equidistantly inside the cell. In eukaryotic cell, this separation is achieved by the mitotic spindle. In prokaryotes like bacteria such a mechanism is lacking. Jacob and Brenner have proposed a model to explain the replication and separation of daughter chromosomes. According to this model, the chromosome gets attached to the mesosome at the point of origin of replication. After the completion of replication, a new membrane attachment site is formed adjacent to the old one. The strand which has acted as a template remains a circle, while the newly formed one is broken. One of the free ends of the newly formed chromosome attaches to the new binding site. During replication, the replicating fork and the enzyme system remains stationary while the DNA molecule moves along and gets replicated. Simultaneously a new membrane is synthesized between the two and the two genomes get separated. The newly synthesized DNA becomes circular eventually.

Carin's Model

At a specific point (origin), the DNA breaks and replication begins and proceeds in two opposite directions. As the replication proceeds, the DNA strand unwinds. This unilateral unwinding (the other cannot unwind freely) creates a pressure and it is transfered to the unreplicated part resulting in supertwisting. This results in the stoppage of replication. To begin replication once again, a 'neck' is made in one of the strands by a swivelling protein. This facilitates rotation of the strands and the break is closed subsequently.

Yoshikawa Model

This is a modification of the Cairn's model. According to Yoshikawa, the newly formed DNA strands covalently join to the ends of original chromosome.

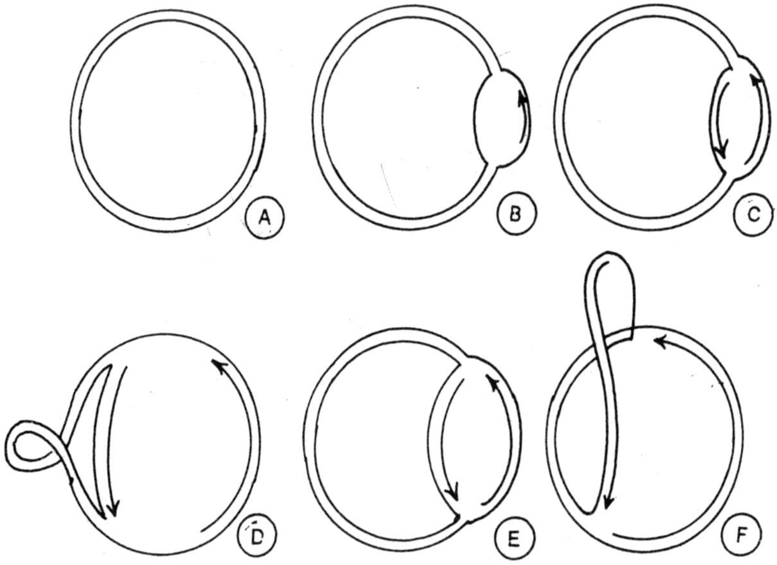

Fig. 5.12 Replication of DNA
Cairn's model for circular DNA replication

Rolling Circle Model

This is the model accepted for the replication of single stranded DNA in viruses (ϕ x174). The single stranded circular DNA becomes double stranded by the synthesis of another strand starting at the origin. One strand of this duplex ring is now cut at a point by an endonuclease. As a result of this, a linear strand with 3' and 5' end is created. The 3' end serves as a primer for the synthesis of a new DNA strand. The unbroken strand is used for this purpose; with the parental strand itself acting as a template. The 5' end of the strand becomes attached to the membrane. Meanwhile the original, unbroken single strand unwinds and

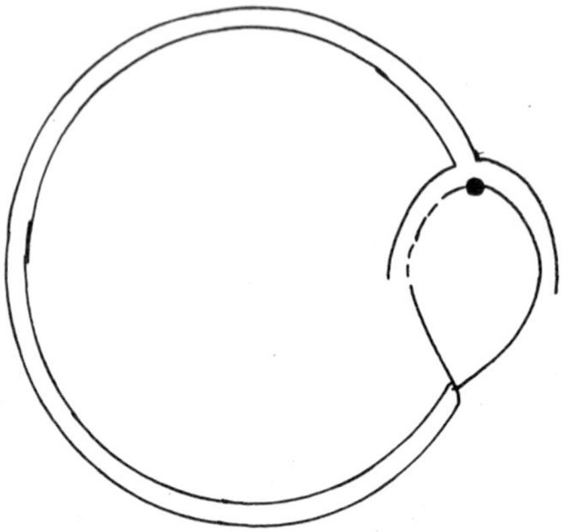

Fig. 5.13 Replication of DNA
Yoshikawa model of a circular DNA replication

separates leaving the tail attached to the membrane. The tail is cut off into a specific length – *Progeny rod.* The rod rolls and forms a circular molecule, with the enzyme ligase binding the two ends. The newly formed circular molecule which is double stranded can in turn become rolling circles. The genetic information is preserved in the single stranded template which remains circular always. New DNA molecules may be formed either from the single stranded original molecule or from the double stranded rolling circles. Evidence for the rolling circle model has been obtained from the replication of several viruses (M13, P2 T4, λ etc).

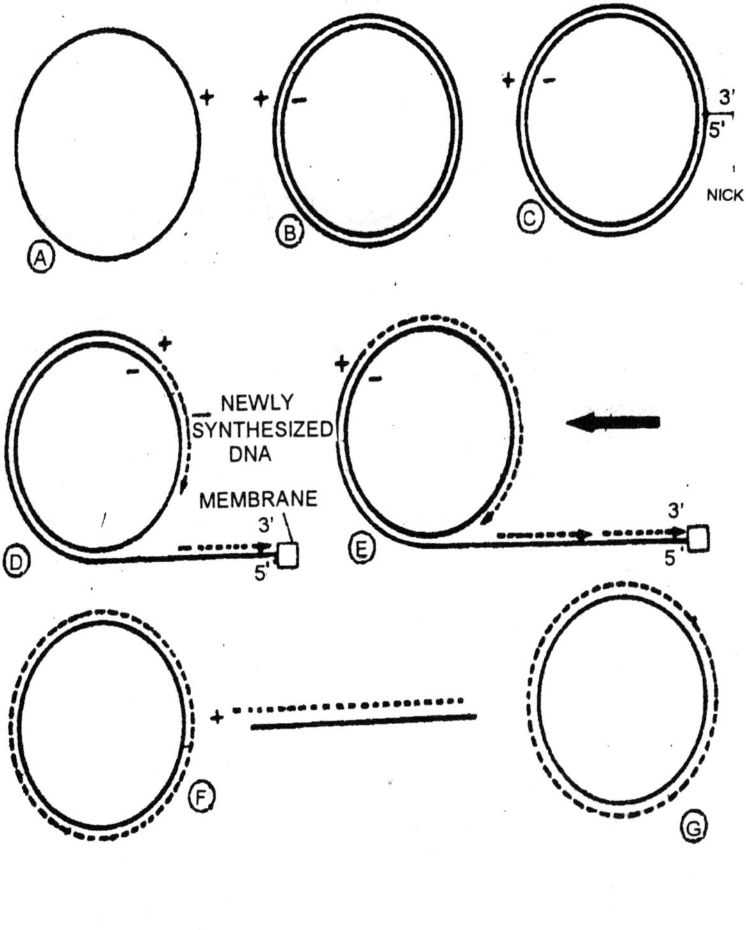

Fig. 5.14 Replication of DNA
Rolling circle model of a circular DNA replication

6

GENETIC CODE

Proteins are the fundamental biological constituents of the protoplasm. They form 15% of the body in living organisms and are important building material of the cells. They play a significiant role in the physical and chemical activities of the cells. They are involved in the structural and functional organisation of the cell. The structural proteins form the various cellular organelles. The functional proteins like enzymes and hormones control the metabolic, biosynthetic, biogenetic, bio-energetic, growth regulating, sensory and reproductive functions of the cell. The 'gene actions' take place by means of enzymes. The enzymes are the important catalysing agents in most biochemical reactions. Thus, proteins through their particular structure, composition and enzyme activity form the most important biochemical compounds through which 'hereditary traits' are expressed in the organism.

Proteins are giant molecules formed of polypeptide chains of several hundred amino acids. About 20 odd number of amino acids are known to occur in the cells. Amino acids are formed of a basic amino (NH_2), and an acidic carboxyl (COOH) groups. The variety of cells in an organism represent myriads of proteins. Different types of proteins are synthesised by different cells of the body. Different combinations of amino acids produce different proteins.

Synthesis of Protein

The process of protein synthesis is the most complicated biological activity. Basically, it is governed by the *genetic information* carried in the genes (DNA) of the chromosomes. Specific amino acids are linked in particular sequence or order to form a polypeptide chain of protein. The various cell organelles and chemical compounds involved in the synthetic proces are –

1. Deoxyribonucleic acid .... DNA
2. Three types of ribonucleic acids .... RNA
3. Amino acids

4. Enzymes and
5. Ribosomes

The discussion above will clearly reveal that proteins are the representative of the master molecule of the cell DNA to carry out and regulate all the metabolic activities. Proteins are important in another way also. They are important for the existence of DNA itself. A number of enzymes (proteins) are necessary for the replication of DNA itself. When proteins are of such fundamental biological importance, no wonder then the cell machinery to synthesize them (proteins) is not only complicated but precisely regulated.
Which molecule(s) present in the cell can regulate the synthesis of proteins? Obviously there cannot be another compound than DNA. The structure and properties of the DNA molecule are eminently suited to exercise such a regulation over the synthesis of proteins.

Having conceded that DNA directly is involved in the regulation of protein synthesis, we may now try to answer the two following questions – which aspect of protein structure is controlled by DNA? And how and by what mechanism DNA regulates the assembly of amino acids to synthesize the proteins?

Proteins are extremely complicated biological macromolecules. There are different dimensions for its structural complexity and integrity. These are theprimary, secondary, teritary and quaternary structures (for details – see the chapter on protein synthesis). DNA molecule specifies only the primary structure, while the other dimensions of protein structure like folding etc. is controlled by proeins themselves.

The primary structure of proteins involves the following aspects – quality, quantity and sequence of amino acids and it is this that is regulated by DNA.

What is the mechanism adopted by DNA to regulate the synthesis of proteins? Obviously, proteins cannot be synthesized *in situie* within the nucleus, because all the components needed for proteins–amino acids, ribosomes etc., are not present in the nucleus. It is not possible also for the DNA itself to migrate to cytoplasm to synthesize the proteins. The only alternative is for DNA to send a messenger or representative which carries the message to specify the quality, quantity and the sequence of amino acids to be asembled to produce the protein. This message is called the genetic code, because it specifies in genetic language the amino acids to be assembled.

The language of genetic code

A code is said to be a cryptic or hidden language, carrying a message and upon deciphering or translation, the message is understood. In the genetic code, the message is translated as the type of aminoacid to be assembled. What is the composition of the genetic code that comes out of DNA? To answer this, we must first understand the manner in which genetic information exists in the DNA molecule.

It is well known that eventhough the DNA molecule has three components namely sugar, phosphate and nitrogen bases, it is the last one that is unique to each DNA molecule in the sequence of their arrangement. The nitrogen bases vary from one DNA segment to another in arrangement, so the information is based upon the sequence of their arrangement. The sequence of nucleotides is then the code in which the DNA sends the message. How is the message sent? The DNA molecule replicates in a particular segment and the replicated segment in the form of RNA now called mRNA carries the genetic message in the form of nucleotide sequence (for details see protein synthesis). The sequence of nitrogen bases have been found to be identical to the linear sequence of amino acids in a peptide chain. In otherwords, there is colinearity between mRNA and polypeptide chain (see later in the chapter for details of colinearity).

It is thus clear that the genetic code is in the language of nitrogen bases and the reading of the code is in their sequence of arrangement.

Amino acids involved in protein synthesis

A large number of amino acids – almost upto 150 are found in nature. Of these, however, quite a number of them remain as free amino acids in the sense they do not join together to form a molecule of protein. It has been found out that only about a little more than 20 amino acids are protein amino acids in that all the known proteins are made up of only these amino acids. Some amino acids, however are found in almost all protein chains while a few others are found only in very few proteins. For example, cystine is present only in proteins which have a disulphide linkage. Hydroxylproline is another amino acid found only in few proteins.

The code is triplet (basis of crypto analysis)

A DNA molecule from every source contains only 4 types of nucleotides– adenine, guanine, cytosine and thymine. On the other hand the basic building blocks of proteins namely amino acids are more than 20 types. A basic problem that one has to analyse before finding about a genetic code is how many bases specify a unit of amino acis. Sevral opinions have been put forward in this regard.

Single code. According to one of the opinions, each nucleotide will code for one amino acid. But this would be insufficient to code for all the 20 amino acids as there are only 4 nucleotides. With 4 nucleotides only 4 amino acids can be coded. What about the rest?

Double code. This asumes that each code is formed of two letters i.e., 2 nucleotides. In other words, 2 nucleotides will specify one amino acid. In this type of a code, all the 20 amino acids cannot be coded but 16 codes (4 x 4) are possible. But there will be a number of ambiguous determination of amino acids.

Possible singlet, doublet and triplet codes of mRNA

Single Code (4 words)	Double (16 words)				Code Triple Code (64 words)			
					AAA	AAG	AAC	AAU
					AGA	AGG	AGC	AGU
					ACA	ACG	ACC	ACU
					AUA	AUG	AUC	AUU
					GAA	GAG	GAC	GAU
					GGA	GGG	GGC	GGU
					GCA	GCG	GCC	GCU
A	AA	AG	AC	AU	GUA	GUG	GUC	GUU
G	GA	GG	GC	GU	CAA	CAG	CAC	CAU
C	CA	CG	CC	CU	CGA	CGG	CGC	CGU
U	UA	UG	UC	UU	CCA	CCG	CCC	CCU
					CUA	CUG	CUC	CUU
					UAA	UAG	UAC	UAU
					UGA	UGG	UGC	UGU
					UCA	UCG	UCC	UCC
					UUA	UUG	UUC	UUC

Triple code. This was first suggested by the physicist, George Gamow (1954). According to this code, 3 bases (nucleotides) will code for one amino acid. In other words, each code will have 3 nucleotides which will specify the particular amino acid. These triplets of codes are given the name codons. With triplet codes, thepossibilities would be 64 (4x4x4) codons. This would be more than the number of amino acids (which are only 20). Obviously then each amino acid will have more than one code. Experimental evidence shows that the code is a triplet and about 61 of the 64 codes are used to specify amino acids during protein synthesis.

The maximum possible number of codons in the singlet, doublet and triplet codes

Type of code	Number of bases in codon	Number of codons	Ambiguous/ degenerate
Single code	1	4	Amgiguous
Double code	2	4x4=16	Amgiguous
Triplet code	3	4x4x4=65	Degenerate (more than required number)

Properties of code

1. The code is non-overlapping

The DNA molecule is a very long chain and has series of nucleotides. There is a gap between the nucleotides. Naturally a question arises that if three nucleotides constitute a codon will there be any gap between two successive codes or do the codes overlap. In otherwords, is the code overlapping or non overlapping? If the code is over lapping from where will the reading begin. Because the beginning of the code reading will alter the nucleotide combination and hence will alter amino acid which is specified by the code. When Gamow proposed the code he assumed it to be overlapping. Many people attempted to rationalise the excess number of codes i.e., 64 for 20 amino acids by assuming that the codes are overlapping and the codons can be reduced to about 20. But this involves a lot of problems. If we take for instance, the codes for lysine (AGG and AAA), and codes for phenyl alanine (UUU and UUC). The code letters of these two amino acids are mutual exclusive i.e., lysine forbidden combination of amino acids and proteins. But, on the other hand these amino acids are found in proteins proving that the genetic code is non-overlapping. Further, the overlapping code assumes that a nucleotide is read twice in different combinations to code for amino acids. If this is true, the entire precision of code would be disturbed. Further, any mutational change in an overlapping code would affect more than one code thus disturbing a number of amino acids.

Mutational studies have shown that the code is non-overlapping. In TMV (Tobacco Mosaic Virus), the change of one base in the DNA due to mutation would alter the composition of protein by a single amino acid proving that the code is non overlapping. If the code is overlapping change of one base should have disturbed more than one amino acid. Studies on normal and sickle cell haemoglobins have shown that single base mutations result in the replacement of only one amino acid.

2. The code is degenerate

(Synonym codons). As it has been mentioned above, there are 64 possible codons considering the code to be a triple one. Of these 64 codes about 61 code for amino acids. As they are only amino acids for exceed the number of amino acids. This leads to the conclusion that each amino acid must have more than one code. Of the 20 amino acids, only tryptophan and methionine have a single codon each, while all the others have more than one codons. *Phenylalanine, tryosine, histidine, glutamine, asparagine, lysine, aspartic acid* and *glutamic acid* have two codons each (see the codon table). *Leucine, argenine* and *serine* have 6 codons each. Thus, there is no specificity as to the number of codons that decide an amino acid. The number of codons (whether one or many) specifying the amino acid possibly have been evolved over a long period of time during the process of evolution. The variability in the number of codons perhaps may explain the differential distribution of amino acids in proteins (those with multiple codons occur more frequently).

When there are multiple codons for an amino acid, it must be noted that these codons are not totally different. Synonym codons specifying an amino acid have the first two bases of the triplet being constant. Whereas only the third will vary. For instance, proline is coded by the following codons – CCU, CCC, CCA and CCG. It is obvious here in these synonym codons, the first two bases are common.

The multiplicity or degeneracy of the code may pose a question as to why nature has provided alternate codons for an amino acid where a single codon would have been more precise and specific. One possible reason for the degeneracy of the codes is that multiple codons protect the organism from mutations which might eliminate an amino acid altogether if there were to be only one code.

Number of codons coding for different amino acids

	Amino acids	Number of codons
1.	Tryptophan, methoionine	1
2.	Phenylalanine, tyrosine, histidine, glutamine, asparagine Lysine, aspartic acid, glutamic acid, cysteine	2
3.	Isoleucine	3
4.	Valine, proline, threonine, alanine, glycine	4
5.	Leucine, agrinine, serine	5

3. Polarity of the code

It is very essential that a particular gene should specify an amino acid always and in all situations unless it undergoes mutational changes. It is necessary therefore, that the code must be read between a fixed beginning point and a fixed end point. These points are referred to as the initiation and termination codons respectively. Another necessity is the reading of the direction of the code because any change in the direction would alter the amino acid. All these features clearly point out that the code must have a polarity i.e., a fixed start, a fixed end point and a predetermined direction. For instance, if the codon direction as given below varies, the composition of the amino acids would completely vary.

	Leucine	Isoleucine	Valine	Serine →
	UUG	AUG	GUG	UCG
←	Valine	Leucine	Leucine	Alanine

It is obvious from the above that the direction of reading the code would alter the sequence of the amino acids and obviously changes the composition of the proteins. Available molecular biological evidences indicate that the message in mRNA is read in 5' →3' direction. Based on this message, the polypeptide chain is synthesised from the direction of amino group to carboxyl group.

4. The code is comaless

One of the most interesting questions with reference to the non-overlapping code is, are there any gaps between the codes in the form of punctuation marks so that one code is separate from the other. In genetic terms this means that there could be bases between codes indicating where one code ends and the other begins. For instance, codes with intervening punctuation marks can be represented as follows –

UUG	X	AUG	X	GUG	X	UCG
Valine		Leucine		Leucine		Alanine

In the above, X can be regarded as a base separating the two codons. If a mutation occurs in any one of the codes, it will not affect other codes and consequently, the disturbance in the protein synthesised would be to the extent of only one amino acid. On the other hand, if the code is commaless (no punctuation mark-base between codes), mutation in one base would result in a drastic change of the genetic message. Researches of Dr.Khorana and others had clearly indicated that the code is commaless i.e., there is no delimiting point between one codon and the other. Khorana and his associates used artifically synthesised long polynucleotide chains which had repeating sequences to produce polypeptide chains. For example, the repeating sequences CUCUCU contains the codons for leucine (CUC), and serine (UCU). When this sequence was used for producing the proteins neither amino acid was incorporated with the protein unless the other one is also present. This clearly explains that the triplet code has to be commaless, and only then there would be alternate translation of the codes into leucine and serine.

5. Universality of the code

The code was first deciphered and worked out in microorganisms but subsequent researches have clearly shown that these codes are same in their amino acid specification from bacteria to man. Thus, the code is said to be universal. Marshall Caskey and Nirenbrg (1967), have shown by their experiments that diverse organisms such as *E.coli* (bacterium), *Xenopus laevis* (amphibian), and Guinea pig which is a mammal used the same code for their amino acyl tRNAs.

Gene mutations provide additional evidence for the universality of codes. These mutations result in the substitution of one amino acid for other. Studies have indicated, gene mutations bringing about amino acid substitutions effect the same amino acid in TMB, *E.coli* and also in man. For instance, a single point mutation, changes the amino acid in the coat protein of TMV, alpha chain of tryptophan synthetase in *E.coli,* and also in the haemoglobin of man.

Another remarkable feature of the code is not only that it universally occurs in all organisms, but it has also remained probably without any change ever since it evolved in some bacteria a few billion years ago. Any mutation that occurs would change the composition of mRNA, and consequently the amino acid composition in the protein. Changes such as these apparently have proved to be harmful to the organisms and as such there will be a strong selection pressure against the occurrence of such mutations. Hence, natural selection has persisted with the codes perhaps ever since they originated.

Ambiguity in the genetic code

There is no evidence to indicate that the genetic code is ambiguous *in vivo.* Ambiguity means that a single codon may code for more than one amino acid. The only exception to the non-ambiguity is the AUG codon in prokaryotes–it codes for formyl methionine at the initiation site while at other positions it specifies methionine. It is not clear as to how this distinction is always met precisely but it is believed that the secondary structure of the mRNA at the initiation site may have got something to do with the coding for formyl methionine. In *in vitro* conditions, however, genetic codes sometimes appear to be ambiguous. For example poly-U codon is known to incorporate some amounts of leucine in addition to phenylalanine. Treatment of ribosomes with streptomycin is known to introduce ambiguity to codons. For example in addition to coding for phenylalanine UUU is also known to code isoleucine, serine and leucine.

Codons and anticodons

During the process of translation, the message carried in the mRNA would be converted into the sequence of amino acids which are picked up by tRNA. This is achieved by the process of binding. For this binding to take place, obviously then there should be a specific relationship between the nucleotide. Sequence of mRNA is called the codon and that of tRNA is called anticodon. The reading of mRNS is in the 5'→ 3' direction and the codons are also written in the same direction. For example, codon AUG is written as 5' AUG 3'. The corresponding anticodon and tRNA should therefore be represented as 5' CAU 3'. In such a configuration, the first nucleotide of both codon and anticodon would be the ones at the 5' end and third nucleotides would be at the 3' end.

Very often, the anticodon is written in the 3' → 5' direction so that it will be easy to correlate between the bases of codon and anticodon.

Initiation codons

In most of the polypeptide chains, the first amino acid which begins the chain is methionine in eukaryotes and N-formlymethionine in prokaryotes. Methionine or N-formylmethionine tRNA complex specifically binds to the initiation sites in mRNA containing the codon AUG which codes for methionine. The AUG code, therefore, is referred to as the initiation codon. Less frequently, however, GUG (code for valine) may also serve as the initiation codon in bacterial protein synthesis. In the bacteriophage MS2, GUG is the initiation codon for the protein referred to as A protein. Researches have indicated that the GUG may replace AUG as initiation codon when the latter is lost due to chromosomal abbreation. It has also been found out that chain initiation by GUG is not efficient as AUG possibly because the former has less affinity for formyl methionyl tRNA.

Termination codons (nonsense codons)

It has been mentioned above that AUG (occasionally GUG) initiates the synthesis of a protein chain. The natural question would be which codon specifies or signals the end of termination of the synthesis of the protein chain. It has been found out that three of the 64 codons do not specify any tRNA and as such they cannot assemble any amino acid. As such, these codons whenever they are found in mRNA would signal the end of the chain. They are called termination codons as they terminate the synthesis of the polypeptide chain.

They are also called *nonsense codons* as they donot specify any amino acid. The nonsense codons are UAG (*amber*), UAA (*ochre*), and UGA (*opal*). Of the three termination codons, UAG was the first termination codon to be discovered. It was named amber after a graduate student in Germany, named Bernstein (means amber in German). This graduate student helped in the discovery of the class of mutations. The other two termination codons were obviously given the name of the colours only to maintain uniformity.

Termination codons found in mRNA will not specify any amino acid and as such stop the process of translation and bring about the release of the peptide chain from ribosomes. No tRNA molecules have anticodons complementary to the termination codons found in mRNA. Termination codons seem to have some affinity to proteins called release factors which bring about the release of the protein chain. In prokaryotes, there are three release factors RF-1, RF-2, and RF-3. RF-1 recognises the amber and ochre codons, while RF-2 recognises ochre and opal codons, RF-3 stimulates RF-1 and RF-2 ultimately releasing the newly synthesised peptide chain. In eukaryotes there is a single factor (RF protein) which is capable of recognising all the three termination codons.

Evidence for genetic code (deciphering the code)

Earlier to the pathbreaking experiments of Nirenberg, Matthaei and H.G. Khorama, work on the genetic code was largely speculative and restricted to development of theoritical concepts. The great physict George Gamow, who proposed the triplet code gave the comprehensive theoretical concept for various aspects of genetic code. The main features of his proposals are –

1. The code is a triplet.
2. Translation is a direct procedure between codon and amino acid.
3. Translation of codes is of overlapping type.
4. Code is degenerate.
5. Nuclei acid and the peptide chain are colinear.
6. Code is universal.

The works of Brenner (1957), showed that the triplet code is of non-overlapping type. Gamow's idea of the direct translation process became unacceptable when Crick proposed adoptor hypothesis. As per this hypothesis, during the translation process some mediator molecules intervene between amino acids and nucleic acids. It is now known that the intervening molecules are tRNA.

Experimental evidence for the existence of genetic code may be obtained by two basic approaches – *in vitro* approach and *in vivo* approach. In *in vitro* method, nucleotide sequence are compared with the amino acid residues of the proteins in a culture medium. In *in vivo* method, proteins and nucleic acids are obtained from intact cells and then compared.

The *in vitro* approach has the following four methods :

1. Genetic analysis of deletion mutatants of a cistron..
2. Incorporation of amino acids by synthetic polyribonucleotides.
3. Binding of known triplets of aminoacyl tRNA.
4. Amino aicd sequence generated by mRNA transcribed from synthetic tRNA of known nucleotides sequences.

1. Genetic analysis

Genetic analysis of deletion and insertion mutants (spontaneous as well as induced) in phage T4 was reported by Crick *et al* in 1961. This deletion or insertion of a nucleotide in a DNA chain leads to a frame shift mutation after the point of deletion or insertion. As a result of this, the reading of the codon subsequent to the mutation gets altered. Obviously then, the amino acid composition of the resultant protein would be different. As a result, the protein may be functionally inactive. All the mutations tested by Crick *et al,* produced a phenotype called mutant II. It was further observed by Crick *et al,* that a series of three mutations was necessary to supress the effects of a previous mutation. In other words, three supressor mutations would nullify the effect of three frame shift mutations and bring back the original phenotype. This is a clear evidence to show that the altered codon reading would return to the normal state after three point mutations thus indicating that the codon must have three nucleotides. The study also indicates that the code is commaless. If there were to be a punctuation mark between the two codes, a deletion or insertion of a base would change the number of bases between two commas adjacent to the point of insertion or deletion.

2. Incorporation of amino acids by synthetic polyribonucleotides

Synthetic polyribo nucleotides were used as mRNA by Nirenberg and Matthai to determine the meaning of codons. They produced a synthetic RNA by using the enzyme polynucleotide phosphorylase discovered in 1955 by Grunberg. Manago and Ochoa. This enzyme does not use DNA as a template and can incorporate ribonucleotides in a random order. The first synthetic RNA used for protein in a self-free system (the *in vitro* system consists of ribosomes, tRNA's and other necessary factors for protein synthesis) was poly U i.e., the mRNA consists of only one nucleotide namely, U in a series. This can be called a homopolymer as it is a polymer of only one nucleotide. The poly U mRNA was able to synthesise a homopolymer of a polypeptide chain containing only phenylalanine residues. This proved that the triplet code UUU specifies phenylaline. Subsequently, many such homopolymer species of mRNA were synthesized and were directed to produce polypeptide chains. For example, AAA was shown to code for lysine and CCC specified proline.

Nirenberg and Ochoa subsequently used heteropolymeres of mRNA to produce polypeptide chains. A heteropolymer includes a mixture of two, three or all the four nucleotides in mRNA. This yielded further information on the possible codons for certain amino acids. For example, poly CG directs the incorporation of arginine and alanine in addition to the proline coded by poly C. Some additional information on the assignment of codons to individual amino acids was obtained by varying the ratios of different nucleotides in the synthetic mRNA. In this experiment, the code assignment was at best ambiguous since all the four bases were not required for the incorporation of any of the amino acids.

3. Binding of known triplets to aminoacyl tRNA

A triribonucleotide cannot direct the synthesis of a polypeptide but it leads to binding of specific aminoacyl tRNA to the ribosomes. Nirenberg *et al,* used this property for assigning specific codons of specific amino acids. Their technique is known as Nirenberg – Leder technique. Conventionally, a trinucleotide with 5' terminal phosphate are more active and those lacking this phosphates in inducing the binding of aminoacyl tRNA to ribosomes. On the otherhand, those with 3' terminal phosphate are inactive. This shows that the codons must be triplet. In the Nirenberg - Leder technique, trinucleotides with 5' terminal phosphate are added to a cell free system containing a mixture of 20 amino acids bound to their respective tRNA. One of the 20 amino acids had radioactive carbon. For different experiments, different labelled amino acids were used but with the same trinucleotide after the synthesis of protein, the cell free mixture containing ribosomes were analyed to detect radioactivity confirming the binding of aminoacyl tRNA to the ribosomes. These studies showed that UUU, CCC and AAA induced the binding of phenylalnine, proline and lycine respectively to ribosomes. Using this technique Nirenberg *et al,* assigned most of the codons to their respective amino acids.

4. Amino acid sequence generated by mRNA transcribed from synthetic tRNA of known nucleotides sequences

H.G. Khorana and his associates used the above technique for codon assignment in *in vitro* conditions. Known as Khorana technique, in this method long specific DNA sequences, as first synthesized. These are then transcribed to mRNA molecules using the DNA directed RNA polymerase enzyme. These mRNA molecules which have known sequences of nucleotide are directed to synthesize polypeptides. The polypeptides were then subjected to sequential degradation to know the amino acid sequences. The nucleotides sequence is then compared with the amino acid sequences. The nucleotides sequence is to deduce specific codons for different amino acids. Using this technique, almost all the codons for amino acids were designated. In this technique, Khorana and commaless nature, non overlapping and triplet nature also.

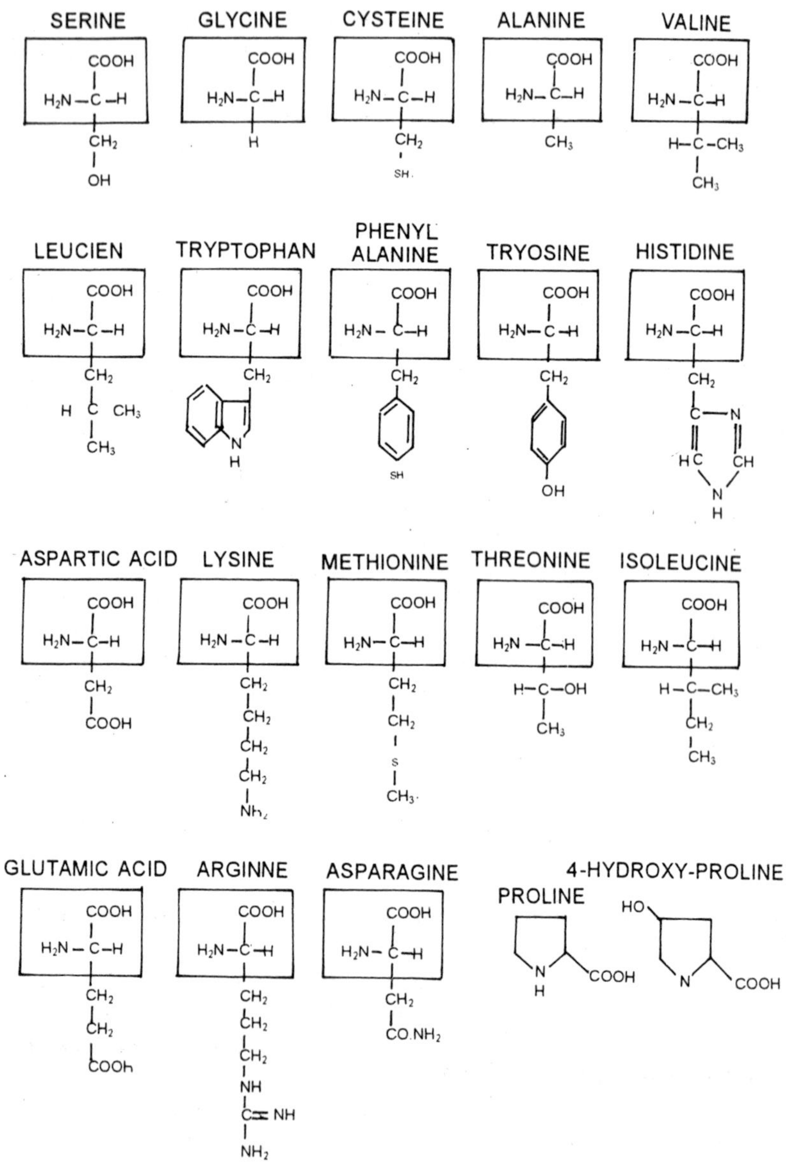

Fig. 6.1 Genetic code
The codon dictionary

In Vivo codon assignment

In the cell free systems, genetic codes have been deciphered with reference to their amino acid affinity but a doubt may arise whether these codes are the same in living systems also. In otherwords, whether the living systems use the same codes for protein synthesis. In order to prove the similarity of codes in both in *vitro* and in *vivo* three different techniques were used by three different molecular biologists.

(i) Amino acid replacement studies
(ii) Frame shift mutation studies
(iii) Comparison of a nucleotide sequence of a cistron and the amino acid sequence of the polypeptide produced by it

Amino acids replacement studies on the synthesis of tryptophan synthetase in *E.coli* (Yanofsky, 1963) have clearly indicated that the codons for the replaced amino acids, and for the amino acid replacing it prove the specificity of a codon for a particular amino acid. Frame shift mutations on lysozol enzyme of T4 bacteriophages have also shown the affinity between a specific codon and a particular amino acid.

The direct in *vivo* evidence for the genetic code comes from a comparison of the base sequence of a cistron and the amino acid sequence in a polypeptide chain produced by it. Such a study has been done with RNA bacteriophage MS2 which has only three cistrons. One of these cistrons produces the coat protein having 129 amino acids. The cistron responsible for it has 387 nucleotides. There is a total agreement between amino acid sequence expected from the base sequence of the cistron and the actual amino acid sequence of the head protein. In othewords, each codon of the cistron precisely designates the same amino acids as expected from the coding dictionary.

Wobble hypothesis - for details see protein synthesis

Colinearity of the gene structure and its product

The gene is an unbranched chain of nucleotide pairs in a particular sequence. The polypeptide chain is also unbranched and is made up of amino acid residues attached in a linear series. Thus, a linear relationship between the gene and the polypeptide chain produced by it can be indicated. Yanofsky *et al* (1964), established this colinearity in their study of tryptophan synthetase enzyme of *E coli.*

The enzyme tryptophan synthease is made up of two chains α and β. A genetic map of the α chain gene was established by determining the recombination percentage of mutations for the amino acid tryptophan. The complete amino acid residues of the α chain of the enzyme tryptophan synthetase was thus established. Subsequently, the location of amino acid substitutions in the chain for 10 mutations was determined. It was then found out that the order of mutations corresponded with the order of amino acid substitutions in the α chain of the polypeptide.

Sarabhai et al (1964), came to the same conclusion in their study of nonsense mutations in bacteriophage T4.

Pattern of the genetic code

A distinct pattern emerges when one analyses the amino acid structure and the type of codons that specify them (fig 13.1). Amino acids with similar structural properties tend to have related codons. For example, codons for aspartic acid (GAU and GAC) are similar to the codons for glutanic acid (GAA and GAG). Same relationship is found between the codons of phenylalanine, tyrosine and tryptophan – all these begin with uracil. It is believed that this feature of the code is evolved to minimise the consequences of mistakes that may arise during translation and base substitutions due to mutation. In otherwords, if an amino acidin a protein is replaced by another with almost similar properties, the protein may still remain functional.

Another pattern that is seen in the code is that for synonym codons (multiple codes for thesame amino acid), the first two bases are always constant whereas only the third one varies. The flexibility of the third nucleotide of the codon is another method to help minimise the harm posed by mutations. Crick (1966) has called this wobbling and proposed a hypothesis called Wobble hypothesis (see protein synthesis for details).

New genetic codes in mitochondria and ciliate protozoa

- It has been mentioned above that the code is universal and ever since it originated it has remained static. However, during 1980s it was observed that the genetic codes of mitochondrial DNA in yeasts may be different from the others. In yeast, mitochondria UGA codes for tryptophan, while in nuclear genes in the same organism UGA is the termination codon. This, however, is not to say that in all mitochondria of all organisms, the UGA codes for tryptophan. For example in maize cytochrome, oxidase tryptophan is coded by CGG. This code in nuclear genes codes for arginine. There are also indications that AUA which codes for isoleucine in nuclear genes codes for methionine in *mit* DNA of mammals, yeast, Drosophila etc.

A different genetic code in DNA is perhaps necessiated because of the small *mit* genome size. T.H. Jukes (1966), presented a concept called *archetypal code* which will allow one anticodon (of tRNA) to pair with 4 different codons due to wobbling at the first base of anticodon and the third base of codon.

In 1986 a different genetic code were shown to be working in a ciliate member of Mycoplasma (*mycoplasma capricolum*). In the genetic code of this organism, UAA and UAG are not nonsense codons but code for glutamine.

Evolution of genetic code

Many of the features of the genetic code like degenerase, non-overlapping commaless nature etc. have led to a suggestion that the present version of the genetic code has evolved from a more primitive version which are perhaps a double than a triplet. It is believed that this shifting from doublet to triplet occurred very earily in the origin of cellular life. It is not certain, however whether all the codes for all the amino acids were evolved simultaneously or the codon dictionary gradually arrived at the present stage of 64 is not certain. One thing however, is certain eversince it came into existence in the present form, the biological necessity of it imposed such a selection pressure on it that it has remained unchanged upto the present day.

7

PROTEIN SYNTHESIS

Protein Structure And Classification

Of the two important macromolecules that mastermind all the activities of the living systems, proteins are one, the nucleic acids being the other. Proteins are present all over the cell and are important both structurally and functionally.

The name protein was first suggested by Berzelius (1838). *Protein* in Greek means 'First rank'. Proteins are highly complex organic molecules with molecular weight ranging upto many millions. Chemically, proteins are madeup of C, H, O and N. In addition to these major elements, some proteins may also haave sulphur, phosphorous, iron zinc, iodine etc. as components. The ratio of major elements in proteins are C= 50-55%, O=20.23%, N=15-18% and S=0-4%.

Proteins are made up of building blocks called amino acids (i.e. on hydrolysis, proteins yield these units). The idea that proteins are build up of amino acids was first proposed by Emil Fischer.
Amino acids are carboxylic acids which have atleast one COOH group and one NH_2 group. Other functional groups may also be present. There are atleast 22 different types of amino acids encountered in plants.

The amino acids are bound to one another in a protein chain by means of linkages or bonds called peptide linkages. Peptide linkages are formed between the COOH group of one amino acid and the NH_2 group of the other, with the elimination of a molecule of water. Several amino acids combine together with the peptide bond and form a peptide chain. Several of these peptides constitute a polypeptide chain. The polypeptide chain folds and refolds in a characteristic fashion to form a protein molecule.

Peptides

These are made up of amino acids linked with CONH (peptide) bonds. Peptides are intermediate compounds produced during the synthesis or degradation of proteins.

Amino acids ——— peptides ——— proteins

Proteins

These are the final productsin the N2 metabolism, and are made up of folded peptide chains which have many additional linkages (depends upon the point of folding), in addition to the CONH bond. Proteins have a three dimensional configuration, and have a very high molecular weight reaching upto 10^4 - 10.

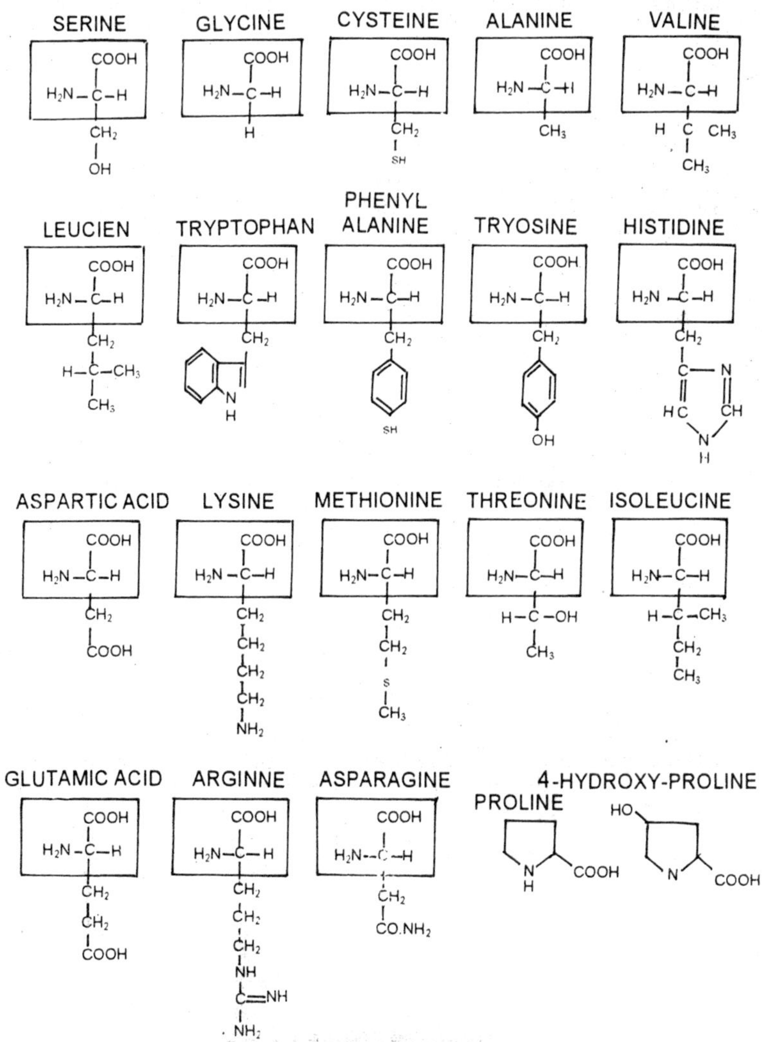

Fig. 7.1 Protein Synthesis
Structure of the amino acids

Although proteins contain not more than 20 amino acids, they are of immense variety and complexity due to variations in combination of amino acids. In general, the three following aspects pertaining to the amino acids decide the type of protein. These are – the quality, quantity and sequence of amino acids in the peptide chains.

In other words, the type of amino acid, its quantity and the sequence in which these acids are arranged forms the basis for the categorization of proteins. Besides, proteins may differ in the number of peptide chains per molecule, type of secondary linkages (SH) etc.

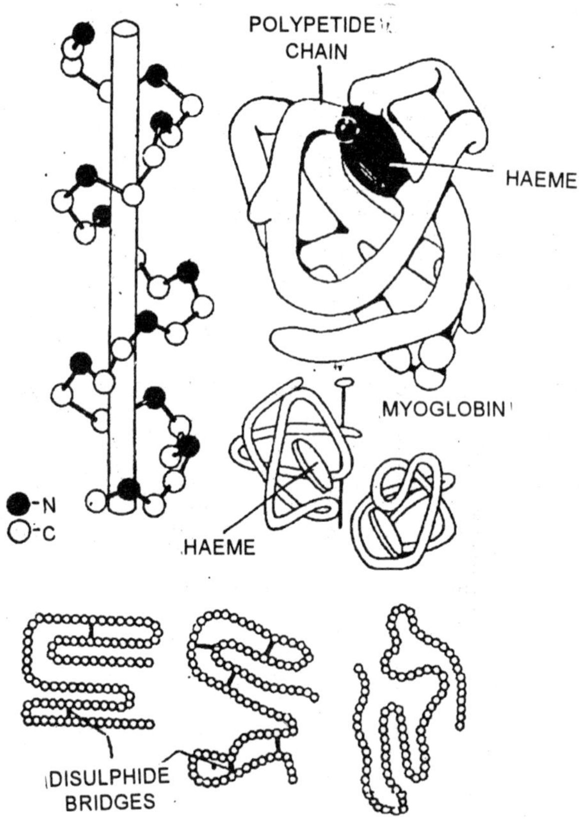

Fig. 7.2 Protein Synthesis
Protein structure. **A.** Polypeptide chain helical configuration, **B.** Tertiary and quaternary structure of myoglobin, **C.** Denaturation of proteins

Structure of proteins

Modern biochemists regard four structural levels in the protein molecule which accounts for its complexity. These levels are – primary, secondary, tertiary and quaternary structure, in the order of increasing complexity.

Primary structure

This is the simplest and the most basic structure and refers to the linear peptide chain. Besides the CONH bond, disulphide bonds (-S-S-) are found among the peptide chain linkage and the sulphur containing amino acids.

Secondary structure

The primary polypeptide chain spirally coils forming the L helical chain. The characteristic bonds of the secondary structures are hydrogen bonds between the carbonyl and amino groups of peptide links, salt linkages and Vander Wall's forces which help maintain the helical structure.

Tertiary structure

Folding and superfolding of the helical chain in a three dimensional orientation constitutes the tertiary structure. Interaction between the amino acid residues situated far apart in the chain brings about the folding.

Quaternary structure

This consists of more than one identical sub units (poly peptide chain) which associate with one another to form a functional unit. Individually, each subunit is non-functional. Such complicated structures are usually seen in enzymes. For example, the enzyme phosphorylase has two sub units, which are inactive individually, but as dimers, they are active. A protein with identical sub units in the quaternary structure is known to be homogenous or if the sub units are dissimilar, it is known as heterogenous. A protein with more than one sub unit (monomer) is referred to as an oligomeric protein.

Classification of Proteins

As the proteins possess, a very complex strucuture, their classification also poses a problem as many categorizations are possible based on structure, function etc. The following is a generalized classification. The major groups are recognized in proteins based on their solubility and chemical properties.

I. Simple proteins
II. Conjugated proteins and
III. Derived proteins.

I. SIMPLE PROTEINS

Chemically, these have only amino acids. No additional compounds are present in the molecule. Simple proteins are of the folowing kinds –

1. Albumins

These are soluble in water, dilute salts, acids and alkalies. Heat denatures the protein structure by coagulation. Some of the commonly occuring albumins are Legumelin (from legume seeds), Amylase (from barely), Leucosin (from cereals) etc.

2. Globulins

Less soluble in water than albumins but soluble in less concentrated salt solutions. Heat coagulation is seen in globulins also. Eg. Legumin (pears), Tuberin (potato) etc.

3. Glutelins

Insoluble in water and salt solutions, but soluble in weak acids and alkalies. Glutelins are generally found in plants and are heat immune. Seed proteins found in cereals like Wheat (glutenin) and rice (oryzenin) are best examples of glutelins..

4. Prolamines

These proteins are water insoluble, but dissolve in 70–80% alcohol. Prolamines have a high concentration of the amino acid proline and other amides. E.g. Gliadin (Wheat), Hordein (barley) and Zein (maize).

5. Scleroproteins

These are fibrous proteins found in integral part of the structural framework of plants. These proteins are insoluble in most of the solvents.

6. Histones

Most of the nucleoproteins, haemoglobin and globins are histones. They are insoluble in water but soluble in dilute ammonia. Histones have a high concentration of lysine and arginine.

7. Protamines

These are soluble in water, dilute acids and ammonia and do not undergo heat coagulation. Basic amino acids are rich in protamines, while tryptophan and tyrosine are completely absent. Eg. Nucleoproteins of sperms in fish.

II. CONJUGATED PROTEINS

In these proteins, non-amino acid components are associated with the protein molecule. The non-amino components are called prosthetic groups. There are seven major categories of conjugated proteins.

1. Nucleoproteins

Proteins are found in combination with nucleic acids. Hydrolysis can separate nucleic acids and simple proteins. Nucleoproteins are water soluble.

2. Glycoproteins

Also called mucoids, these are conjugated with carbohydrates like macromolecular polysaccharides, sugars, sugar phosphate etc. Glycoproteins are found in cell membranes.

3. Chromoproteins

These constitute the pigments and are coloured, the coloured appearance is due to the presence of metals such as Cu, Fe, Mg etc. Chlorophylls, flavoproteins, carotenoids, cytochromes etc. are typical examples of chromoproteins.

4. Lipoproteins

Proteins in conjugation with lipids are called lipoproteins. Lecithin, Cephalin etc. are the lipids found in combination with proteins. Lipoproteins are an integral part of all cell membranes.

5. Metalloproteins

Certain types of enzymes which require metals as activators belong to this group. Many of the respiratory enzymes belong to this group.

6. Phosphoproteins

These are proteins with phosphoric acid as the prosthetic group. Phosphoproteins are insoluble in water, but soluble in alkalies. E.g. milk and egg protein (casein and viteline).

7. Leeithoprotein

Proteins with lecithin as the prosthetic group (E.g. white of egg) belong to this group.

III. DERIVED PROTEINS

These are degradation products of proteins. Primary proteins on hydrolysis with acids or alkalies yield derived proteins.

(A) Primary derived proteins and

(B) Secondary derived proteins

A. Primary derived Proteins

There are two categories of primary derived proteins. These are Metaproteins and Coagulated proteins.

(i) Metaproteins : These are insoluble in water but dissolve in dilute salt

solutions, acids and alkalies. Hydrolysis of natural proteins on long treatment with alkalies or acids yield metaproteins.

(ii) Coagulated proteins: Treatment with heat or alcohol of natural proteins yields these proteins.

B. Secondary derived proteins

Three types of secondary derived proteins have been identified. These are proteoses, peptones and peptides.

(i) Proteoses : Prolonged hydrolysis of metaproteins yield proteoses (eg. Albuminoses from albumin). These are water soluble but immune to heat treatment.

(ii) Peptones : Peptones are obtained by continued hydrolysis of proteoses. Like proteoses, these are also water soluble and heat insensitive.

(iii) Peptides : Prolonged hydrolysis of natural proteins yields peptides. These are also water soluble and not coagulated by heat.

Functions of proteins

Most of the vital activities of the cell are controlled by the two master molecules, namely proteins and nucleic acids. Proteins are important both from the point of view of structure and function of the cell. Structurally, proteins constitute in integral part of membranes, pigments etc. Functionally, in the form of enzymes, proteins play such a vital role in cell physiology, that without proteins (enzymes), no metabolic reaction is possible. Even the synthesis of DNA is regulated by proteins. In the form of Chromoproteins (Chlorophyll, Cytochromes etc), proteins mediate all energy transformation reaction. It is because proteins are so important that the nucleic acid directly regulates the protein synthesis.

Protein structure is highly complex (as has been pointed out earlier), as well as specific. The primary structure is determined by the quality, quantity and the sequence of amino acids. Even an alteration of the position of one amino acid would alter the structure of the protein, and if it happens to be an enzyme regulating an important reaction, the entire reaction is disturbed leading to several abnormalities. Hence it is but natural that there is direct control of DNA in the synthesis of proteins. DNA, RNA and ribosomes of the cell are involved in the synthesis of proteins.

The central dogma : The basic mechanism of protein synthesis is that the nuclear DNA regulates the assembly of amino acids in the ribosomes to produce a polypeptide chain. Since DNA cannot move out of the nucleus it sends a messager – mRNA which carries the message in the form of sequence of nucleotides. This mRNA forms a template on the ribosomes. Another type of RNA, tRNA found in cytoplasm (whose nucleotide sequence is complementary to that of mRNA), picks up the amino acids and assembles them on the ribosomes to produce a peptide chain.

This may be briefly expressed as follows :
DNA Replication → DNA Transcription → RNA Translation → Proteins translation. Transcription is the copying of a complementary messenger RNA strand on a DNA strand. In translation, the genetic information present in mRNA (in the form of sequence of nucleotides) directs the sequence of amino acids to be assembled. We shall study these events in detail.

Transcription

Copying of a complementary strand of mRNA requires the following – a template, activated precursors, a divalent metal ion and RNA polymerase.

DNA molecules serves as a template for the production of mRNA. Of the two strands of DNA, only one strand called the 'sense' or the 'coding' strand transcribes mRNA. Single stranded DNA also can serve as a template.

The activated precursors required for mRNA synthesis are – Adenine triphosphate, Guanine triphosphate, Uracil triphosphate and Cytosine triphosphate (ATP, GTP, UTP and CTP)

Mg^{++} or Mn^{++} are the divalent metal ions required for mRNA synthesis.

The enzyme RNA polymerase is necessary for transcription. The nucleotides of mRNA are assembled together with the help of RNA polymerase. The enzyme consists of two parts - a core enzyme and a sigma factor. The sigma factor initiates transcription of mRNA on the DNA template and the core enzyme continues transcription. The DNA base sequence are thus transcribed into mRNA base sequence.

Genetic Code

This is the message in mRNA in the form of nucleotide sequence. Three nucleotides code for one amino acid. It is called a triplet code or codon. There are 64 codons for the 20 odd amino acids. i.e., there is more than one code for each amino acid. These alternative codes are helpful in the cae of random mutations which may alter a code (nucleotide).

The nucleotides of mRNA are complementary to the nucleotides of tRNA (present in cytoplasm). The triplet nucleotides of tRNA are called anticodons. Each tRNA has three nucleotides. The pairing between the codon and anticodon (mRNA and tRNA), follows the A-U, G-C combination. For example, if mRNA consists of the first code UUU, it pairs with a tRNA having an anticodon. AAA. Thus the series of codons of mRNA determine the series of anticodons of different molecules of tRNA, and hence the amino acids (Each tRNA molecule is specific to a particular amino acid). Since the triplets of mRNA are determined by the base sequence of DNA, it follows that the DNA molecule decides the sequence of the amino acids and hence the structure of the protein molecule.

Some triplets do not code for any amino acid (UAA and UAG). These are called nonsense codes. The so called nonsense codes are actually terminator codons and are now called ochre and amber (UAA and AUG). Code AUG is called initiator codon since it initiates the synthesis of a polypeptide chain.

The triplet codes for all amino acids collected together is called a codon dictionary. The dictionary was first proposed by Nirenberg (1965), and subsequently worked out in further detail by Khorana H.G.
The mRNA is released out of the nucleus into the cytoplasm where it is attached to the smaller sub unit of the ribosomes. Many ribosomes are connected by a single mRNA molecule, constituting a polysome or polyribosome. The mRNA attaches itself to the ribosomes in such a way, that its bases remain free (to be able to undergo pairing with the bases of tRNA). The whole structure (polyribosomes + mRNA), is called mRNA complex.

tRNA (transfer RNA) : This is present freely in cytoplasm. Also called sRNA (soluble RNA), it is the smallest among the various types of RNA molecules found in the cell. Although the tRNA is single stranded like other types of RNA, it appears double stranded because of coiling. Complementary base pairs of the same strand form pseudo bondings of hydrogen due to folding. There are two regions which are free without bonding – one at the acceptor and which always has CCA base pair sequence, and the other in the middle region of the molecule called the anticodon region, where there bases turn out of the folded portion and remain free. The types of tRNA is based on the types of these free bases. tRNA has the following functions.

(i) To attach to a particular amino acid at the acceptor, and (CCA) with adenine nucleotide. The choice of amino acid, however, depends on the type of free bases present at the anticodon end.
(ii) to attach itself at a specific site on the mRNA, and
(iii) to free the amino acid after it gets attached to another amino acid by means of peptide linkage.

The Wobble hypothesis
The triplet code or codon is degenerate i.e., there are many more codons than there are amino acids. Only 20 amino acids are involved in protein synthesis, while there are 64 possibilities or types of codons (43). Hence, more than one codon can code for an amino acid. It is interesting to find out as to why the

codes are degenerate? There could have been only so many codons as there are amino acids. But there is a problem here. If there were to be only one code for an amino acid, what happens if there is a mutation at that site in DNA? Obviously the code changes and the amino acid cannot be incorporated in the protein leading to abnormalities. On the other hand, if there are alternate codes, they act as a protection against mutation in a specific base. Hence, degeneracy of codes is a must.

Regarding tRNA present in cytoplasm, since they have to bind themselves to a specific codon, there should have been as many tRNA's as there are codons. But the number of tRNA is only as many as there are amino acids. This means, the anticodons of the tRNA's must be able to 'read' more than one codon of mRNA. How is this reading possible when bonding between base pairs is highly specific?

Crick (1966), proposed **'the Wobble hypthesis'** in order to solve this apparent dilemma. According to this hypothesis, only the first two bases of the codon have a precise pairing with the bases of the anticodon of mRNA, while the pairing between the third basis of the codn and anticodon may Wobble (non-specific). The pairing in the third base is ambigous. Thus a single tRNA can pair (bind) with more than one mRNA codon differing in only the third base. For example, the anticodon UCG of tRNA can read codons AGC and AGU of mRNA. The pairing between UGC and AGU is a perfect pairing as per the Crick and Watson pattern. But in the second case i.e., between UGC and AGU, the pairing between the first two bases is normal, while between G and U is against the normal pairing pattern. This unsual bonding G and U is called Wobble pairing.

The degeneracy of the code is not totally randon. Multiple codons for an amino acid always carry the same bases in the first two positions in a triplet with only the third 'Wobbling'. For example all the four codons for Valine have the first two bases - GU and for alanine it is GC.

Valine - GUU, GUC, GUA and GUG.

Alanine - GCU, GCC, GCA and GCG.

Translation

The process of translation consists of activation of amino acids, transfer of activated amino acid to tRNA, initiation of synthesis of polypeptide chain, chain elongation and chain termination.

Activation of amino acids

Before amino acids (20 of them) get themselves attached to the specific type of tRNA they have to be activated (energized). The activation is carried out by the ATP molecule in the presence of the specific enzyme called aminoacyl synthetase with Mg as the cofactor. The reaction produces aminoacyl adenylate (AAA) or aminoacyl AMP. Pyrophosphates are released in the process.

Amino acid + ATP + Enzyme ———— Enzyme AA> AMP complex + Ppi

Transfer of activated amino acids to tRNA : The activated amino acid (aminoacyl AMP) is transferred on to a specific tRNA, during which AMP and the enzyme (aminoacyl synthetase) are released. The amino acid gets attached to the CCA and tRNA.

Enzyme AA. AMP + tRNA ———— AA. tRNA + AMP + Enzyme

Initiation of Polypeptide chain synthesis

This involves two steps viz., preparation of the site of protein synthesis and initiation of the polypeptide chain.

(i) Preparation of the site of synthesis

Ribosomes in cytoplasm are the actual sites of protein synthesis. These are usually into their subunits (50s and 30s), when they are not participating in protein synthesis. The mRNA usually attaches to the 30s subunit through its 5' end carrying the AUG codon. In the process of attachment certain substances called initiation factors (IF) help. These factors are of three types in prokaryotes (IF1, IF2, IF3) and are of six types in eukaryotes (eIF2, eIF2al, eIF2a2, eIF2a3, eIFa4 and eIF3).

The attachment of the mRNA to the 30s submit requires the help of IF3 and Mg^{++}. The IF3 probably unwinds the rRNA (ribosomal RNA), so that it can bind to mRNA. After the attachment of mRNA, an mRNA - 30s complex is formed.

The starting amino acid in a polypeptide is methionine *(met)* in the case of eukaryotes, and N-formly methionine *(f-met)* in the case of prokaryotes. The amino acid (met or f-met) tRNA complex attaches to the initiation codon (AUG) on the mRNA 30-s complex through its anticodon (UAC). This forms the 30s initiation complex and requires the participation of IF2, IF3 and GTP (energy donor).

IF3

(i) 30s + mRNA ———— mRNA - 30s complex

(ii) mRNA - 30s complex + met tRNA ———— 70s - mRNA - met - tRNA

During the formation of the initiation complex , the 50s subunit combines with the 30s subunit to form the 70s unit. This union is facilitated by the participation of IF1. With the formation of the 70s ribosome), the initiation of the polypeptide chain begins.

(ii) Initiation of the polypeptide chain

Each ribosome has two cavities - P site (polypeptide site) and met-tRNA (aminoacyl sites). AA tRNA complex occupies the A site, while the met - tRNA (tRNA with the initiator aminoacid - methionine) will be at the P site. This occupation requires a molecule of GTP and the participation of IF_1. The energy of GTP is utilized for the formation of the peptide linkage.

Peptide linkage occurs when the carboxyl (COOH) group of peptidyl tRNA (met. tRNA) present at the P site reacts with the amino group (NH_2) of AA tRNA at the A site. The enzyme peptidyl transfers and catalyses the formation of peptide bond. Later the AA+tRNA moves from A site to the P site in the presence of GTP and IF_2

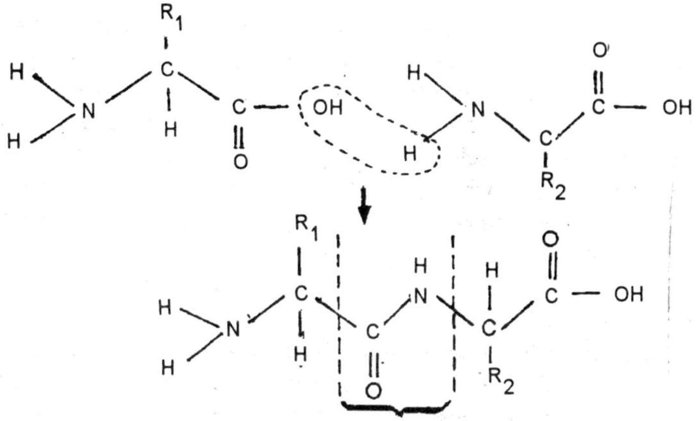

Fig. 7.3 Protein Synthesis
Formation of peptide linkages

Formation of the polypeptide chain

A specific type of tRNA (which has bases complementary to the codon of mRNA), linked to a particular amino acid gets attached to a specific site on mRNA when its anticodons can form a pairing with the first codon of the mRNA. Soon after, a second tRNA with its amino acid joins the second site (triplet) on mRNA. The first and the second amino acids attached to different tRNA form a peptide linkage as both are linked to mRNA *via* the anticodons of tRNA. A molecule of water is released during the formation of the peptide linkage.

$$\mathrm{COOH} + \mathrm{NH_2} \xrightarrow[\text{Peptide bond}]{} \mathrm{CONH} + \mathrm{H_2O}$$

As soon as the peptide linkage is formed, the tRNA is made free and comes back to cytoplasm. This whole process in which a specific tRNA carries its amino acid to a specific site is called translation i.e. the code (triplet) is translated into an amino acid. Here the sequence of bases in tRNA will be identical with the base sequence of that part of DNA coding for mRNA.

Chain elongation

Elongation of the chain requires the participation of certain factors called elongation factors (EF). These are EF-Tu, EF-TS, and EF-G in prokaryotes, and EF-1 and EF-2 in eukaryotes. After the first amino acid tRNA complex has moved from the A site of the ribosome to the P site, a second AA. tRNA complex now occupies the A site. The codons are recognized enzymatically. This requires the participation of Ef-Tu, GTP and Mg^{++}.

Peptide linkage takes place between the two amino acids and the AA. tRNA complex migrates to the P site rendering the A site vacant. Now a third amino acid tRNA complex occupies the A site and the process continues. The shifting of th AA. tRNA complex from A site (releasing site) is called translocation. At the P site the, tRNA is released from the ribosome while A site has a peptide chain liked to a tRNA which has brought in the last amino acid of the peptide chain. During translocation, the ribosome moves relative to the mRNA in the 5'-3' direction.

Chain termination

The elongation of the chain continues by the alternate occupation and vacancy at the A site by the AA. tRNA complexes until a termination codon (UAA, UAG or UGA) reaches the ribosomes. The termination codon sends signals to the ribosomes to attach release factors (RF-1, RF-2 and RF-3 in prokaryotes and RF in eukaryotes).

The release factors interact with the enzyme peptidyl transferase (which catalyses the formation of peptide bond) resulting in the hydrolysis of the (i)bond between the polypeptide chain and the tRNA (linked to the last amino acid of the peptide chain at the site) and the former is released from the ribosome.

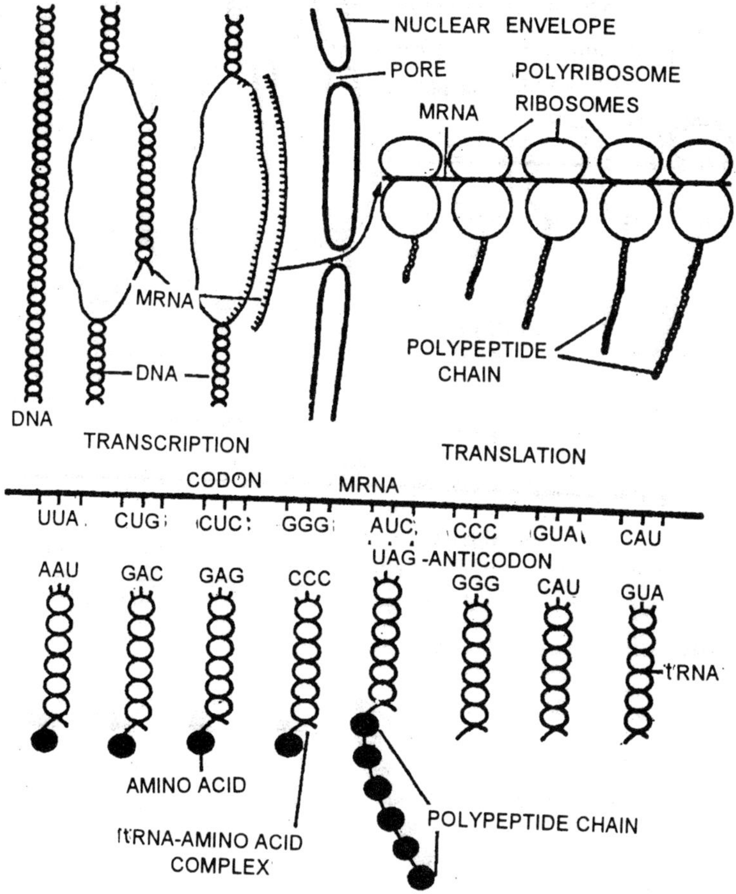

Fig. 7.4 Protein Synthesis
General scheme of protein synthesis

Other events that occur at the termination of the polypeptide chain are –

Release factors get disassociated from the ribosomes due to the hydrolysis of GTP.

(ii) tRNA is also released.

(iii) Ribosomal sub units (30s and 50s) get disunited.

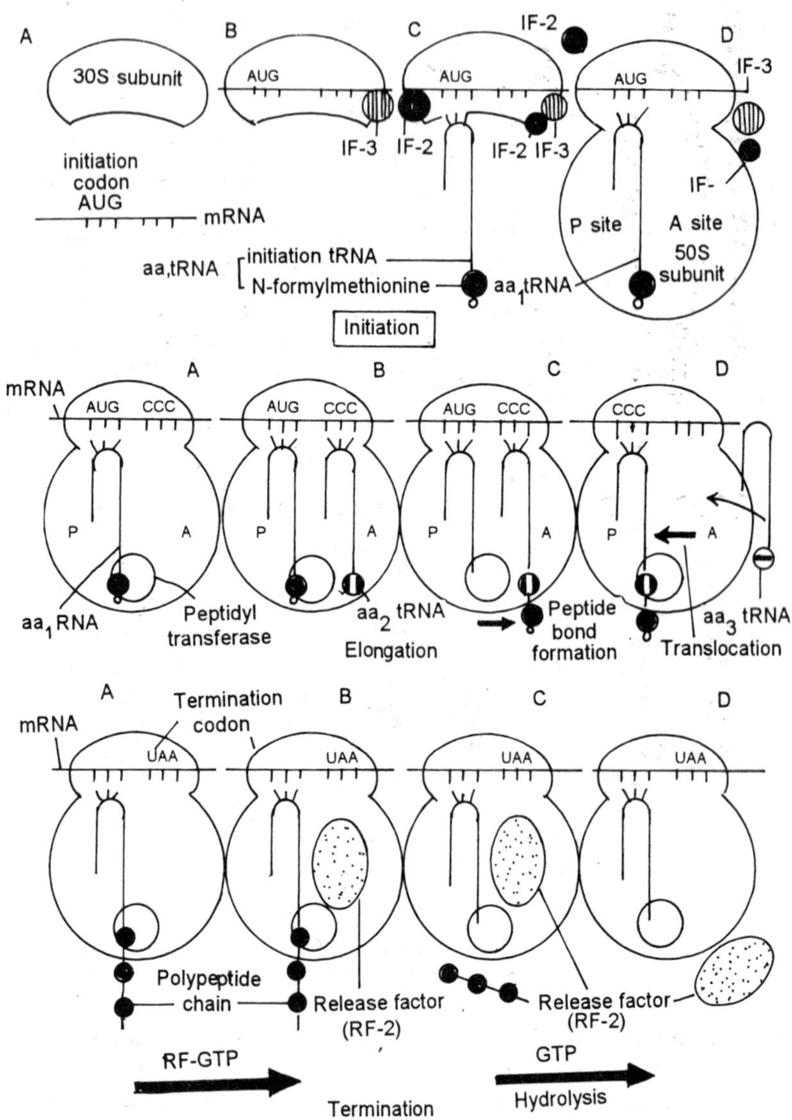

Fig. 7.5 Protein Synthesis
Different stages of translation

(iv) The polypeptide chain is processed by the removal of the methionine residue from the chain. Depending upon the circumstances, the chain may get cleaved into two or more pieces.

In eukaryotes, the polypeptide chain passes through a tunnel between the two subunits of the ribosome into the cavity of the cisterna of the endoplasmic reticulum.

Rate of Protein synthesis

The rate of transcription in *Escherichia coli* DNA for the enzyme controlling trypotophan synthesis is 28 nucleotides per second at 39°C. The speed of translation is 7 amino acids per second. For the haemoglobin molecule, the rate of translation is 4 aminoacids per second.

DIFFERENCE IN PROTEINS SYNTHESIS BETWEEN PROKARYOTES AND EUKARYOTES.

No.	Character	Prokaryotes	Eukaryotes
1.	Ribosomes involved in protein synthesis	70s, with 30s and 50s sub units.	80s with 40s and 60s sub units.
2.	Occurrence of transcription and translation	Translation begins even while transcription is transcription and still going on in mRNA	Translation begins after complete mRNA comes out of the nucleus.
3.	Migration of mRNA	Ribosomes pull out mRNA from the DNA	mRNA moves out of the nucleus through the pores in the nuclear membrane.
4.	Sites of initiation of protein synthesis in mRNA	many	one
5.	Initiator amino acid for protein synthesis.	N-formyul-methionine	Methionine
6.	Initiator factors	IF1, IF21 and IF3	eIF2, eIFwa1, eIF2a2 eIF2a3 and eIF3
7.	Elongation factors	EF-Tu, Ef-Ts and Ef-G	Ef-1, Ef-2
8.	Release factors	Rf-1, Rf-2 and RF-3	Rf (single factor)

8

GENETIC REGULATION

All the activities of an organism, be it unicellular or multicellular, are ultimately controlled by the genes. In multicellular organisms, where cells differ widely in structure and function all of them carry the same genes, since all of them are traceable to a single cell viz., zygote. This poses an interesting questioen as to the function of genes. How do similar genes function differently? The answer to active at all times. The genes are *switched* on and *switched off* at different times. This differential gene action is also responsible for the synthesis of different types of proteins (enzymes) at different times. Not all enzymes are required at all times. Enzymes required during early stages of development may not be required at later stages in the life cycle and *vice versa.* The metabolic products produced may induce or repress the enzyme synthesis.

Genetic control on protein synthesis is exercised at three stages – transcription stage, translation stage and at the release of polypeptide chain (post translation stage).

The control mechanism ensures that the cell is not unnecessarily flooded by the enzymes which are required only at particular times and only in particular quantity. Depending on the requirement the enzyme synthesis can either be induced or suppressed. Genetic regulation is seen both in prokaryotes and in eukaryotes. While the purpose of genetic regulation is same in both prokaryotes and eukaryotes, the mechanisms differ. In prokaryotes, the mechanism of genetic regulation is simple when compared with the eukaryotic organisms. The reasons for this are not far to understand. As the eukaryote organisms are far more complex they do require a very sophisticated mechanism of genetic regulation. In this chapter we will discuss genetic regulation in both prokaryotes and eukaryotes.

GENETIC REGULATION IN PROKARYOTES

Induction and Repression

In *E.coli,* the synthesis of the enzyme β *galactosidase* has been studied in considerable detail. This enzyme breaks up lactose into glucose and galactose. If lactose is not supplied to *E.coli* cells, the presence of the enzyme β *galactosidise* is hardly detectable, but as soon as the substrate (lactose) is added, the production of the enzyme β *galactosidase* increases by as much as 10,000 times. The production of enzyme again gets reduced if lactose is removed from the system. Such enzymes whose sythesis can be increased many times by the presence of the substrate are called **inducible enzymes.** The genetic system responsible for synthesis of such enzymes is called the *inducible system.* The inducible enzyme is an example of adaptive enzyme in the sense its (enzyme) synthesis gets increased due to the presence of the substrate.

In some other systems, the reverse of the inducible system occurs. For instance, in a culture of *E.coli,* if no amino acids are supplied from outside, the cells can produce all the enzymes required for the production of amino acids. On the otherhand, if a particular amino acid like histidine is added to the system, the production of histidine synthesising enzymes will fall down. In these systems the addition of a product of a synthetic pathway will check the production of enzymes responsible for the synthetic pathway. In other words increased presence of a substrate would reduce the synthesis of an enzyme which has the enhancing effect of the production of that substate. Systems such as these are called *repressible systems*, and the enzyme whose synthesis gets checked is known as the **repressible enzyme.**

Inducer and Co-repressor

It has been mentioned above that an enzyme gets induced or represented by the presence of a particular substrate. The substance which induces the synthesis of an enzyme is called the inducer and the substance which supresses the synthesis of the enzyme is called the repressor. In the synthesis of β *galactosidase,* lactose acts as the inducer and in the supression of histidine synthesising enzyme, histidine acts as the repressor.

Enzyme induction or repression is generally brought about by the substrate on which an enzyme can act. But in some instances, there will be molecules which resemble the substrate which also can act as the inducers. A typical example of this kind of induction is IPTG (isopropyl thiogalactoside). IPTG resembles lactose and can induce the production of the enzyme β *galactosidase.* Such inducers which are not substrates themselves but can bring about induction because of similarities with the substrate are known as **gratuitious inducers.**

In the above paragraphs, we have discussed as to how in the absence of lactose there is no synthesis of β *galacotosidase.* This means then in the absence of the inducer, the genes responsible for the production of the enzyme protein do not function. It will be interesting to find out as to why the enzymes do not function in the absence of lactose, and how in the presence of lactose the genes regulating the enzyme synthesis get activated. It has been found out that a group of molecules called *repressors* are present in the cell and they check the activity of the enzymes. In the presence of the inducers, the repressors get inactivated. In some instances, ordinarily the repressors are inactive but they become active in the presence of another group of molecules called co-repressors. The induction and repression is supposed to work in the following manner :

1. Inducible system : active repressor + inducer = inactive repressor
2. Repressible system : inactive repressor + corepressor = active repressor

The operon concept

F.Jacob and J.monod (1961), the Nobel laureates have studied in detail the inducible system in the synthesis of the enzyme β *galactosidase* in *E.coli.* In order to explain the induction and repression of enzyme synthesis, they have proposed the involvement of a group of genes along with the presence of cytoplasmic inducers and repressors. Their model of gene action is popularly called the operon model. This scheme of gene action proposed by Jacob and Monod as originally conceived by them consisted of an operator gene and a number of structural genes. As per the current understanding, however the operon system consists of structural gene, operator gene, promoter gene, regulator gene and repressor, inducer and co-repressor. The following is an account of the components of the operon model.

Structural genes

There are many structural genes associated in an operon system. These direct the synthesis of the mRNA and govern the sequence of amino acids in a protein molecule. Each structural gene might produce a particular kind of protein or all structural genes might regulate the production of a single protein. The activities of the structural gene(s) are controlled by the promoter and operator site of the operon system. The most well studied structural genes (z,y and *ac*), are those of the lactose operon system in *E.coli.*

Operator gene

The operator gene is situated adjacent to the first structural gene. It switches on or switches off the functioning of the structural gene (protein synthesis). In case a structural gene has to be suppressed, a repressor attaches itself to the operator to form an Operator-repressor complex. In the case of protein synthesis, the operator-repressor complex prevents the transcription by blocking the movement of RNA polymerase.

Promoter gene

The promoter gene is continuous with the operator gene and is believed to lie left to it. It is suggested that RNA polymerase binds to the promoter site during transcription. Three regions have been recognised in the promoter site. These are (i) recognition site, (ii) initial binding site and (iii) the mRNA initiation site (operator site).

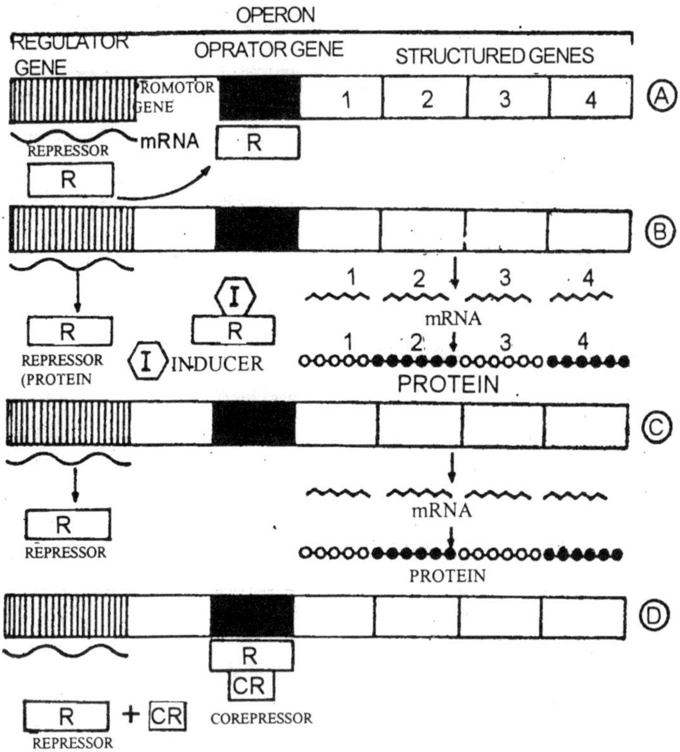

Fig. 8.1 Genetic regulation

Genetic regulation of protein synthesis in prokaryotes (the Operon system of Jacob and Monod, 1961)

(i) Recognition site

Also called the *cga* site (catabolic gene activator site), it consists of certain palindromic sequence of DNA. These symmetrical sections of DNA are recognised by proteins having symmetrically placed sub units. This site, also called the CRP site (cyclic AMP receptor protein site) binds a Carotenoids protein to the promoter gene and thus facilitiates the binding of the enzyme RNA polymerase. It has been found that in *E.coli*, CRP combines with cAMP (cyclic adenosine monophosphate), forming a CRP + cAMP complex which binds to the promoter enhancing the binding of RNA polymerase and activates transcription. This regulation is called positive control.

(ii) Initial binding site

This consists of seven bases (DNA) to which the RNA polymerase binds.

(iii) RNA initiation site

The site where transcription begins is called initiation site. This is the region overlapping with the operator region.

Regular gene

The regulator gene directs the activity of the operator gene by producing inhibitor proteins called repressors. This repressor protein binds to the operator gene and blocks the path of RNA polymerase, thus preventing transcription. If an inducer is present in the system, it binds to the repressor which undergoes conformational change and becomes inactive. As the inactive repressor cannot bind to the operator, the structural genes get activated and protein synthesis continues.

In a repressible system, if only a repressor is present, it is inactive by itself and cannot attach to the operator gene and hence transcription continues. However, when another protein called co repressor is present, it attaches to the repressor forming a *repressor-corepressor* complex, which blocks the operator gene thus preventing transcription.

Types of regulatory mechanism

Earlier in the chapter it has been mentioned as to how the presence or absence of a substrate can induce or suppress the synthesis of the enzyme protein. Here we shall study in some more detail the various regulatory mechanisms that are involved in the synthesis of enzymes.

Induction. It has already been mentioned that in inducible systems, the inducible enzymes are normally present in small quantities and their synthesis can be increased manifold by the presence of the inducer. Such as example of inducer in *E.coli* is lactose and the inducible enzyme is β *galactosidase.* In the absence of lactose, only one ore two molecules of the enzymes β *galactosidase* may be present in the system. When the inducer is added, within three minutes about 3,000 molecules of the enzyme are synthesised by the gene.

When the inducer is absent, the repressor protein will block the production of the enzyme by not allowing transcription of mRNA by the structural genes.

The repressor protein actually blocks the operator gene. When the inducer is present the active repressor binds to the inducer to form a repressor inducer complex. As a result, the repressor gets inactivated. An inactivated repressor cannot bind to the operator and hence the operator will allow the structural genes to transcribe mRNA and the enzyme is synthesised.

Repression. The presence of a metabolite inhibiting the synthesis of an enzyme in the system is caled repression. It has been mentioned already that histidine acts as a repressor and blocks the production of the enzyme responsible for histidine synthesis. This is called enzyme repression. Another example of repression is the blocking of the production of the enzyme tryptophan synthetase by the substrate tryptophan. The enzyme tryptophan synthetase is produced only when the bacterium (*E.coli*) is cultured in a tryptophan free medium.

In the repressible system, the protein produced by the regulator gene is an inactive repressor (*aporepressor*). On combining with a corepressor (effector), the aporepressor molecules get activated due to a conformational change. As a result of this the aporepressor becomes active and binds to the operator gene. As a result of the operator genes getting blocked the structural genes cannot by itself bind to the operator gene. The operator gene being free will allow the structural genes to transcribe mRNA. The following is a summary of the regulation of enzyme synthesis in inducible and repressive systems.

I. Inducible system

1. ***Inducer present :*** Regulator gene → Active repressor + inducer → repressor inactive → operator gene free → structural genes transcribe mRNA → synthesis of enzyme protein.
2. ***Inducer absent :*** Regulator gene → active repressor → binds to the operator gene which becomes inactive → no mRNA transcription → inhibition of enzyme synthesis.

II. Repressible system

1. ***Corepressor present :*** Regulator gene → inactive repressor + corepressor → active repressor → binds and inactivates operator gene → structural genes do not transcribe mRNA → inhibition of enzyme synthesis.
2. ***Corepressor absent :*** Regulator gene → inactive repressor → operator gene not blocked → structural genes transcribe mRNA → enzyme protein synthesised.

Negative control

In the inducible and repressive systems, protein synthesis is allowed to take place when the operator gene is free and it (protein synthesis) stops if the operator gene is blocked. In other words, gene expression is possible only when the operator is free. Such a controlled mechanism of protein synthesis is said to be of a negative type. A typical example of negative control is the production of the enzyme protein in the presence of lactose and in the absence of glucose. The presence of glucose actually inhibits the production of the enzyme β *galactosidase* while the presence of lactose promotes the production of the eyzyme.

In negative control, the regulator protein acting as the repressor prevents the transcription of mRNA by structural genes. The main controlling factor is the operator gene.

Positive control

In this control, the regulator protein functions as an activator and increases the enzyme synthesis. The activator binds itself to DNA in a location called the initiator site. Here the initiator site need not necessarily be the operator gene. As the regulator protein enhances the transcription of mRNA by structural genes this kind of regulation is called a positive control. A typical example of a positive control is the arabinose operon seen in *E. coli* (see later for details).

Replication of DNA is also an example of positive control. The bacterial genome is made up of replicating units called replicons which are capable of independent duplication. One of the structural genes of a replicon is known to produce an activator protein which in turn is induced by some cytoplasmic factors. The activator binds to a specific site on the replicon called the replicator and initiates replication from this point.

Catabolic repression

The works of Magasaine and others have shown that the presence of glucose inhibits the synthesis of enzymes involved in the breakdown of sugars such as lactose, arabinose or galactose. This phenomenon of repression is known as catabolic repression. This has been in many microorganisms including *E.coli,* yeasts etc. For instance, when a culture of *E.coli* maintained on lactose is supplemented with glucose, it has been seen that the synthesis of β *galactosidase* enzyme becomes suppressed as long as glucose is present in the medium. The inhibitory effect of glucose is due to a marked drop in the intercellular levels of a nucleotide cyclic AMP. The studies of Zubay and Beckwith (1970), have shown that glucose suppresses the level of cyclic AMP which in turn inhibits the transcription of mRNA.
In vitro studies of Eron, et al (1971), have indicated that lactose operon transcription requires not only cyclic AMP but also a protein called CAP

(catabolic activator protein). CAP binds to the cyclic AMP. Thus, catabolic repression has two steps at the first stage CAP binds to cyclic AMP and at the second stage, this complex (CAP cyclic AMP) exerts its effect on the transcription process.

Feed back inhibition

In some instances, the final product of a particular biosynthetic pathway gets accumulated if it is not utilised for metabolic activities. Such a kind of accumulation many a time will cause the stoppage of further synthesis of the substance. For instance, if the normally growing *E.coli* cells are supplied with isoleucine (an amino acid), it has the effect of blocking the isoleucine synthesising enzyme. This inhibition is due tothe binding of isoleucine to enzyme *threonine deaminase.* This kind of inhibition is called feed back inhibition. Feed back inhibition may be defined as the inhibitory effect of the final product on the enzyme involved in the first step of production when a synthetic pathway is made up of several steps and is controlled by the activity of specific enzymes. Feed back inhibition has an advantage in the sense if the enzyme of the first step is blocked, there is no necessity to block the intermediary enzymes as in any case they would not get the substrate.

Feed back inhibition is mostly through the allosteric transformation of the enzyme so that its active site gets inactivated. Feed back inhibition is different from repression in that in the latter there is no mRNA synthesis, and hence no enzyme synthesis. Feed back inhibition is actually a case of control of gene expression after translation.

Role of Leader sequences in genetic regulation

Gene expression in prokaryotes is controlled both at the level of transcription as well as after the transcription. In the mechanisms which operate after transcription, a prominent role is played by leader sequences and attenuators. For example, in the tryptophan operon of *E.coli,* the mRNA of gene trpE (gene for anthranilate synthetase) has a sequence of 162 bases between initiator codon and the promoter operator region of this gene. This sequence of 162 bases is called leader sequence and is thought to exercise post transcriptional control. This leader sequence of the gene trp E has a mechanism which allows transcription when synthesis is needed, and stop the transcription whenever synthesis is not needed.

Regulation by antisense RNA

Some of the recent researches have shown that regulation of protein synthesis can be controlled at the translational level by using RNA which has complementary bases to mRNA. When such an RNA is introduced, it binds to mRNA forming RNA mRNA hybrid and prevents the translation, it (RNA) which is this RNA is complementary to mRNA and interferes with the translation, it (RNA) is called antisense RNA. Antisense RNA can be used in experimental conditions to stop the translation of mRNA. In *in vivo* conditions also, antisense RNAs are known to regulate the translation of mRNA. In transposon Tn 10, the production of the enzyme *transposase* is prohibited by antisense RNA. The production of antisense RNA is controlled by a promoter which directs transcription in a direction opposite to the transposase gene, thus producing antisense RNA.

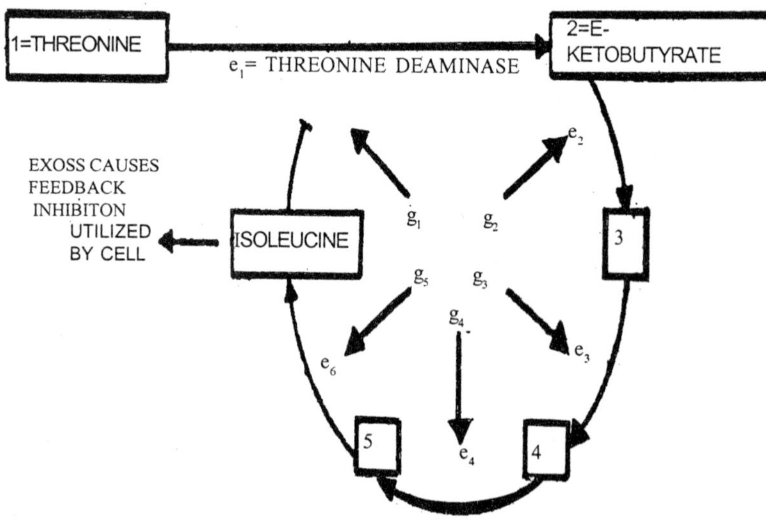

Fig. 8.2 Genetic regulation
Feedback inhibition in *E.coli* (after Stansfield, 1969)

Role of antisense RNA in regulating gene expression has also been demonstrated in the case of the two genes involved for the synthesis of outer membrane protein.

Examples of a few operon systems

The lac operon of E.coli. The lac operator consists of three structural genes.

1. Gene *z.* This is made up of 3,063 base pairs and codes for enzyme β *galactosidase,* which is active as a tetramer and breaks lactose into glucose and galactose.
2. Gene *y.* This is made up of 800 base pairs and codes for the enzyme β *galactoside permease* which is a membrane bound protein and helps in the transportation of metabolites.
3. Gene *ac.* This consists of 800 base pairs and codes for β *galactoside*

transacetylase, an enzyme that transfers an acetyl group from acetyl coA to β *galactosides*.

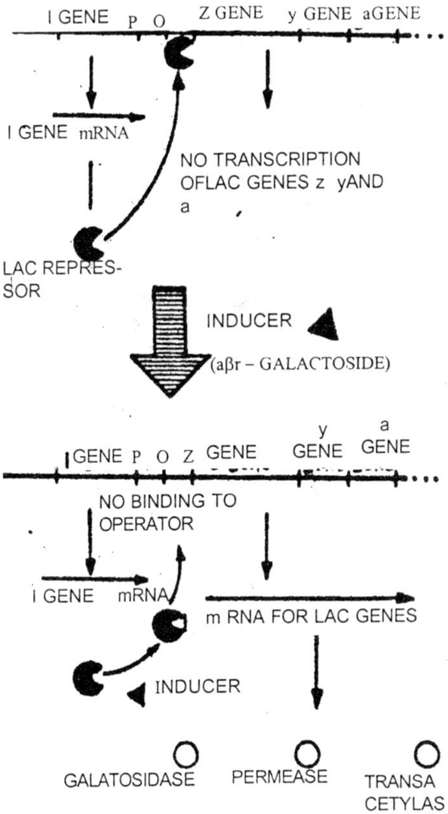

Fig. 8.3 Genetic regulation
Lac-operon model

In addition to the structural genes, the lac operon consists of a regulator gene *i*, an operator gene *o* and a promoter gene *p*. These six adjacent cistrons (genes) constitute the lac operon system. An operon can be defined as an integrated genetic unit.

In lac operon, the gene *i* produces a repressor (protein) which inactivates the operator by binding with it. Subsequently, an inducer such as lactose binds with the repressor molecule there by inactiviting it. As a result, the operon gene *o* is free from the repressor and becomes active. When the operator gene is active mRNA is produced from the structural genes *z, y* and *ac*. The production of mRNA starts from the promoter gene. It has been found that all the three structural genes produce a single mRNA molecule which codes for the sequence amino acids for all the three enzymes – β *galactosidase*, *permease* and *transacetylase*. Thus equal amounts of all the three enzymes are produced and the first message translated is for β *galactosidase;* this is followed by the synthesis of permease and then followed by transacetylase. Thus the operons show polarity. This has been proved by mutational studies. Mutations in the *z* region will reduce the quantities of all the three enzymes. While mutations in *y* locus will have no effect on β *galactosidase* but will reduce the quantities of other two enzymes. Such mutations are said to be **polar mutations.**

The gal operon system

In the gal operon system in *E.coli*, the sequence of three structural genes are gal E, gal T and gal K. These genes are directly responsible for the sequence of the enzymes produced namely galactokinase, galactose-1-phosphate uridyl transferase, and uridinediphospho-galactose-4- epimerase. These enzymes are responsible for the degradation of galactose into glucose-1-phosphate and UDP glucose.

Tryptophan operon system (trp operon)

The trp operon is also seen a *E.coli,* and consists of five structural genes ABCD and E. These genes code for the following enzymes :

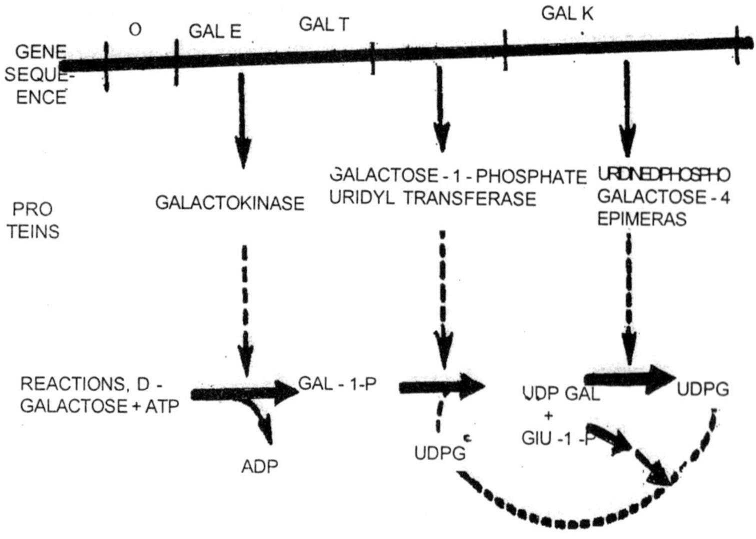

Fig. 8.4 Genetic regulation

The galctose (gal) operon of E.coli. Abbreviations : O-operator, Gal-1-galactose-1-phosphate, UDPG-uridine diphosphoglucose; UDP GA1-uridine diphosphogalactose; Glu-1-P=glucose-1-phosphage (after Goodenough and Levine, 1974)

1. Genes A and B code for tryptophan synthetase.
2. Gene C codes for PRA isomerase and IGP synthetase.
3. Gene D codes for phosphoribosyl pyrophosphate anthranilate transferase.
4. Gene E codes for anthranilate synthetase.

The levels of these enzymes is low when the amount of tryptophan is more and the level high when the amount of tryptophan is less. All these enzymes are regulated in a coordinated fashion. The operator and promoter are overlapping and are located next to *trp* E gene. The regulator gene (R) is located away from the operon and the repressor produced by it is inactive until it combines with tryptophan. All the enzymes are produced constitutively if the repressor is either absent or fails to combine with tryptophan. All the structural genes are translated together and form a polycistronic mRNA. This polycistronic mRNA besides the transcription of the bases of structural genes has a 160 nucleotide long leader sequence. This contains the attenuator site to which the substrate tryptophan can get attached and thus reduce the rate of transcription. Thus in *trp* operon, there are two regulation – the regulator gene and the attenuator site in the RNA itself.

The *trp* operon thus seems to have a dual control–a control by regulator and also by attenuator. In addition to this, feed back regulation is also seen in *trp* operon.

Arabinose operon

This is also seen in *E.coli* and is an example of positive control. Like the lac operon, the arabinose operon also consists of three strcutural genes. These are –

(i) *ara* A gene codes for arabinose isomerase
(ii) *ara* B gene codes for L-ribulokinase
(iii) *ara* B gene codes for L-ribulose-5-phosphate epimerase

These structural genes are located together on the *E.coli* chromosome. Their expression is under the regulation of a regulator gene C. This gene C called activator must be present for transcription of the arabinose operon. The product of C gene binds to the operator in the presence of arabinose and forms the activator protein. As a result of this, transcription takes place in the arabinose genes. If the activator is not bound to the arabinose or if the arabinose is absent, the operon is not expressed because the regulator protein binds to the operator gene (*ara* o) and functions as a repressor. In arabinose operon, the gene C acts both as an activator and repressor. In this way, arabinose operon is different from lac operon.

Gal genes in yeasts

In yeasts, galactose is fermented by the activity of three enzymes–kinase, transferase and epimerase. These are coded by the three structural genes GAL 1, GAL 7 and GAL 10. The regulator gene is located away from the structural genes. If a mutatkon occurs in a regulator gene, all the three enzymes are synthesised without control indicating thereby that the product of the regulator gene is a repressor. The regulator gene of gail in yeasts is very much similar to the gals in *E.coli*. A detailed genetic analysis has revealed that the product of the regulatory gene controls the functioning of an unlinked gene GAL 4. The product of GAL 4 is a positive regulator for the three linked genes. The GAL 4 gene has an operator gene *(c)* also. The regulator gene is *i,* the product of this gene exerts a negative control over gene c (operator). As a result, transcription of GAL 4 is reduced. When GAL 4 is free it produces a product which stimulates the transcription by the structural genes.

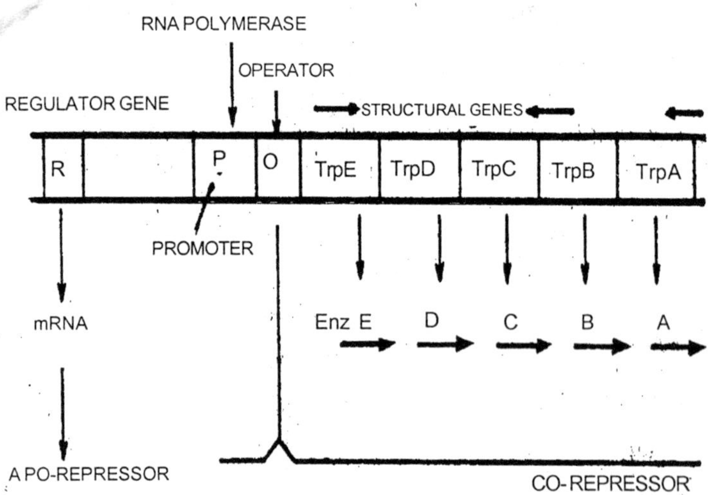

Fig. 8.5 Genetic regulation
Tryptophan operon

Regulation of transcription in λ phage DNA

Regulation of the λ phage DNA has both negative and positive control mechanisms. The phage exists in a lytic or lysogenic state and this is controlled by some regulatory proteins. Depending on when the genes get activated during infection, multiplication or maturation phase, the λ genes can be classified into *early genes* and *late genes.* The early genes as the name indicates are required at the beginning of the infection cycle, while the late genes are required for the assembly and maturation of phages. The following table gives a list of genes identified on the λ genome along with their functions.

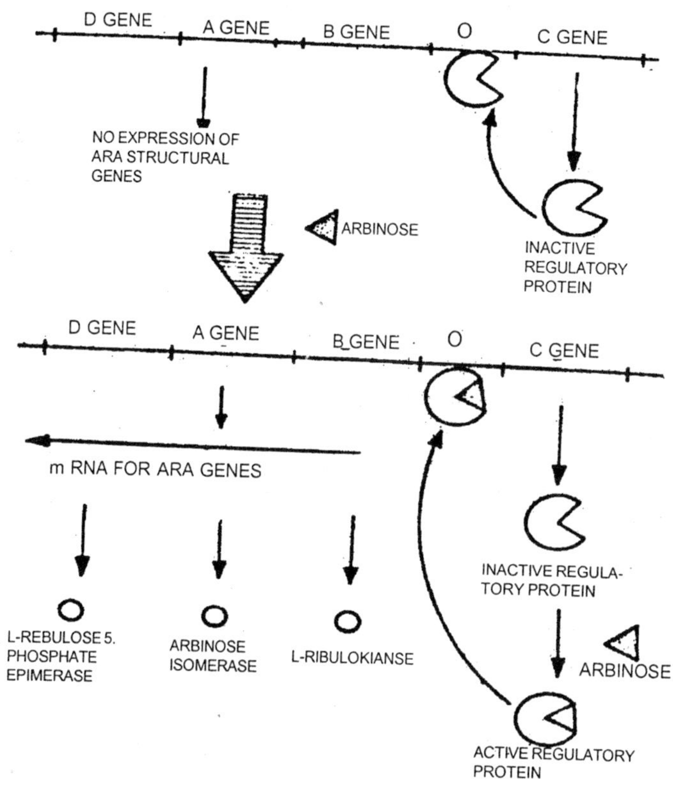

Fig. 8.6 Genetic regulation
The L-arabinose operon of E.coli

List of genes of the λ phage genome

A,W,B,C,D,E and F	-	Head genes
Z,U,V,G,T,H,L,K,Iand J	-	Tail genes
att λ of the	-	site of attachment and insertion phage DNA into the host DNA

Gene		Function
int	-	integration of λ DNA e
xis	-	excision of λ DNA
red α	-	exonuclease
red β	-	λ recombination system
c III	-	required for the expression of the c I gene
N	-	required for the expression of early gene
OL	-	operator of a set of early genes
rex	-	restriction of growth of T4 R II
c I	-	λ repressor
OR	-	operator of another set of early genes
tof	-	negative regulator of c I transcription
c II	-	required for the expressions of the c I gene
ori	-	point of origin of replication
O	-	required for λ replication
P	-	required for λ replication
Q	-	stimulates transcription of late genes
pl	-	promoter of late genes
S	-	required for host lysis
R	-	required for host lysis

The cI gene codes for the repressor which binds to the operator genes thereby blocking the transcription of both the early operons. The repressor can bind to the operons blocking its integration and replication resulting in immunity due to a negative control exerted by a repressor. The product of gene N, however, acts as a positive regulator and allows transcription. The product of the N gene is known as antiterminator because instead of stopping, it allows transcription. Another example of positive regulation is the influence of Q gene product on late genes. This product combines with the coat transcriptase enzyme of *E.coli* and acts favourably on the promoter of the late genes. This is the λ phage, there are both negative and positive regulatory mechanisms to control the initiation and completion of lytic or lysogenic cycle. This is very necessary for an organism which has to survive in a continuously changing environment such as the bacterial cell.

GENETIC REGULATION IN EUKARYOTES

The mechanism of gene regulation of protein synthesis explained above illustrates the phenomenon in prokaryotes, whereby a repressor protein transcribed by regulator genes may activate or prevent the protein synthesis depending on the presence of inducer or corepressor.

In eukaryotes, genetic regulation takes place by a slightly different mechanism. Two types of proteins associated with the DNA of the chromosomes possibly play a role in gene regulation. These are histones and non-histone chromosomal proteins (NHC proteins). Since histones of different organisms have similar amino acid composition it is unlikely that they can selectively repress gene action. At best, histones may probably bring about a general (non specific) repression. It is believed that histones under certain situations bring about supercoiling of DNA preventing the movement of RNA-polymerase and thus blocking transcription. This is known by the fact RNA precursors are incorporated only in those regions where chromatin is diffuse (no super coiling).

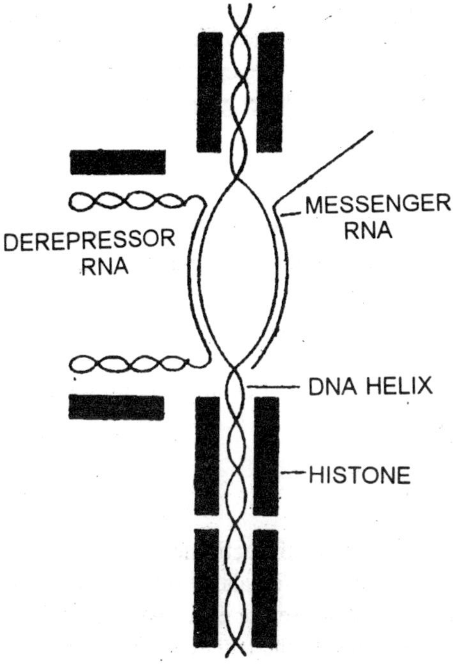

Fig. 8.7 Genetic regulation

Genetic regulation of protein synthesis in Eukaryotes-Frenster's model of gene regulation

Three models have been proposed by various scientists to account for gene regulation in eukaryotes. These are - 1. Frenster's model, 2. Non-histone depressor model and Britten and Davidson's model.

Frenster's model

This is also called gene specific derepressor RNA model. As has been pointed out earlier, histones bring about non-specific repression of transcription. Derepression (initiation of transcription) is brought about when nuclear polyions separate histones from DNA at certain specific parts (of DNA), and form complexes with histones.

Non-histones derepressors

The model proposed by Paul et al (1971), is similar to Frenster's model except the derepression is brought about by NHC acid proteins instead gene specific RNA. The fact that NHC proteins are of immense variety in different organisms, and also at different stages of growth even in the same individual goes to show that they can undertake selective derepression at specific gene loci. Derepression is brought about when NHC proteins bind themselves at specific loci to histones forming complexes. This complex separates itself from the DNA and allows it to transcribe, only at those specific loci.

Gene regulation by NHC proteins

(a) Non specific repression by histones.
(b) Specific binding of NHC proteins to histones.
(c) Removal of histone by NHC proteins.
(d) De repression of DNA, leading to transcription.

Britten Davidson Model

Also known as operon operator model, this is somewhat different from the other two models. As per this model, four types of genes are involved in the regulation process. These are sensor genes, producer genes, integrator genes and receptor genes.

1. Sensor genes

Comparable to the promoter genes, these are sensitive to the metabolic status of the cell. The sensor gene is stimulated by various metabolic substances.

2. Producer genes

These regulate the output of metabolism and control the formation of enzymes, cell organelle, membranes etc.

3. Integrator gene

As the name suggests, the gene integrates the activities of other genes. The integrator gene sends specific molecular signals to other genes as to whether the sensor gene is activated or not. Each sensor gene may be associated with a single or a group of integrator genes. In the latter case, a sensor gene may send a variety of molecular signals to other genes through its various integrator genes.

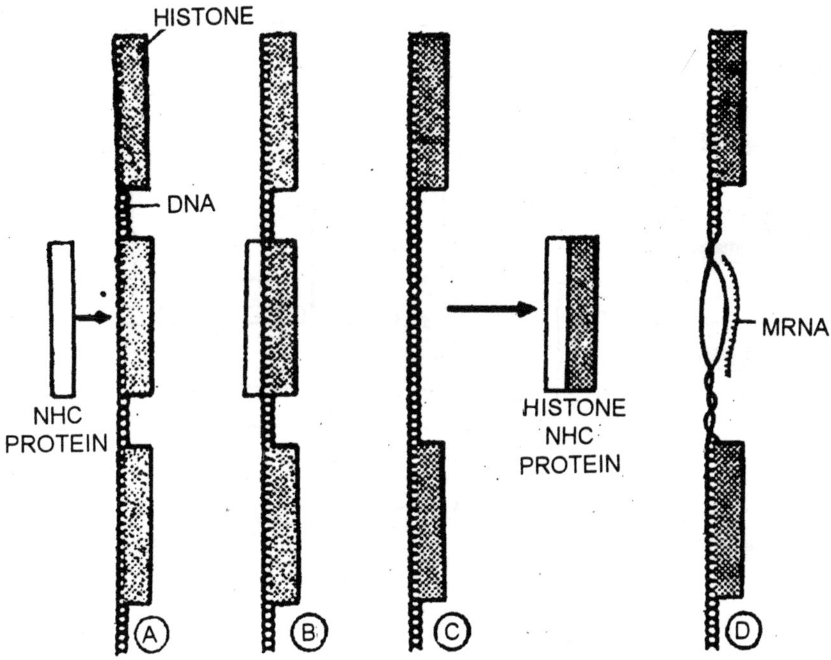

Fig. 8.8 Genetic regulation
Gene regulation by NHC histones

A. Non specific repression by histones
B. Specific bindng of NHC proteins to histones
C. Removal of histone by NHC proteins
D. De repression of DNA, leading to transcription

4. Receptor gene

This acts as an intermediate between the integrators and the producers. When the molecular signal reaches the receptor gene, it inturn activates the producer gene. A producer may have one or a group of receptor genes associated with it in the same way as sensors have seveal integrators. This helps, not only in the variety of origin of molecular signals, (from sensors) but also variety in reception of these signals (by receptors). Perhaps, multiintegrator gene complex and multireceptor gene complex may have a specific multichannel relationship.

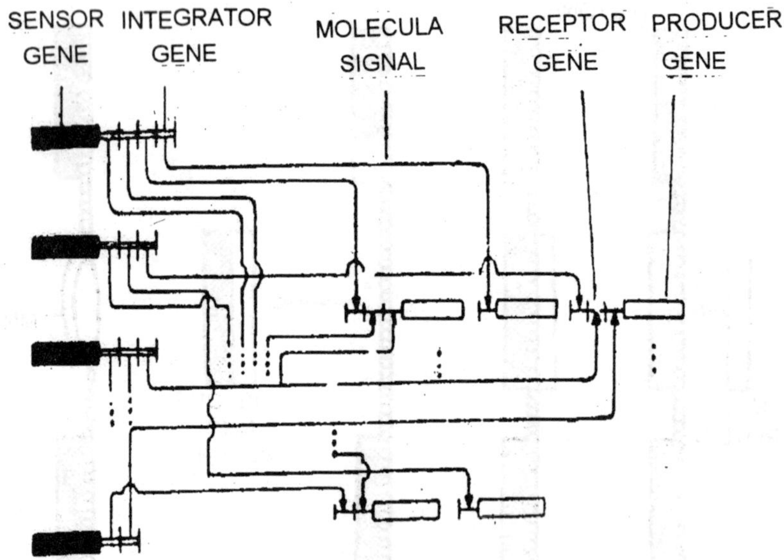

Fig. 8.9 Genetic regulation
Britten Davidson's Operon-Operator model

9

GENETIC RECOMBINATION

Genetical studies in microorganisms is extremely difficult, as they are very minute in size and do not possess any variability in their phenotype which can be subjected to genetic analysis. Another difficulty of microbial genetical studies is that microorganisms reproduce asexually to multiply their number and microbiologists thought that microbes like bacteria do not reproduce sexually. Extensive and repeated reseaches were made to explore the possibility of sexual reproduction in bacteria. Most of the earlier attempts were unsuccessful, because they were based purely on morphological observation of cell fusion. When Luria and Delburck established the assay procedure for obtaining the meaningful results, bacteriologists knew that their approach towards bacterial sexuality should change and a new technique is necessary for the analysis of sexuality in bacteria. The first successful experiment in bacterial genetics was performed by Joshua Lederberg and E.L. Tatum. These two collaborated in 1946, and using K_{12} strains of *E.coli* conducted what can now be called a classical experiment in bacterial crossing. They were able to demonstrate, that in *E.coli,* exchange of biochemical traits is possible. For their work on bacterial genetics, lederberg and Tatum shared with G.W. Beadle, the Nobel prize in medicine, for the year 1958.

With the establishment of the fact, that bacteria do exhibit sexuality, a vast area of microbial genetics opened up for geneticists. Prokaryotic organisms like bacteria were favoured for genetic experiments because –

1. Large amount of material could be handled as they are small.
2. Several generations could be obtained within a short period.
3. Pure cultures could be easily obtained for analysis.

With the advantages mentioned above, microbial genetics has made rapid strides in the last few decades.

Types of genetic material transfer

As in any other organism in bacteria also, sexuality involves transfer and recombination of genetic material from one bacterial cell to the other. Three methods of genetic recombination have been demonstrated in bacteria. These are conjugation, transformation and transduction.

CONJUGATION

Genetic recombination

Lederberg and Tatum (1944), demonstrated this in experimenting with the biochemical mutants of *Escherichia coli.* These bacteria require the following four nutrients 1. Biotin 2. Methionine 3. Leucine 4. Thereonine. But they can be grown on a minimal medium (a medium containing only agar and carbohydrates) because, by using carbohydrates they can synthesize all the above four nutrients. They are called 'Paratrophs' and can be genotypically designated as follows :

$$B^+ M^+ L^+ T^+$$

In this, letter stands for the nutrient (B = biotin etc) and the sign + stands for the capacity to synthesize that particular nutrient.

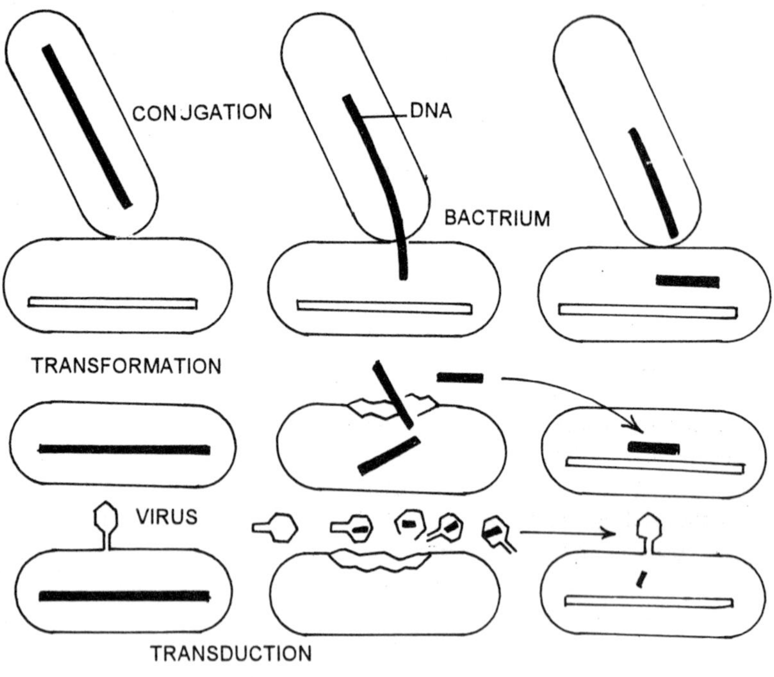

Fig. 9.1 Genetic Recombination
Genetic recombination in bacteria

By subculturing in the laboratory, the paratrophs undergo mutation and lose the capacity to synthesize one, or more nutrients. The mutants are called *Auxotrophs*. By careful isolation, Lederberg and Tatum obtained two complementary polyauxotrophic varieties of *E.coli*. They can be designated as follows

1. $B^+ M^+ L^+ T^+$
2. $B^- M^- L^- T^-$

(The sign–stands for the loss of the synthesizing capacity. For ex : L means leucine cannot be synthesized, and hence it has to be added to the culture).

It is obvious that both strains 1 and 2 cannot be cultured on a minimal medium. Strain 1 cannot be cultured because it cannot synthesize leucine and theronine. Strain 2 cannot be cultured because it cannot synthesize biotin and methionine.

But when both of them are kept on a minimal medium, curiously they survive (normal expectation is, they should not live). On nutritional analysis, it was revealed that they were Paratrophs. The formation of paratrophs from two polyauxotrophic varieties can be explained only on the possibility of sexual recombination between strain 1 and 2.

$B^+M^+L^+T^+$ $B^-M^-L^-T^-$
1 2

Because of recombination of 1 and 2 the following combinations may be obtained.

$B^-M^-L^-T^-$ and $B^+M^+L^+T^+$
(a) (b)

Of the above, (a) will not survive because it has lost the capacity to synthesize all the nutrients, but (b) will survive because it has become a paratroph.

Mechanism of conjugation

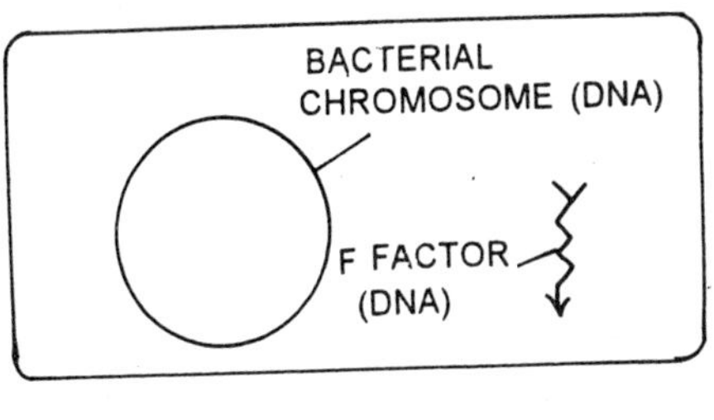

Fig. 9.2 Genetic Recombination
A bacterial cell showing circular DNA and F factor

The exact mechanism of genetic exchange was worked out by three groups of scientists led by Lederberg in Wisconsin USA, Hayes in London and Woolman in Paris. Genetic transformation was not involved in the process, was confirmed when cell free extracts of one bacterium failed to transformer other cells. It was demonstrated that the two bacterial cells physically come together and DNA from one passes into the other through the conjugation tube.

It was observed that in bacterial cultures, conjugation did not occur randomly between cells; rather it was selective. Genetic material seemed to migrate selectively only from particular cells to certain other specific cells. What is this specificity? What determines that only certain cells can conjugate while the others cannot. Careful studies indicated that the K_{12} *E.coli* has two strains – *Donor strain* and *recipient strain.* The donor strain possesses an episome (plasmid), while the recipient does not have this episome, as F+. The recipient strain does not have F factor and can be designated as F-. The donor strain may also be called male and the recipient female.

The fertility factor is free of the main genome of the bacterium and lies freely in cytoplasm. It is made up of a small DNA molecule having a molecular weight of 2,50,000. Normally, only one F factor is present in a cell and it replicates along with the main genome during cell division.

When conjugation occurs between F and F cells, the main genome is not involved. Consequently, chromosomal genes are not transferred from donor to recipient cell. In normal conjugation, the F factor of donor replicates and one copy of the F factor migrates along the conjugation tube to the recipient cell. As a result, both the cells now have a F factor. Consequent to this transfer, even the original recipient cell now becomes a donor as it has got an F factor. In other words, both the cells are now F^+. This means that the female cell has become male. The newly converted donor cell can now transfer its F factor to other recipient cells by *infecting* them with F factor.

If the entire conjugation process were to be limited to only transfer of the F factor with chromosome genes not being involved, it would have been perhaps of little genetic interest. But the discovery of another K_{12} *E.coli* strain by Cavalli- the *Hrf* (High fertility) strain changed the entire approach to the phenomenon of bacterial conjugation.

What is a *Hfr* strain? The F factor in the donor cell exists freely in cytoplasm – a point that has already been stressed. Occassionally, however, the F factor gets integrated with the chromosome (main genome). The F^+ cell that has an integrated F factor (with the chromosome) is now called a *Hfr* strain, as it can increase the rate of recombination many fold. So a F^+ male that brings about recombination has now been converted into a supermale (High fertility male). Integration of F factor into the chromosome involves a break (in the circular chromosome) and attachment of F factor. As a result of this integration, the chromosome along with its F factor becomes linear with two free ends. It is this chromosome. As the F factor becomes a part of the chromosome and as the chromosome has free ends, one end enters the other bacterium through the conjugation tube.

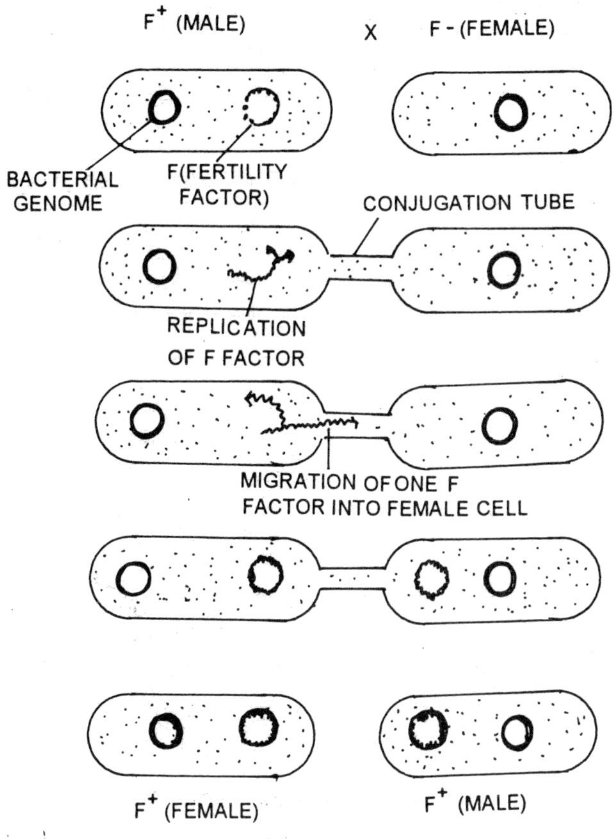

Fig. 9.3 Genetic Recombination
Conjugation between F+ (male) and F- (female) resulting in the conversion of female into male due to transfer of the fertility factor, F.

When a *Hfr* cell conjugates with a F cell, the latter receives the chromosomal genes. Which end of the chromosome of Hfr cell enters the F-cell? It has been found out that when F+ joins the chromosome with a single nick, it will be at one end. This end can be called posterior and the other end, anterior. Usually, chromosome movement from *Hfr* to F begins at the anterior end and the F factor enters the F strain only when conjugation is complete. This, however, rarely happens as conjugation and the F factor remains in the Hfr due to partial conjugation. Two possibilities exist here.

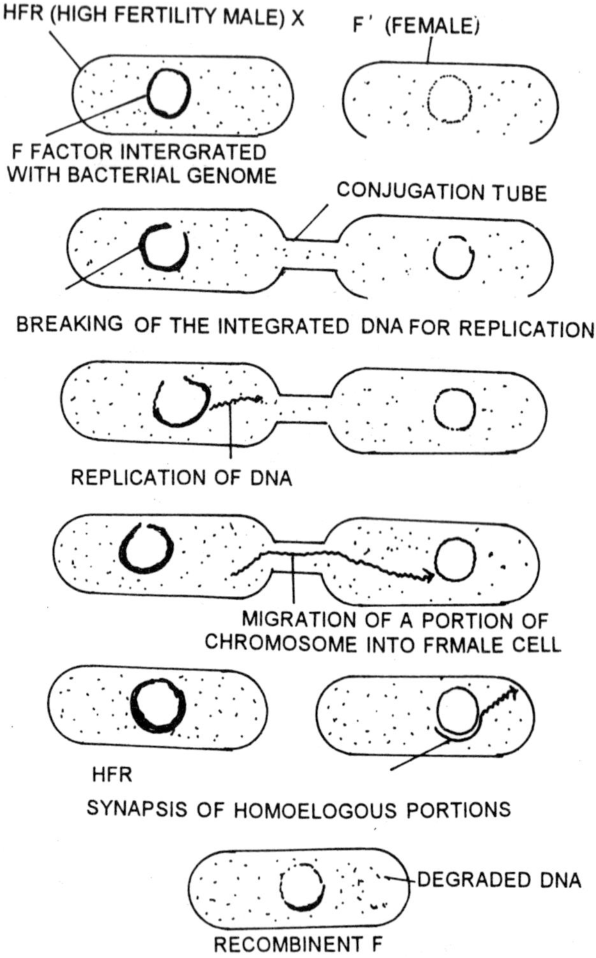

Fig. 9.4 Genetic Recombination

Conjugation between high fertility male (Hfr) and female (F-) cells, resulting in production of recombinant female

1. Complete conjugation

Total transfer of genome along with F factor from *Hfr* to F. As a result, not only the F acquires the genetic traits of *Hfr*, it becomes a Hfr. This is total recombination.

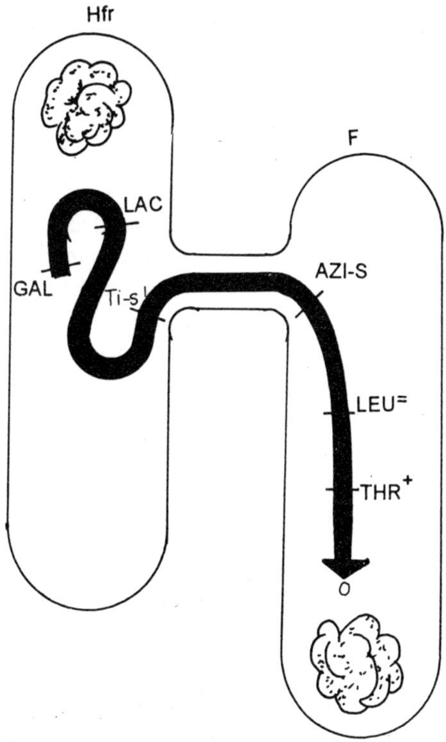

Fig. 9.5 Genetic Recombination
Transfer of genes from Hfr to F in a definite linear order

2. Interrupted or partial recombination

Here, chromosome entry from *Hfr* to F stops in the middle. As a result, F remains female (F) only, but acquires new chromosomal genes as a result of partial recombination. Whichever genes have entered from *Hfr* to F, they express themselves i.e., the F acquires new phenotype.

The F factor attachment to the chromosome is not at any fixed locus. There are several loci at which the chromosome can have a single break and the F factor can integrate to change the cell into a Hfr from F+. In otherwords, the integration of F factor into the chromosome may take place at any one of the possible loci. This interesting observation has led to the conclusion of the existence of several Hfr strains ategorized on the basis of locus of attachment of F factor to the chromosome. This can be further clarified as follows–let us suppose that the bacterial genome has genes A to Z on it. Theoretically, the F factor can attach between A and B or Z and A, C and A and soon. The point of attachment of F factor can be proved by the fact, that if gene A appears first in the recombinant followed by B, then the attachment is between A and B. The F factor always forms the rear end of the genome. For different Hfr strains, the recombination values of various genes in a population varies. But when once, the recombination frequency for a gene is known, the frequency of other genes is always fixed. In otherwords, in a chromosome from Hfr whichever gene may enter first, its distance from other genes will not vary. This aspect has been exploited for mapping the genes on bacterial chromosome.

Interrupted conjugation and chromosome mapping

It has been pointed above, that in a mating between *Hfr* and F, the process of conjugation is rarely completed. More often, it is interrupted by several environmental factors and as a result only few genes of the *Hfr* chromosome get transferred (after duplication) to F. This can be tested as the recombinant acquires a new phenotype (as it has genes obtained from *Hfr*). This interruption of mating can be artificially induced in a culture medium and the recombinant tested for acquiring the new trait. By carefully controlling the duration of mating one can observe, the series of new phenotype acquired by recombinant. Let us suppose, after 5 minutes of conjugation *gal* - gene enters the recombinant, and after 10 minutes *lac* - gene enters followed by leu-gene after 15 minutes, it is obvious that the genes are located in the following sequence.

gal- ———————— lac- ———————— leu-

By following the method of controlling conjugation time, one can map all the genes on a bacterial chromosome.

Jacob and Woman (1957), used the technique of "interrupted mating" between a *Hfr* strain and a F strain. At different time intervals, the mating mixture was diluted and agitated on a waring blender to separate the mating cells forcibly, thus interrupting conjugation and then plated on selective agar media. They demonstrated a lag period of 8 minutes before the first recombinant appeared. The recombinants then appear for other genes in a linear order in relation to the time allowed for conjugation. From this, Jacob and Wolman concluded that the latent period of 8 minutes is required for the conjugating cells to make contact,

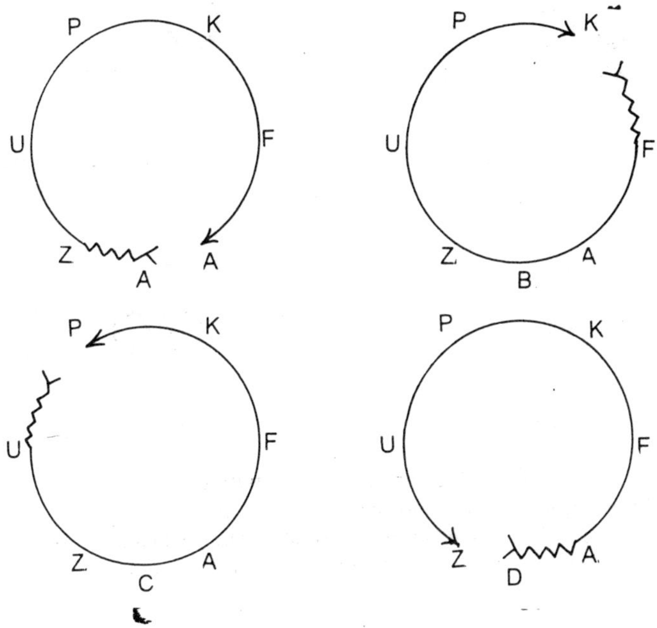

Fig. 9.6 Genetic Recombination
(A,B,C,D). Four hypothetical Hfr strains having different genes at their proximal ends

whereafter, the genetic material from *Hfr* is transferred to F in a linear order. They showed that 89 minutes are required for the complete transfer of the chromosome form *Hfr* to F. The bacterial chromosome is therefore, 89 time units long. Wolman and Jacob in 1956, produced the first genetic map of a bacterial chromosome in a *Hfr* strain.

The transfer of genes in linear order from male bacterium (tagged with virus) to female; the length of transferred segment is dependent the time taken for transfer, and the chromosome is, therefore, mapped in terms of time units.

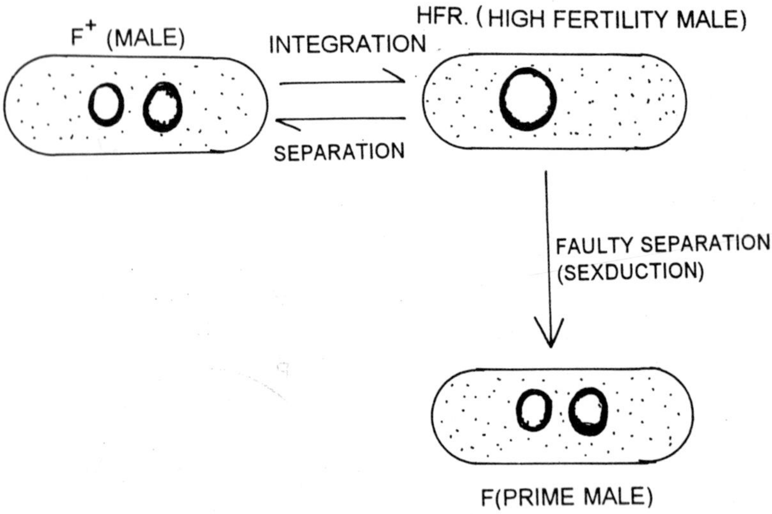

Fig. 9.7 Genetic Recombination
Showing interconversion between male and high fertility male. Faulty release of the fertility factor results in Sexduction and production of F (F prime) mal

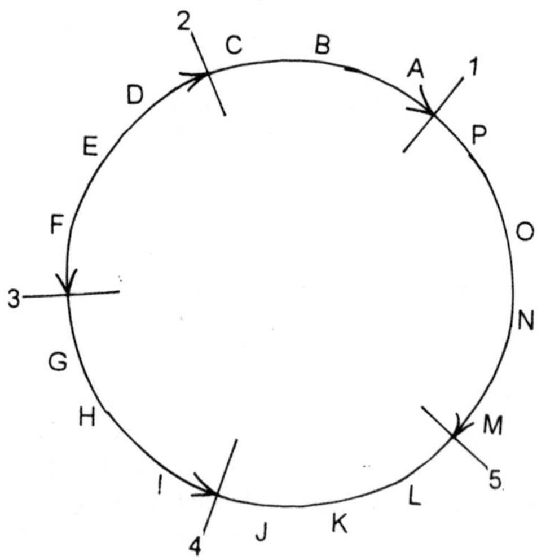

Fig. 9.8 Genetic Recombination
The linear order of genes in five hypothetical Hfr strains, resulting due to breakage of bacterial strain at different sites and subsequent orientation acquired by the resulting linear chromosome

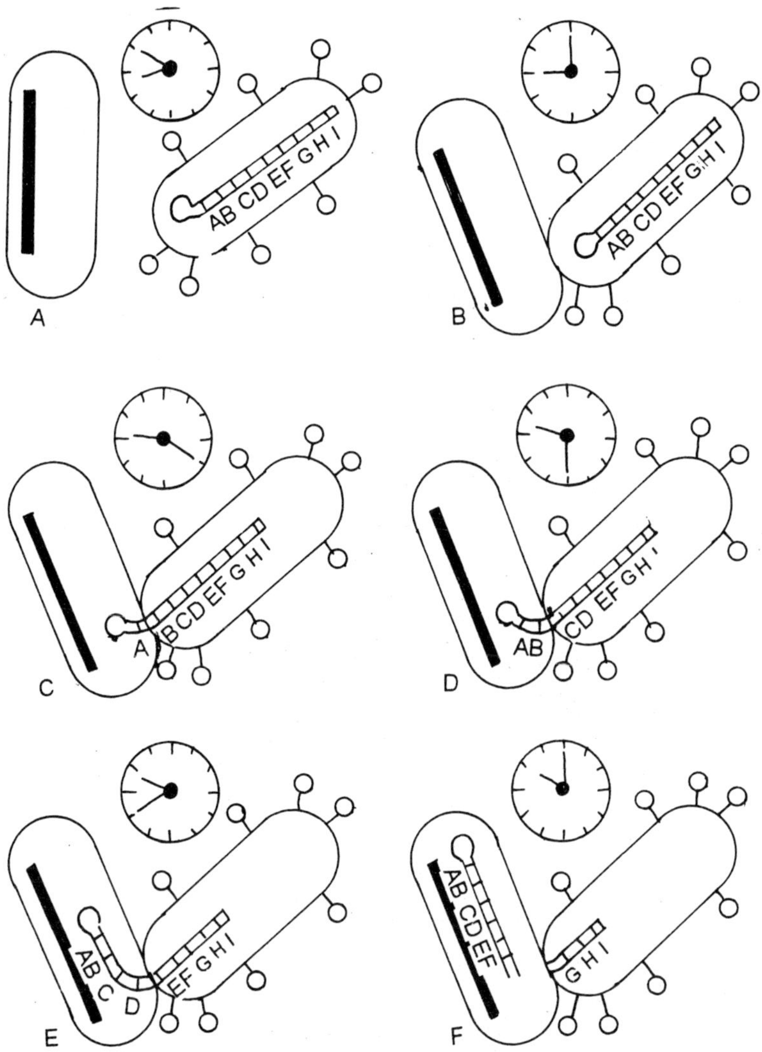

Fig. 9.9 Genetic Recombination
The transfer of genes in linear order from male bacterium (tagged with virus) to female; the length of transferred segment is dependent upon the time taken for transfer and the chromosome is, therefore, mapped in tems of time units

Sex duction

A *Hfr* strain originates as a result of integration of F factor with the bacterial chromosome. After integration into the chromosome, the F factor might get separated again and remain as an independent entity in the cytoplasm. If this happens, the *Hfr* strain would change into a F^+ strain. In otherwords, *Hfr* and F^+ strains are interconvertible. (Fig. 9.7)

In some unusual cases, when an integrated F factor gets detached from the chromosome, it may carry along with it some of the chromosomal genes (some porition of the chromosome gets torn and remains attached to the F factor).

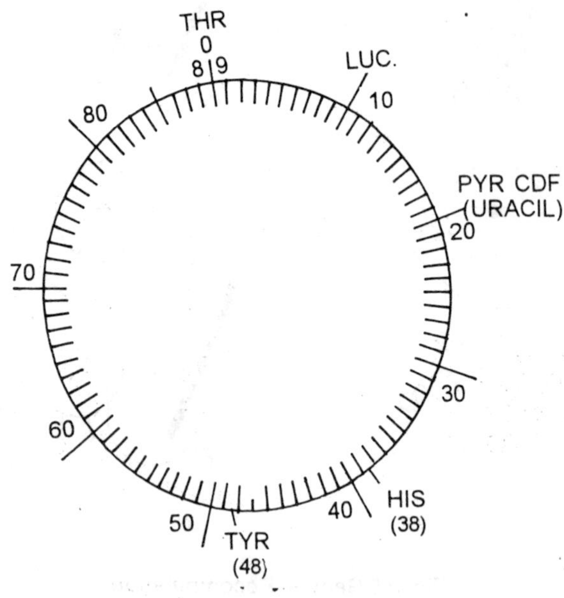

Fig. 9.10 Genetic Recombination
Circular linkage map of *Escherichia coli* chromosome, the units of recombination shown as minute required for transfer

Such an independent F^+ factor is given the name F^1 (F prime). An F^1 cell is male and like the F^+ except that the F factor has some genes of the chromosome.

When a F^1 (male) cell conjugates with a F cell, the recipient cell on getting the F^1 becomes a donor. In addition, it becomes diploid for certain genes of the chromosome which are present in the F^1. Because these genes are present on the chromosome as well as an F^1, this phenomenon is known as *sex duction* (Jacob and Alderberg, 1959). The recipient cells become *paritial diploids* for the genes introduced by the F^1.

GENETIC TRANSFORMATION

Every individual born out of sexual reproduction will have the mixture of traits of both the parents. In genetic terms, this means that there is genetic recombination in the off spring. Genetic recombination or transformation which is an index of genetic change has also been reported in bacteria.
Genetic transformation i.e., change of one genetic type to other was first reported by British bacteriologist F.Griffith (1928), while working with pneumococci (*Diplococcus pneumoniae*).

There are two strains in *Diplococcus pneumoniae.* One is capsulated, and the other one is non-capsulated. Non-capsulated ones form glistening smooth (S) colonies on the culture medium, while the non capsulated ones form dull, rought (R) colonies. Further, the capsulated (S) strains are virulent while the non-capsulated (R) strains are non virulent i.e., do not produce toxins. The ability of producing or not producing a capsule is a genetic character.
Griffith injected experimental mice with R strains of pneumococci and found that they survived, but when S strains are injected, the mice died. When heat killed (killing the bacteria by applying heat when DNA gets denatured and will not produce toxin) 'S' strains were injected the mice survived. Griffith next injected both R strain and heat killed 'S' strain to the same experimental mice and they (mice) died. Blood analysis of the dead mice (which were infected with R strain and heat killed S strain) showed the presence of 'live' S strains. There are three possibilities how the heat killed 'S' strains can become live again.

1. Heat killed S strain might have become alive in the body of mice–this is an impossibility.
2. A mutation might have occured in the R strain converting them to S strain. This is also not possible because mutations will not occur frequently, but everytime the experiment showed the presence of capsulated S strains in the blood of the dead mice.

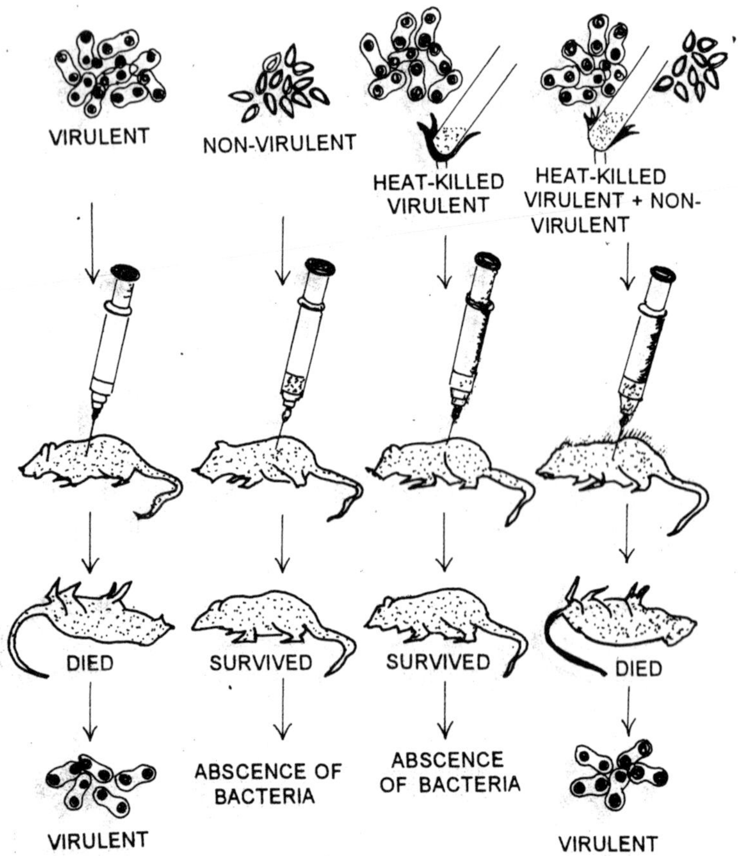

Fig. 9.11 Genetic Recombination
Schematic representation of Griffith's experiment to demonstrate transformation in *Diplococcus pneumoniae*

3. Some factor in the heat killed strain has transferred the genetic information for capsule synthesis to R strain making them produce the capsule. In other words, they become S strain. This is the only way a live S strain could be found in the blood of mice. Griffith (1928), thought that the transforming principle was a bacterial polysaccharide. But, Avery *et al* (1944), demonstrated that the transforming principle was DNA.

This experiment clearly demonstrates that a piece of DNA responsible for capsule synthesis recombined with the DNA of the R strain inducing it to produce a capsule. This can be called a natural method of genetic engineering.

Mechanism of transformation

This involves the following steps – competence, binding, penetration, synapsis and integration.

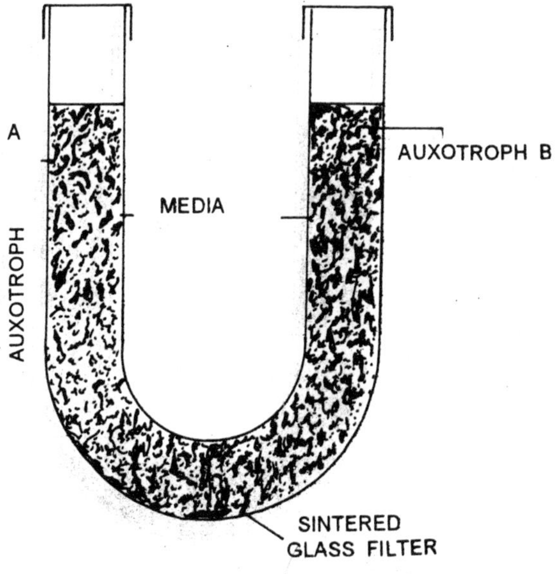

Fig. 9.12 Genetic Recombination
The Transduction (U tube) experiment of Zinder and Lederberg

Competence

The recipient bacteria must be competent if they are to accept the foreign DNA and undergo transformation. Competence of recipient bacteria depends on several factors such as age, physiological condition etc. The duration and time of occurrence of competence varies with species also. During growth, competence is observed during the end of logarthmic phase and just at the beginning of stationary phase. Experiments of Euphrissi–Taylor and Freed (1964), have shown that in *Pneumococcus,* competence developed in 90–100 minutes old culture and lasted for just 7–15 minutes. In *Bacillus subtilis* competence lasts for several hours, but only a few cells in the culture become competent. Several external factors are known to induce competence in an otherwise non competent cells.

Binding

DNA of the transforming bacterium comes into contact with the cell surface of recipient cells as a result of random collision. Subsequently, the DNA binds to the surface at places called *receptor sites* which appear to be associated with mesosomes.

Penetration

After binding itself to the cell surface, the DNA penetrates the bacterial cell. It is believed that the double strand DNA (dsDNA) becomes single strand DNA (ssDNA) on entering into the recipient cell in *pneumococcus.* According to one of the theories when one end of dsDNA enters the cell, an endonuclease enzyme of the recipient digests one of the strands, while the other strand is pulled into the cell. Labelling experiments have shown, half the labelling is present in one strand of DNA, while the other half is in the degraded strand. The ssDNA of the donor migrates from the periphery to the chromosome of recipient. In *Bacillus subtilis,* this takes 15–60 minutes. Mesosomes seem to play a role in carrying the ssDNA on to the chromosome.

Synapsis

The dsDNA of the recipient unwinds and initiates pairing with the donor ssDNA. The donor ssDNA is identical, but not exactly homologous with the recipient DNA. This is necessary as otherwise there would be no variation.

Integration

The unpaired recipient DNA strand is cut in two places by an endonuclease enzyme. An exonuclease enzyme also degrades the unpaired portion of ssDNA. Integration of donor DNA is brought about when *repair synthesis* of DNA takes place to fill any gaps. These are joined by ligase. The transformed heteroduplex undergoes replication to produce copies when the recipient cell divides and expresses its phenotype ie., virulence in the case of *Pneumococci.*

TRANSDUCTION

Transfer of genetic material from one bacterium to another mediated by bacteriophages is known as transduction. Transduction was first discovered by Zinder and Lederberg (1952), in *Salmonella typhimurium,* and subsequently, by other workers in *Escherichia, Shigella, Staphylococcus* etc.

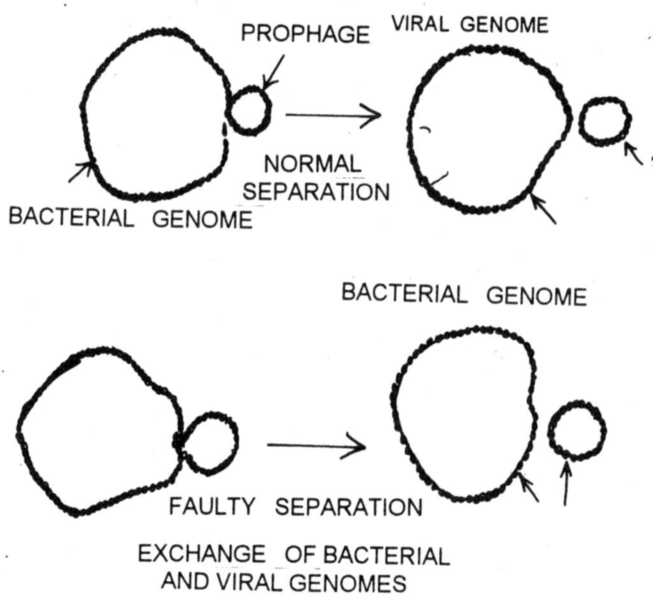

Fig. 9.13 Genetic Recombination
Showing normal and faulty separation of phage DNA from bacterial genome. A piece of bacterial DNA is incorporated into the phage DNA and vice-versa.

In transduction, when bacteriophages infect a bacterium they often incorporate a portion of the genetic material of bacterium into their genetic system. If such a bacteriophage infects another bacterium (of the same species), the bacterium genes (of the first infection) may get incorporated into these bacteria. These bacteria now exhibit some of the traits of the first strain. This shows that bacteriophages have picked up some genes from the first bacterium and transferred it on to the second bacterium.

In order to prove bacteriophage mediated genetic transfer. Zinder and Lederberg performed a 'U' tube experiment. The experimental set up consists of a 'U' tube where the two arms are separated at the base by means of a sintered glass filter (Bacterial filter), Zinder and Lederberg (1952), grew two different strains of bacteria *Salmonella typhimurium* in the two arms. One strain was infected with bacteriophages. A little later, when the bacteria in both the arms of the 'U' tube were analysed, it was noticed that the strains in both the arms had become recombinants. Since the bacterial filter does not allow the migration of bacteria, the only way for transfer of genetic material was through bacteriophages since they can move across bacterial filters.

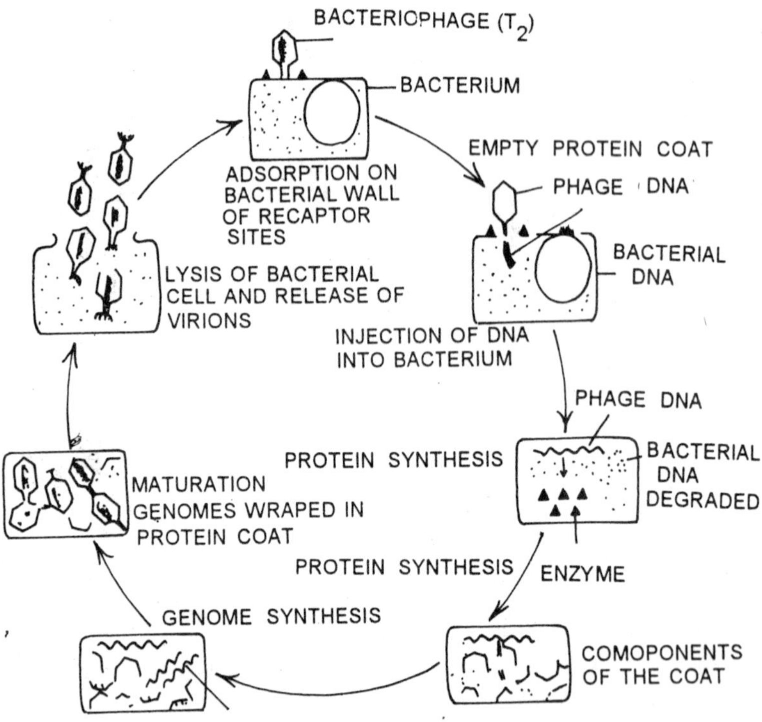

Fig. 9.14 Genetic Recombination
Lytic cycle of a bacteriophage (T_2).

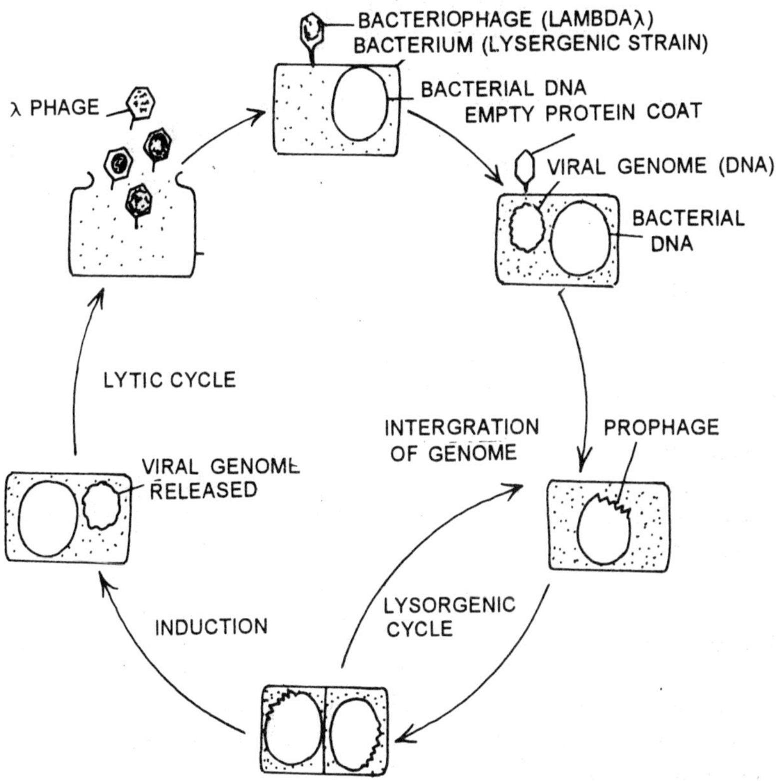

Fig. 9.15 Genetic Recombination
Lysogenic cycle of λ bacteriophage

Types of transduction
There are two types of transduction – Generalized transduction and special or restricted transduction.

Generalized transduction

Phage mediated transfer of small segment of DNA from any part of bacterial chromosome is called generalized transduction. Bacteriophages of P1 of *E.coli*, P22 of *Salmonella* and SP10 and PBSI of *Bacillus subtilis* sometimes make mistakes during their lytic infection cycles. At the end of the lytic cycle, when phage particles are packaged into protein coats and the host chromosome is getting degraded, a small piece of host DNA, usually the same length as the phage chromosome, gets incorporated into the phage and becomes a part of phage DNA. The virus particle now infects a second bacterium. In this bacterial cell, the fragment of DNA (which has been picked up by the virus), can be inserted into the bacterial chromosome. The first bacterium from which the phage picked out a piece of DNA can be called a donor, and the second in which the DNA fragment gets integrated can be called a recipient. The recipient will be a recombinant. This is called generalized transduction because random fragments of bacterial DNA get incorporated with the viral genome.

Specialized transduction

Also called restricted transduction, in this, the phage is able to transduce or transfer only certain segment of the bacterial genome. Specialized transduction takes place as follows.

When prophages are induced to leave the bacterial chromosome, the viral genome carries with it only some bacterial genes covalently linked to the viral genes. This hybrid DNA packed in the phage coat is called *special transducing particle* because it can transfer from one bacterium to another only certain genes – genes that are adjacent to the prophage attachment site. Specialized transduction is induced by λ 80 prophage, which usually occupy a fixed site on the bacterial chromosome. The site is called prophage site. In *E.coli,* the prophage site is flanked on one side by bacterial genes gal K, T and ε genes of the galactose gene cluster, and on the other side by block of genes involved with the synthesis of biotin–bio A,B,F,C and D. Thus, λ prophage can transduce only *gal* or *bio* genes.

The total length of hybrid DNA in a transducing prophage is exactly equivalent in length to the normal viral genome (as otherwise it cannot be packed into the coat), which means to accomodate the bacterial genes, a corresponding length of viral genome should be missing. As such, these phages are called defective paste (they lack some viral genes).

Abortive transduction

In this, the foreign DNA (of the donor tagged to a viral genome) does not get incorporated into the genome of recipient bacterium. The DNA fragment remains free in the cytoplasm and cannot undergo replication. When the bacterial cell divides, only one cell in each generation will have the foreign DNA (exogenote), while all the others are non-transducted. The non -transduced cells might have the gene product (produced by the exogenote before cell division), but progressively it gets reduced as it is distributed among the cells. This phenomenon which brings about incomplete transfer of genes is called *abortive* or *incomplete transduction.*

Growth phases of transducing phages

Bacteriophages which a bring about transduction can be divided into two groups based on their growth cycles. These are **lytic cycle** and **lysogenic cycle.**

Lytic cycle

This is seen in phages called virulent phages typified by T series phages. Here, soon after infection, the host cell lyses releasing the newly formed viral particles. The lytic cycle seen in Tphages infecting *E.coli* bacterium has the following stages.

Adsorption

The phage attaches to specific receptor sites on the bacterial surface.

Injection

The tail fibres hold the phage in position and a pore is formed in the bacterial wall, through which the viral DNA is injected inside the host cell.

Protein synthesis

Phage DNAm 'takes over' the synthetic machinery of the bacterial cell, and forms viral mRNA and through this the viral proteins.

Genome synthesis

Viral DNA replicates and produces several copies of the viral genome.

Maturation and release

Viral genome gets packed into the newly formed coats, thus assembling new phage particles. At a particular time, the host cell lyses (breaks up) releasing the phage particles.

Lysogenic cycle

This is seen in λ phage. The phages are also called temperate phages (as they show lysogenic cycle). Lysogenic cycle was first discovered by Andre Lwoff.

The characteristic feature of this cycle is that the host cell do not lyse as a result of infection by the phages. But the phages will remain in the host cell in a sort of an association. The lysogenic cycle is as follows.

The DNA of the phage, after adsorption and injection, enters the bacterial cell. But, instead of starting its degrading activity, the phage genome gets integrated with the host genome. This integrated condition of the viral genome is said to be **prophage.** When the bacterial cell divides, the integrated genome replicates in the normal way. The viral genome may be said to be in a hibernating state as it is not trying to express itself. Occasionally, the viral genome gets separated from the bacterial genome and is released into the cytoplasm where it starts multiplying and causes the lysis of cells. The dissociation of viral genome from bacterial genome (with, causes lysogeny to become lytic), is called *induction.*

10

GENE MUTATION

Evolution in the biological world is primarily dependent on two events to bring about hereditary variations. These are - **mutation** and **recombination**; while mutation provides the raw material for the changed potentiality of inheritance, recombination permits permutation combinations in the gene pool, thus stabilizing the mutational change in the environment.

What are mutations? To understand this we will briefly analyse how characters are trasmitted from parents to offspring. We know that the characters are trasmitted from parents to offspring through the genes. But the offspring will not have the same genes as those of the parents, rather they have copies of them. In other words during reproduction, parental genes produce copies of themselves to be passed on to the offspring. Gene copying or gene reproduction is a precise process and generally there are no mistakes. Occasionally however, mistakes may occur while copying, so that the newly formed gene gets an altered structure. This modified gene then goes on multiplying and produces its own copies just as the original gene does. This alteration in the changed gene structure is called **gene mutation.**

Broadly speaking, mutation includes all those phenotypic variations, which arise due to altered gene structure and are heritable. Hence the term encompasses, small and minute changes brought about by altered individual genes (gene mutation) and large changes brought about by the altered chromosome structure or number. These latter types of changes can also be called chromosome mutations, and they have been described under the chapter chromosomal aberrations. In this chapter we will study gene mutations. Gene mutations are also called point mutations as they occur at a specific gene locus altering its molecular structure. Sometimes chromosomal aberrations also can bring about a slight phenotypic change, in which case it is difficult to distinguish between these two.

History : The term mutation was first coined by Hugo devries in the early 1900 to designate sudden and drastic changes that occured in plants. Devries was conducting breeding experiments on the plant *Oenothera lamarckiana* (of the family Onagraceae) commonly called the *Evening primrose*. He noticed that some of the progeny possessed characters which were not to be found in either of the parental types. He further observed these changed types bred true. He therefore described these sudden changes as mutations and ascribed them to some changes occuring in the germplasm of parents.

Even Charles Darwin was aware of mutations, even though he did not use the term (mhtations). He was aware of sudden changes occuring in organisms both plants and animals and had used the term sports, to describe them.

The first recorded instance of mutation dates back to 1791, to the pen of Seth Wright, a New England farmer, where appeared a male lamb with short bowed legs. Wright reared this lamb and it bred true. A lamb with short legs was of great advantage as it could not jump even a low level fence. This short legged breed was called **Ancon breed.** This breed however became extinct after about eight years. But the same Ancon sheep appeared some fifty years later in the flock of a Norwegian farmer, representing probably the repetition of the same mutation. Since then a number of mutant types have been discovered. Hornless individuals in cattle, double toed cats, albino rats, mule footed swine, dwarf cupid sweet pea, double flowered, white flowered varities of many plants are all mutants.

The scientific study of mutations began with the work of Morgan on *Drosophila* in 1910. He observed that among a population of red eyed flies, some individuals with white eye appeared and they were predominant among the male individuals. He further found out that the white eyed gene is recessive to red eye and located on the X chromosome. Since the male individuals possessed only one X chromosome and the females had two X chromosomes, the recessive character appeared more frequently among males than among females.

After the original discovery of the white eyed mutant in *Drosophila* by Morgan and his co-workers, a detailed analysis of millions of these fruit flies for over 17 years was carried out by different geneticists and nearly 500 gene mutations have been discovered. Simultaneously other organisms like maize, rodents, pea, snap dragons, poultry, man etc. also have been investigated and scores of mutations have been discovered. Even in microorganisms also mutations have been detected.

Size or range of mutations : Size of mutation varies in different organisms and in different genes. These may be very minute and may not be detectable at all or as in some cases (Ancon breed), so striking that they can be easily made out. In some microorganisms like Bacteria, Fungi *(Neurospora),* Blue green Algae etc, mutations may not bring out any morphological effect but may bring about some changes in their nutritional requirements. In some cases mutations may be so large as to even result in lethality to the individual.

Fig. 10.1 Gene Mutation
Ancon sheep with ram and ewe

Frequency of mutations : The frequency is generally less unless there is a drastic change of the surroundings. According to Muller, in *Drosophila,* the mutation rate for any gene is less than one in one million. It has also been calculated that a particular gene in *Drosophila* may mutate only once in every 40,000 years.

The rate of mutation also varies between different individuals, organisms and sometimes even under different ages of the same organism. J.B.S. Haldane, a renowned geneticist has estimated that in a comparison between man and *Drosophila,* the former has less chances of mutation as most of them are harmful so there is a selection pressure in man against mutations. If the mutations in man were to take place at the same rate as in *Drosophila,* the whole genetic system of the human race would be impaired. As against the unaltered life expectancy of 40,000 years for a *Drosophila* gene, the human gene has a life expectacy of 25,000,00. In one of the case studies of Chondrodystrophic dwarfness (in these human beings, arms and legs are short, while other parts are normal), it was estimated that one in every 11,500 births haye this genetic defect (the data obtained from a Denmark hospital). This gives a rate of one mutation for every 23,000 genes at this locus. If we assume, the average age of the parents to be 30, who give birth to dwarf children, then one can estimate that the gene mutates once every 6,900,00 years. The following two tables give the frequency of mutations in man and Corn.

Table : Rates of mutation in some human genes

Sl. No.	Phenotypic effect	Genetic condition	Frequency in %	Occurrence oncein every gametes
1.	Colour cannot be distinguished	Colour blindness	0.0028	35700
2.	Blood clots very slowly	Hemophilia	0.0032	31250
3.	No pigmentation in hair, skin etc.	Albinism idiocy	0.0028	35700
4.	Mental faculties dull after birth	Amaurotic	0.0011	90909
5.	Limbs and arms dwarf	Chondrodystrophic dwarfism	0.0070	14300
6.	Face with red patches, tumours in Kidney, brains etc., at later stages	Epiloia	0.0012	83333
7.	Iris in the eyes absent	Aniridia	0.0005	2,00,000
8.	Abnormal WBC resulting in reduced immunity	anomaly	0.0080	12,500

Table : Rates of mutation in maize (After Stadler 1943)

Sl. No.	Gene	Phenotype	Frequency (1 in one million gametes)
1.	R	Plant colour	492
2.	Cl	Chlorophyll	106
3.	Dr	Red aleurone	11
4.	WX	Normal endosperm	0
5.	Su	Sugary endosperm	2
6.	Sh	Full endosperm (not shrunken)	1
7.	Ci	Aleurone colour inhibitor	2
8.	Y	Yellow endosperm	2

Mutational Unit : Earlier it was not known as to what constitutes a mutation in a gene. But recent researches have shown that the smallest amount of DNA within which mutation can take place is called a *muton.* At the molecular level mutation involves alteration in the base pair sequence.

Mutator genes and mutable genes : It is difficult to assess (among plants and animals) as to which gene frequently mutates and which gene is comparatively stable. But in some of the orgamisms like maize (Emerson), it has been observed that some genes change quite frequently, w-hile others are not so. Barbara McClintock (Nobel laureate, discovery of jumping genes) has shown in maize that some genes are not only stable, but influence others to mutate. The genes that influence others (to mutate) are called mutator genes and the genes that are so influenced are called mutable genes. These genes, mutator and mutable are more frequent in plants than in animals.

Classification of mutations : Broadly speaking, mutations are of two types - Chromosomal and gene. Chromosomal mutations actually do not change the gene structure, they bring about change in the phenotype either due to loss, addition, altered seqeuence or trasferred position of the genes. Gene mutation however changes the phenotype by basically altering the gene structure itself at the molecular level.

Gene mutations are basically of two categories - **Spontaneous** and **induced**. Spontaneous mutations apparently arise without the help of any identifiable external agency. They arise due to some internal changes. Induced mutations on the other hand are traceable to some external agency like radiation, temperature, chemicals, gas etc.

Mutations may also be further classified into various categories based on the effect, cell type inwhich it occurs, direction of mutation etc. The following is a brief classification of such mutations.

Gene mutations (both spontaneous and induced)

- Visible mutations
  - Germinal
    - Gametic
    - Zygotic
  - Somatic
  - Reverse
  - Spurious
- Biochemical mutations
- Lethal mutations

We shall briefly study a few examples of spontaneous and induced mutations and then study a few instances of the specific types of mutations above.

SPONTANEOUS MUTATIONS

As has already been pointed out, the agency which brings about spontaneous mutations is not known. They apparently occur due to natural causes. The first recorded instance of a spontaneous mutation is that of the Ancon breed sheep. A detailed study of the spontaneous mutation in *Drosophila melanogaster,* is that of T.H. Morgan, where he described the sudden appearance of white eyed flies in a population of red eyed individuals. Some of the spontaneous mutations noticed in plants and animals are as follows.

Spontaneous mutations in plants : 1) Double petunia, 2)Double rose, 3) Shirley poppy, 4) Single flower mutant in sunflower, 5) Multileaf formation in tobacco reported by Hayes, 6) Reduction in ray size in the ray florets in *Bidens pilosa* of Compositae(reported by the the author).

Spontaneous mutations in Animals : 1) Polydactyly in man, sheep, cats etc. 2) Multinippled condition in sheep, 3) Silky fowls, 4) Albinism in man, pig, pigeon, rodents etc. 5) Double cared cattle, 6) Mule footed swine, 7) White canaries, 8) Hornless cattle, 9) Hares' lip in man, 10) Widows peak in man, II) Lobster claws in man, 12) Cleft palate in man etc.

Spontaneous mutations are generally rare and may be of two types dominant and recessive. Usually recessive mutations are more frequent than dominant mutations.

Factors affecting spontaneous mutations : The rate of spontaneous mutations are known to change under the influence of various external and internal factors. One of the external factors which might bring about spontaneous mutations is a change in the temperature range (normally a higher temperature makes the genes susceptible for change). Disturbed metabolic rates may also bring about mutational changes. Age and sex may also have a bearing on the mutational rate. Increased age may render the genes prone for mutational change. *In Drosophila,* it has been observed that males are more susceptible to mutational alterations than the female.

One of the important factors bringing about spontaneous mutations may be, specific mutation inducing and mutation succeptible genes. As has already been pointed out, these are called mutator and mutable genes.

The mutator genes have been discovered by Me Clintock in maize. She described some genes which had the capacity to influence other genes to mutate. Rhodes has described the role of a mutator gene which influences the functioning of a colour gene. In maize the colour is governed by three recessive genes *a, c,* and *r,* while the dominant alleles of these govern the colour other than green. In addition to these there is a mutator gene)((dotted colour) which controls the variegated colour in the grain kernel. This gene influences the gene *a* to mutate so that red patches are produced. Usually the plants with the genotype *aa cc RR* will be green (all the three dominant gene pairs are required to produce colour other than green), but the presence of *Dt* makes it variegated.

Mutable genes other than the one mentioned above (a) have been reported in maize by Emerson. He investigated a variety of maize withwhite pericarp having several red spots of varying sizes; the homozygous recessive allele producing the white colour. According to Emerson, these red patches are produced by the unstable white gene undergoing spontaneous mutation. Each red spot in the endosperm indicates an area where a mutation has occurred. Jones further continued the studies of Emerson and found that there are six red spots, each having 10 - 45 coloured cells. These coloured cells were exactly identical with those produced by the pure red gene. Jones was conducting an experiment with a homozygous plant and he beleived that the six red spots indicated that the gene for white had mutated six times producing a red cell every time. These cells further multiplied and produced the red patches. The gene for white endosperm colour is thus a typical example of a mutable gene.

Several such mutable genes have been reported *in Drosophila,* sweeet pea, 4 o' clock plant, snapdragons etc. The genes for miniature wings, gene for purple eye colour in *Drosophila* are known to be mutable genes. In maize plant itself, Stadler has observed that the gene R for aleuron colour mutated some 492 times in about a million gametes that he studied.

INDUCED MUTATIONS

Various external agencies are known to bring about mutational changes in genes. All these are included under the general category of induced mutations. H.J. Muller (1927) may be regarded as pioneer in radiation genetics. In his studies on *Drosophila,* he showed that flies subjected to X radiation undergo mutations. Similarly Stadler (1928) demonstraded the mutational effects of X rays in barley and maize. Various kinds of agents causing mutational changes in organisms are known. These are of the following types, a) Radiation b) Temperature changes and c) Chemicals.

Radiation induced mutations : Various types of radiation (excluding visible light) are known to cause mutations in different intensity. Radiations are of two categories depending upon their capacity of penetration into a substance -*Ionizing* and *non ionizing.* Generally speaking, non ionizing radiations are not of a high potentiality to bring about mutations at a rate which is brought about by ionizing radiation. The Alpha rays, Beta rays and the gamma rays and X rays come under the category of ionizing radiations while low penetration rays lik U.V. rays belong to non ionizing radiation.

Mayor (1920) experimented on the effects of X radiation *in Drosophila* (and similar other radiations) to study the effects on germplasm. He demonstrated that as a result of radiation, there was non disjunction of the X chromosome resulting in a XXY individual, which was female phenotypically.

H.J. Muller also worked on the effects of X rays on *Drosophila* and published a classical paper 'The artificial transmutation of the gene'. He also devised a technique to detect induced mutations (see later in the chapter, for detailed explanation).

Stadler studied the effects of X rays on barley plant, while Gager and Blackslee induced mutants in the jimson weed *Datura stramonium*. It should be noted however that X rays are lethal and the dosage of the rays given to the individual has to be carefully monitored. On an average it has been found out that 80% of the X ray induced mutations are lethal.

Oliver, Tirnofeeff Ressovsky and many others have worked out the relationship between the frequency of mutation and the amount of X radiation absorbed by the organism. The intensity of the radiation is measured in terms of Roentgen units or *r*- units. The frequency of mutation seems to be proportional to the amount of *r*- units absorbed by the plant. The duration of exposure or the distance between the radiation source and the experimental organism, however do not seem to affect the frequency of mutations. The following table gives the rates of mutation in *Drosophila melanogaster* due to two different doses of X rays. In the table below t4 dose is twice as large as the t2 dose. (The data in the table below are based on Muller's experiment, quoted from Principles of Genetics, Sinnott Dunn and Dobzansky).

Experiment	Number of chromosomes observed	Number of mutations		
		visible	semilethal	lethal
Control	198	0	0	0
Xrays t_2	676	1	44	9
Xrays t_4	772	3	12	89

Some of the mutations on exposure to X radiation in *Drosophila* are as follows.

Gene wt to W (white), W^e (eosin), and W^a (apricot). Genes coral (W^{co}) buff (W^b), cherry (W^c) and apricot W^a) can mutate into white (W).

Chlorophyll deficiency has been induced in barley due to X radiation (Stadler). X rays can also induce mutations in the wasp *Habrobracon*. Among the other types of ionizing radiation Alpha, Beta, and Gamma rays are important. The First two however do not seem to have a profound mutational role as they cannot penetrate the body cells as deeply as gamma rays.

Mechanism of action of ionizing radiation : There are two theories which account for the mechanism of mutagenic effect of ionizing radiation. These are the **direct hit theory** and the **indirect hit theory**. According to the direct hit theory, the radiation directly strikes the genes bringing about a change in the chemical composition. The electrons that. strike the atoms of genes bring about a change in the chemical composotion, thus altering the gene. But in several experiments, it has been seen that , the same dose of radiation does not result in the same frequency of mutation indicating that direct hit theory cannot account for all radiation induced mutations.

According to the indirect hit theory, gene alteration is brought about by changing the molecular environment of the surroundings. Muller is a supporter of this theory. According to him when the high speed protons of X ray collide with the molecules of the cell, electrons are ejected out of the atoms rendering them positively charged. The electrons that come out move at high speeds and knock out other electrons that come in their way resulting in an environment which is ionized (positively charged). The electrons moving freely may attach themselves to other atoms rendering them negatively charged. Various chemical reactions and shifts take place to bring both the positive and negatively charged atoms to their normal state. During this process, alteration in the chemical composition takes place, thus mutating the gene. At the molecular level, base pair sequences are altered in the DNA molecule (for more details see mutation at the molecular level, in the same chapter).

Effect of non ionizing radiation : Ultra violet rays are non ionizing because they have a long wave length and therefore have less energy. They are generally non penetrative. They may bring about many changes in the traits, but are not as effective as X rays. Altenberg was able to accelerate the rates of mutations by using U.V. rays. Stadler also achieved pollen mutations induced by U.V. rays in maize; the same is the case in *Antirrhinum* as reported by Noethling and Stubbs. The U.V. rays are known to cause alterations in the bond characters of purines and pyrimidines. Between the two, the latter are more prone to mutational changes.

Chemical mutagens : A number of chemicals are known to have mutagenic properties. It was C.Auerbach who discovered during second world war that chemicals could be used to induce mutations. Thorn and Steinberg (1939) had also reported the chemical induction of mutations. They treated the Fungus *Aspergillus* with nitrous acid. Since then a long list of mutangenic chemicals have been identified. Some of the common chemicals used as mutagens are given in the following table.

Table : Some mutagenic chemicals

Class of chemicals	Name of individual compound	Common Abbreviation
1 Aziridine	1 Ethyleneimine	EI
2 Mustards	2 Nitrogen mustard	
	3 Sulphur mustard	
3 Nitrosamines	4 Dimethylnitrosamine	DMN
	5 Nitrosoguanidine	NG
	6 Nitrosomethyl urea	NMU
4 Epoxides	7 Ethylene oxide	Eo
	8 Diepoxybutane	DEB
5 Alkyl sulphonates	9 Diethyl sulphate	dES
	10 Methyl methane sulphonate	MMS
	11 Ethyl methanal sulphonate	EMS
6 Others	12 Nitrous acid	
	13 Colchicine	
	14 Maleic hydrazide	
	15 Hydrazine	
	16 Hydroxylamine	
	17 Caffeine	
	18 Hydrogen peroxide	
	19 Formaldehyde	
	20 Phenol	
	21 Urithane	

Hydrogen peroxide is known to bring about mutations in *Neurospora.* Urithane and phenol are capable of inducing mutations in higher plants and *Drosophila.* Formaldehyde may induce mutations during spermatogenesis in *Drosophila.*

Colchicine is an alkaloid extracted from the bulbs of *Colchicum autmnale* of the family Liliaceae. This is known to induce polyploidy by suppressing spindle formation. Colchicine is not known to bring about any point mutation. Mustard gas is another chemical widely employed as a mutagen. Its mutagenic effect has been tested with reference to *Drosophila, Neurospora,* bacteria, barley, mice etc.

Temperature induced, mutations : Child, Plough and Ires have studied the effect of temperature as a mutagenic agent in *Drosophila.* A range of temperature from 4^{0}C to lethal range was used to find out the possible effect of temperature on the rate of mutation. A higher range of temperature is known to accelerate the rate of mutation. But it has to be borne in mind that a gradual increase in temperature might not be an effective mutagen. Any sudden increase or decrease in temperature might alter the gene copying mechanism thus altering the gene. The temperature effects largely remain as a somatic influence.

Vegetable oils as mutagenes : In one of their experiments concerning oil yielding plants, Swaminathan and Natarajan (1959) noticed that seeds with high oil content are generally resistant to mutagens. They soaked the cereal grains in oils of ground nut, mustard, and castor and wanted to know whether these oils provide any mutation resistance to the grains. Much to their surprise, the results were startling. Instead of the oils providing resistance to the cereals, they (oils) themselves had acted as mutagens as noticed by marked reduction in rates of germination, chromosome irregularities and many ear head variation mutations in wheat, one of the cereals used in the experiment. It was assessed that in wheat *(Triticum astivum)*, the rate of mutation, induced was much higher than that of X rays.

Detection of mutations : A variety of methods are available for detecting mutations, both in plants and animals. Some of these are described below.

CLB method for detecting mutations : This method perfected by Muller has been used to detect mutations in X chromosomes of *Drosophila.* Even though it is primarily meant to detect radiation induced mutations, it can be used in the detection of spontaneous mutations as well. Three characters on the X chromosome are chosen to study the mutation. These are -

C = Cross over suppressor. This represents a long inversion in the X chromosome, suppressing the cross over in the inverted portion. Cross over may still occur, but cross over gametes rarely survive due to deficiences and duplications.

L = Lethal gene. This is a recessive character. Hence homozygous females (l/l) and hemizygous (l/y) males do not survive.

B = Bar eye. This is a dominant gene. This is used as a phenotypic marker for the flies required to mate in the F_1 generation. The working mechanism of this is illustrated in the Fig. A CLB male is mated to a normal male which is subject to radiation. The idea is to find out whether there are any lethal mutations in the X chromosome of the male. Y chromosome is regarded as neutral as for as sex linked characters are concerned, hence the male will be hemizygous as for as genes on X chromosomes are concerned. Therefore any X chromosome of male which is irradiated can be followed in subsequent generations and lethal mutations if any can be detected.

In the X_1 generation (F_1 generation; called X_1 because the individuals are X radiated) the irradiated X chromosome of the male goes to the daughters, who have another X from the mother; hence in this generation any mutation induced can not be detected. The X_1 progeny will be as follows.

Females : 50% CLB; 50% Normal

Males : 50% CLB (die); 50% Normal.

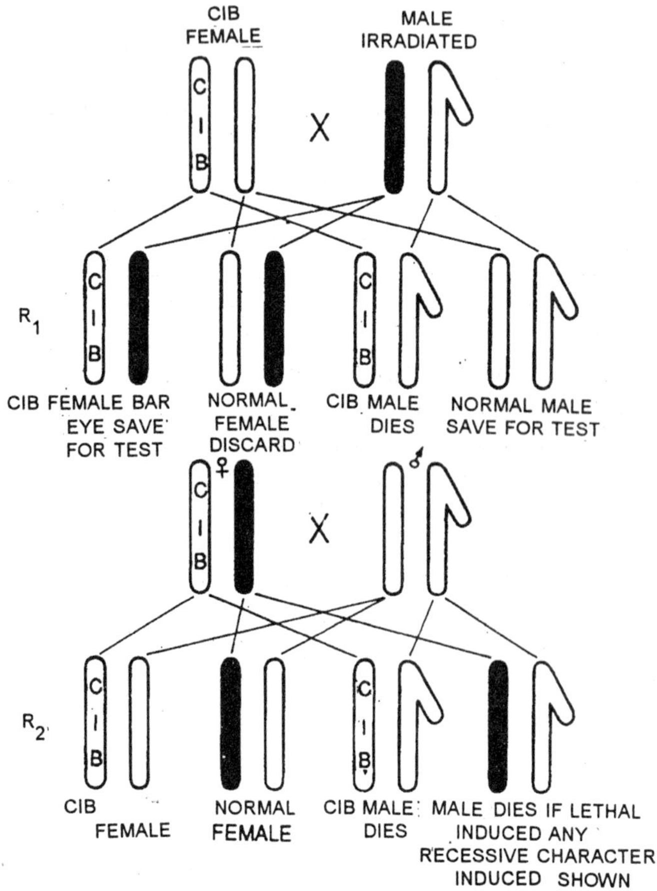

Fig. 10.2 Gene Mutation

Muller's CIB method to detect induced mutations drosophila melanogaster

A cross is now made between the normal male (with X also normal because it is obtained from female) and the CLB female. This CLB female has one of its X chromosome from the male parent which is irradiated and the other is CLB. In this (X_2 or F_2) generation all the males get one of the X chromosome, they die (50%) because there is already a lethal gene. If they get the irradiated X chromosome, and if a new lethal mutation is induced then the other 50% also will die. Hence if there is a lethal mutation on the X chromosome in the F_2, no male progeny will survive and all will be females. This is a clear cut evidence for the presence of a lethal mutation on the irradiated X chromosome. In the X_2 generation the progeny will be as follows.

Females : 50% CLB; 50% Normal (with an irradiated X chromosome)

Males : 50% CLB (die); 50% carry irradiated X chromosome; die if there is a lethal mutation.

The Muller 5 method : .Like the CLB method this is also devised by Muller. The experiment involves crossing a homozygous Muller 5 female *Drosophila* with an irradiated male. The Muller 5 female *Drosophila* has two marker genes - apricot eye (w^a/w^a) and Bar eye (B/B) in a long inversion so that there is no cross over. A homozygous apricot (eyed) Bar eyed female is crossed with an irradiated male. As shown in the figure, the F_1 generation female offspring which gets the irradiated X from the male parent is crossed with a wild male. If a lethal mutation has occurred in the X chromosome, no wild types of male would survive, because they would get the irradiated X.

The absence of wild males in the F_2 is an indication of lethal mutation. Muller was awared Noble prize in Medicine in 1946 for this and many other discoveries that he had made in Genetics.

Detection ofAutosomal mutations (Balanced lethals)

Autosomal mutations can be detected by making use of balanced lethal genes in organisms. A balanced lethal stock is one which carries a lethal gene in one of the homologues and its wild allele on the other homologue. Homozygous individuals will die, so what survives is always heterozygous. Balanced lethals can be studied by using a specific strain of *Drosophila melanogaster* called the **curly lobe plum.**

In this, a stock of flies carry on one of the autosomes two dominant genes *C* (curly wing) and *L* (lobed eye) and a recessive lethal gene. An individual with two *CYL* chromosomes will not survive because the recessive lethals will be in double dose and its homologue carried the *Pm* (plum coloured eye) gene which is also dominant.

When two individuals with the genotype *{(CYL/Pm)* X *(CyllpM)}* are crossed, the heterozygous individuals only would survive, as homozygous for *(CYL/CYL)* would carry lethal genes in double dose, because *CYL* chromosome carries a lethal gene. There is a cross over suppressor to prevent the transfer of the gene *CY to* the chromsome carrying *Pm* gene.

A female *CYL/Pm* is crossed to another normal male which is irradiated.

Four types of progeny are produced in the F_1. These are -

(a) Curly lobe (with an irradiated autosome A)

(b) Curly lobe (with another irradiated autosome B)
(c) Plum (with an irradiated autosome A)
(d) Plum (with another irradiated autosome B)

One of these flies is selected (because each one has one irradiated chromosome) and back crossed to curly lobe plum.

The progenies will be of the following types.

(a) homozygous curly lobe (dies)
(b) Curly lobe plum.
(c) Curly lobe with an irradiated chromsome.
(d) Plum with an irradiated chromsome.

Among the above progeny, (c) flies are crossed among themselves to obtain -

(a) homozygous curly lobe (dies)
(b) Heterzygous curly lobe (with an irradiated chromosome)
(c) same as b.
(d) A fly with both irradiated chromosomes.

It is in the (d) category that any mutation induced would appear. If the induced mutation is lethal, it also dies and only the curly lobed individuals survive. If the mutation induced is not lethal, but a recessive trait, that would be expressed, since it is homozygous. The same test can be done by selecting the plum with an irradiated chromosome and crossing it with a curly lobe plum with an irradiated chromosome.

The significant aspect of this cross is to bring together the same irradiated chromosome in the homozygous condition to detect all the recessive mutations.

Detection of induced mutations in plants

L.J. Stadler of the University of Missouri and Ralph Singleton of the University of Virginia have worked on the methods to detect mutations in maize and barley.

Stadler (1942) has made a detailed study of spontaneous mutations in maize. The following are the important steps.

(a) A genetic stock dominant for a number of genes is grown as the female parent. The tassels are removed before pollen discharge (to avoid self pollination).
(b) A multiple recessive plant in every fifth row is used as the pollen providing parent for (a) plants.
(c) Seeds were observed in large numbers for endosperm characters. Most of them show dominant characters. If any recessive character (of the male parent) appeared it is due to mutation in the female gametes.

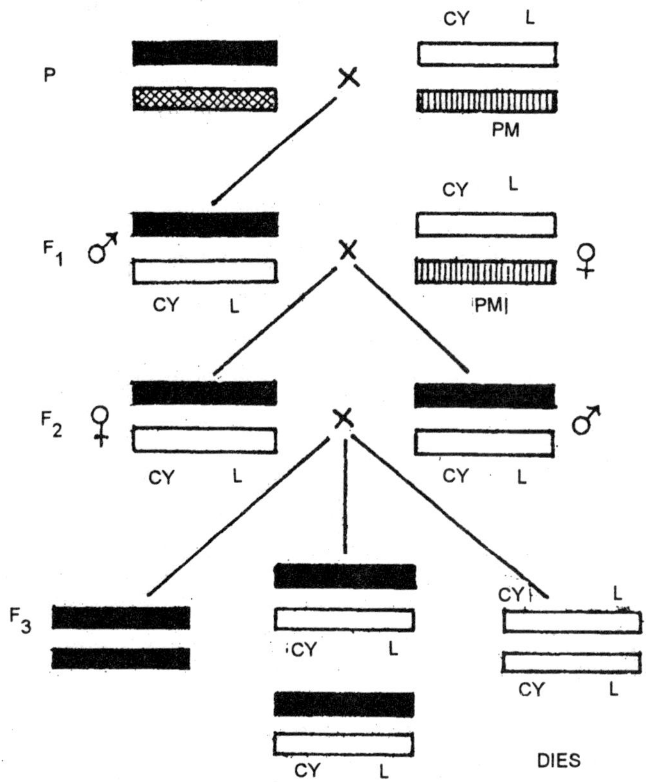

Fig. 10.3 Gene Mutation

Detection of lethal mutations in a balanced lethal system

Singleton (1962) used a techinque similar to that of Stadler do detect both spontaneous and induced mutations in the pollen of maize. His method involves collecting pollen from plants dominant for many genes and cultivated in a field irradiated with cobalt 60 and transferring them (pollen) on to the silks of plants with all recessive traits and cultivated in a normal field. The seeds of the progeny would normally show dominant characters. If any mutation has been induced in a dominant gene, the recessive trait would appear in the progeny. In this way Singleton studied four genes *Su -su,* on chromosome 4, *Pr - pr* on chromosome 5, *Y-y* on chromsome 6 and *Sh -sh* on chromosome 9. The phenotypes of these genes are *Su* - normal endosperm, *su* - sugary endosperm; .Pr- Purple aleurone,pr-red aleurone; *Y*- yellow endosperm, *y* - white endosperm and *Sh* - full endosperm, *sh* - shrunken endosperm.

Apart from spontaneous and induced, mutations may also be classifed (see the beginning of this chapter) into various categores based on different criteria. Some of the examples are discussed briefly below.

Somatic mutations : These are sudden alterations appearing in the somatic characters of an individual. Such mutations normally do not affect the entire plant and express themselves only in those regions where a cell has undergone change. Certain varieties of apples and navel orange are examples of this.

Somatic mutations also occur in Snapdragons, when a white flowered variety produced red flowers in a single blossom.

Another kind of mutation, which has much significance from the poin of view of Horticulture is the *bud mutation.* In this, few cells of the bud or the entire bud (if mutation has occured quite early) shows mutated characters, when grown into a flower or a vegetative branch. If the mutation has occurred a little late during the development of the bud, the flower or the vegetative branch will be a mosaic of normal and mutant characters. Such a combination of two genetically non identical tissues is called a **Chimaera.**

Chimaeras have been seen in *Geranium (Pelargonium zonale)* in which a branch bears both green and white leaves. Another example of bud mutation is that of nectarine mutation in which the skin of the fruits is smooth instead of being rough. Smooth skin is recessive to rough skin.

There are numerous examples of somatic mutations in animals also; for e.g. white eyes, vestigial wings etc. of *Drosophila.*

Germinal mutations : Mutations occuring in the reproductive cells are called germinal mutations. They are heritable and consequently they affect the whole of the organism in the next generation. Germinal mutations are called gametic when they occur in gametes and zygotic when they occur in the zygote. The gametic mutation will express itself readily when either the mutation is dominant or if recessive two mutant gametes should unite. Zygote mutations on the other hand express themselves readily. The ancon breed sheep is a typical example of germinal mutation.

Biochemical mutations : Mutations bringing about alterations in metabolic pathways, nutritional requirements etc. are called biochemical mutations. These are mostly noticed in microorganisms. Biochemical mutations in *Escherichia coli* can be induced by repeated sub culturing. In *E.coli,* the normal strains are called *Paratrophs* while the mutants are called *Auxotrophs.* These mutations are for some of the nutritional requirements like Biotene, Methionine, Leucine etc. The paratroph is called genetically B+, M+ and L^+ because it has the capacity to synthesize all these. On mutation the auxotroph loses the capacity to synthesize these and consequently it is known as $B^- M^- L^-$

Several metabolic disorders due to biochemical mutations have been reported in man also. They are commonly called **Inborn errors in metabolism'**. The first such biochemical disorder to be reported was Alkaptoneuria (Garrod 1902). This relates to a metabolic disorder concerning the amino acid phenyl alanine. **Alkaptaneuric** individuals have a mutated gene which does not produce the .enzyme homogentisate oxidase, resulting in the accumulation of homogentisic acid in the cell, which is excreted along with the urine. Homogenticic acid is a strong reducing agent and gets oxidised in the presence of air into a black pigment, resulting in the black colouration of the urine.

Another biochemical mutation in man is phenyle ketoneuria. It is caused by a recessive mutant gene in which due to failure of conversion of phenylalanine to tyrosine, the former accumulates in the blood and excreted through the urine. This leads to mental retardation. In addition to the above, Tyrosinosis, Albinism and Cretinism are the other metabolic disorders due to biochemical mutations.

Reverse mutations : The back mutation of a gene to come back to its wild state is known as reverse mutation. In *Drosophila,* the wings are normally long and due to a mutation, the wing becomes miniature. In some instances, all of a sudden, a few long winged flies appear in a population of miniature winged flies. This is due to the reverse mutation of the miniature winged gene back to the wild state. Another reverse mutation in *Drosophila* is the mutant, fork bristles back to normal bristles. There are many such reverse mutations in bacteria and fungi.

Lethal mutations : Mutations of gene, from normal to lethal leads to the death of the individual. Normally lethal genes are recessive and nothing happens to the organism in the heterozygous condition. But occassionally there may be dominant lethals in which case individuals do not survive even in the heterozygous condition. The latinum colour in Fox, yellow coat colour (see chapter on Interaction of factors for a detailed discussion) in mice are such examples.

Many lethal mutations are also known in plants *likeAntirrhinum majus* (Snapdragon). The golden plant stock has variegated leaves. On self pollination this variety produces Golden and green in the ratio 2:1 (not 3:1), obviously the golden is heterozygous and green breeds true. The other one (golden) is missing because pure golden can not exist due to homozygosity of lethal genes. Lethal genes are also known in paddy and maize plants.

MUTATIONS AT THE MOLECULAR LEVEL

The basic hereditary material namely DNA is a highly stable biological molecule. The hereditary information is present in the form of sequence of the four nucleotides Adenine, Thymine, Guanine and Cytosine. During the inheritance of characters from one individual to the other, the sequence is duplicated and transmitted in a precise and accurate manner.

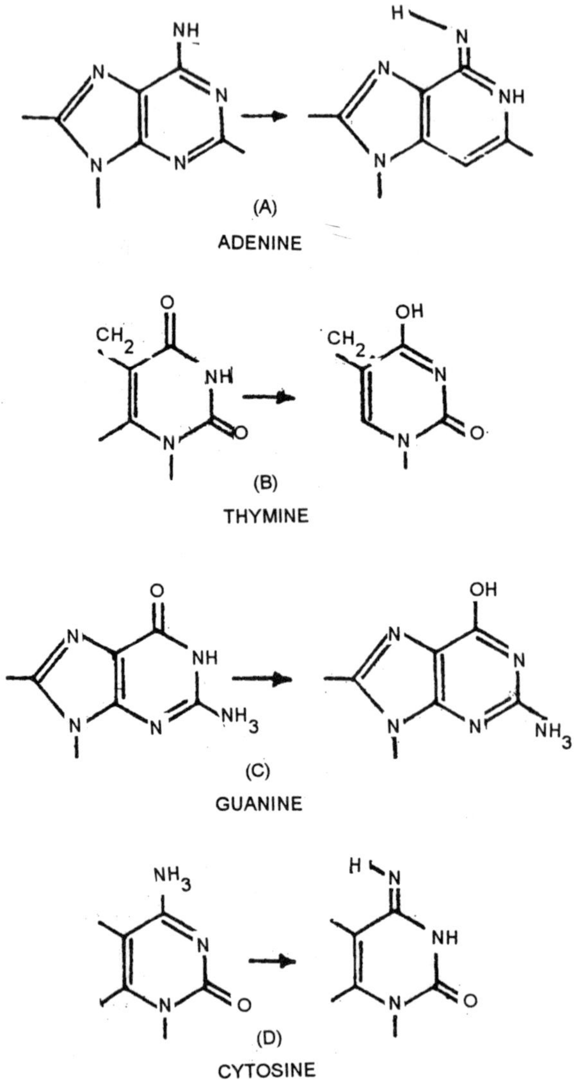

Fig. 10.4 Gene Mutation
Uncommon forms of DNA bases

A slightest mistake at this stage will alter the base pair sequence and results in an entirely different type of genetic code resulting in the formation of an altered protein which may have far reaching effects on the organism. Such mistakes or faults occuring in the DNA molecule during duplication are ultimately the basis of mutation at the molecular level. The following two basic types of sub nucleotide changes are noticed in the DNA molecule - Frame shift mutation and substitution mutation.

Frame shift mutations : Addition or deletions of nitrogen bases either from DNA or mRNA are called frame shifts, because, from the site of the alteration, the reading frame of the message alters. In some cases frame shifts are neutralized ie. deletion may be neutralized by the addition at the same site. Frame shift mutations are of two categories. These are deletion mutations and insertion mutations. In deletion a single nucleotide in a triplet in deleted. These are found frequently in T_4 bacterophages. In insertion mutations an extra nucleotide is added to a gene. These insertions can be induced by chemicals such as acirdine, Proflavin etc.

Substitution mutations : This generally occurs during the replication of DNA, when one nitrogenous base is replaced by another changing the triplet codon. The altered codon now codes for a different amino acid resulting in an altered protein. There are two kinds of substitutions - Transitions and Transversions.

Transitions : The replacement of one purine by another purine or one pyrimidine by another pyrimidine is called trasition. In other words A is replaced by G and T by C or *vice versa.*

Transitions may be brought about by tautomerization, ionization, base analogs or by deamination. In tautomerization, a normal base, pairs with its tautomer partner [A tautomer is an alternate state of existence in a molecule. For e.g. In cytosine and adenine tautomers, the normal amino group NH_2 is converted into the NH (imino) group and the keto group (C = 0) of thymine and Guanine is converted into the OH group in their tautomers. Normally this does not happen]. Such pairing is called *forbidden base pair.* Watson and Crick (1953) opined that the existence of a tautomer to DNA provides a mechanism for mutation.

Ionization is another mechanism by which transition can be brought about. Ionization of a base at the time of DNA replication involves the loss of H from a nitrogen base. Such ionized bases pair with normal bases.

Another method by which transition may be brought about is by the replacement of normal bases by their analogs. These compounds called *base analogs* are derivatives of the natural bases. For eg. cytosine may be replaced by *5 hydroxyl methyl cytosine* or by 5 *glucosyl hydroxy methyl cytosine.*

There are naturally occuring base analogs of cytosine found in T even phages of *E.coli*. *Uracil* has a naturally occuring analog - 5 *hydroxy methly uracil* found in some Viruses. Besides the naturally occuring analogs, artificial analogs like 5 *bromouracil, 5 iodouracil, 5 bromocytosine, 5 methyl cytosine* may be used as analogs of thymine (first two) and cytosine (last two) which may be used to induce mutations.

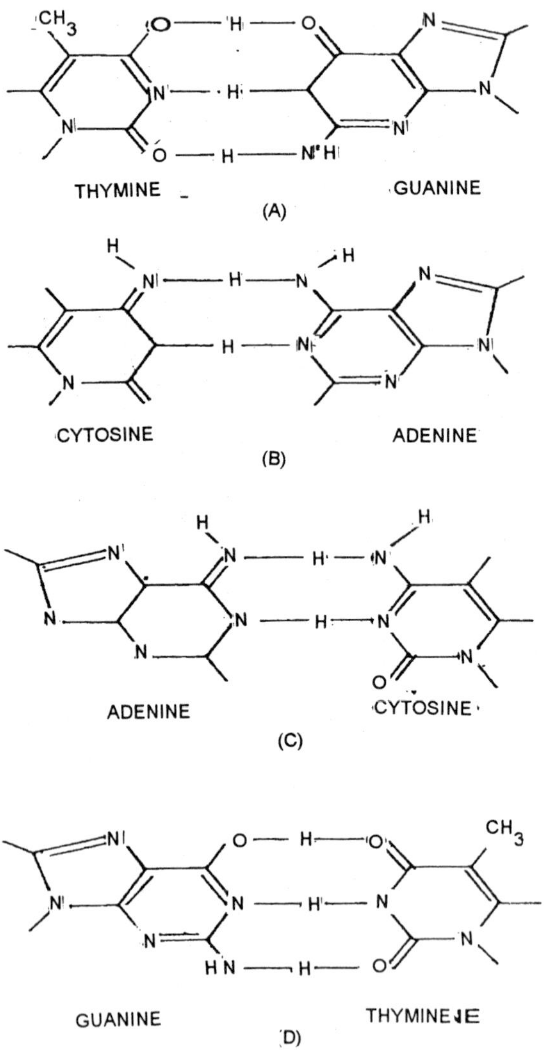

Fig. 10.5 Gene Mutation
Unusual pairing of tautomeic bases in DNA

Deamination is another method of induction of transition, by replacing the NH_2 group of bases with OH group. This may be induced by chemials such as nitrous acid, hydroxylamine, diethyl sulphate etc. Deamination of cytosine leads to the formation of uracil, and that of adenine forms hypoxanthine. During replication U pairs with A and hypoxanthine pairs with C. As a result A = T is substituted by G = C and G=C byA=T.

Trasversions : Unlike m transition, a purine is replaced by a pyrimidine and *vice versa* ie. CG replaced by GC and AT by TA. This change can be induced by alkylating agents like ethylmethane sulphate (EMS) and Methylmethane sulphate (MMS). The alkylating agents alkylate the nitrogen in the 7th position in a purine and lead to its separation from the DNA strand. This gap can be filled by any of the four bases at the time of replication. Trasversions were first postulated by E.Freese in 1959.

PRACTICAL APPLICATION OF MUTATIONS

Generally mutations are lethal and recessive and possibly of no help except as a scientific curiosity in understanding the basic funcitoning of the hereditary material. But there are specific instances where both spontaneous and induced mutations are of selective value from the point of view of Evolution and also from the point of view of human beings. Ancon breed sheep is one such example of a spontaneous mutation. Mutations are also of value in plant breeding, when mutant varieties can be stabilized in the gene pool by cross breeding. Gustaffson has estimated that about one in thousand mutations may be used in breeding experiments. Several mutations in wheat, like branched ears, resistance to lodging, high lysine content, amber seed colour, awned spikelets have been useful inbreeding. The amber seed colour which is originally seen in the mexican variety has been induced in wheat of Indian varieties.

Many useful mutations in rice are also known. Some of these are *Reirnei* variety mutant, for reduced duration of the crop. In barley also mutants like *erectoids* and *eceriferum* have been induced for increased yield. The *Todds Mitchan* variety of mint plant *and-Aruna* variety of castor serve as good examples of the role of mutations in crop improvement. The above mentioned variety of mint has disease resistance as well as high oil content. In the castor variety duration of the crop was reduced by 130 days (originally 270 days). In a survey conducted by FAO in 1974 it has been estimated that 98 varieties of crops have been improved by induced mutations.

11

GENETIC ENGINEERING

Recombinant DNA technology or Genetic engineering as it is popularly known, is the newest tool that is available in the human arsenal of knowledge, in man's quest for a better human life. What is genetic engineering or recombinant DNA technology?

Imagine a situation of chronic diabetic patients being able to get back the capacity of synthesis of insulin; patients suffering from incurable diseases getting an injection of fene therapy–lo and behold they are cured of their diseases; consider a situation which is at present in the realms of fantasy of making man producing his own food just like the green plants do from natural sources; or visualize a situation wherein revolution in agricultural technology would make the use of chemical fertilizers completely redundant or going to the extreme, culturing a human being of the desired trait on a medium of nutrients i.e., man made to order. The scope is fantastic; possibilities are immense; it is as if the role of creator has been for once given to human beings. Well, this is genetic engineering or recombinant DNA technology, the most powerful biological technique which the human beings can use either to make or mar the society either for progress or for destruction. Using the techniques of genetic engineering, you can create deadly pathogenic bacteria and unleash them on human beings; you can produce varieties of agricultural crops which are disease free or you can introduce pathogens into the crops and totally destroy them. Genetic engineering used judiciously can be a harbinger of new revolution but used injudiciously may produce many prototype of criminals, antisocial elements, boot leggers etc. In this chapter, we shall briefly discuss some of the fundamental aspects of genetic engineering and how to use them for betterment of human life.

Genetic engineering may be defined as the method of artificial synthesis of new genes and their subsequent transplantation in the new genome of an organism, by molecular biological techniques. Recombinant DNA is achieved by constructing molecules outside the living cells by joining natural or synthetic DNA molecules that can replicate inside a living cell. Recombinant DNA technology involves several stages or steps.

The main aspects of genetic engineering (gene cloning) are the following –

1. Identification and isolation of the gene to be cloned.
2. Insertion on the gene into another piece of DNA called vector which will allow it to be taken up by microorganisms like bacteria and replicate within them as the cells divide.
3. Transfer of the recombinant vectors into bacterial cells.
4. Selection of those cells which have the desired recombinant vectors.
5. Growth of the microorganisms to get increased quantity of cloned DNA.
6. Expression of the desired gene to obtain the desired product.

Outline procedure of genetic engineering

Genetic engineering of a desired gene and gene product involves the following procedure. First of all, the desired gene has to be obtained from a source which then has to be identified and isolated.

(i) The living cells from which the genetic material is to be obtained has to be broken open. One popular method is to shear the cells in a blender and treat them with a suitable detergent.

(ii) The genetic material from the cell is to be removed. Several techniques are available to remove the DNA. The DNA molecules can even be spooled on to a glass rod. The glass rod bearing the DNA molecules is then lifted up from the debris of the cells.

(iii) The next step is to cut away the specific genes of interest from the DNA molecule. Restriction enzymes popularly called molecular scissors are used to cut and separate the desired gene from the DNA helix.

(iv) In the next step, the specific sections of DNA are incorporated into cloning vectors or vehicles such as plasmids, phages etc. These cloning vehicles are then introduced into a rapidly multiplying living system such as bacteria where multiple copies of the cloned DNA can be obtained. The cloned DNA also called chimeric DNA, contains the desired foreign gene. The coloned DNA is also called recombinant DNA.

(v) In the final step which is a very delicate process, the cloned gene is allowed to express itself in the bacterial cell. Gene expression here means production of a desired gene product. For instance, if a human insulin gene is cloned in a plasmid and is inserted into a bacterial cell, the bacterial cell along with its metabolic products also will produce insulin. This insulin may then be separated by the normal methods of chemical purification and isolation.

RESTRICTION ENZYMES (restriction endonucleases)

The first account of restriction enzymes was given by Werner Arber while studying the bacteriophages. He found that in some instances as soon as the viral DNA entered into a bacterial cell, it (viral DNA) was cut into bits and destroyed. Based on this observation, Arber opined that the bacterial cell must possess certain specific DNA destroying enzymes that could infect only the viral DNA but not the host DNA. In otherwords, the enzyme should be such that it should be able to distinguish between foreign DNA and its own DNA. He gave the name restriction enzymes and hypothesized that these restriction enzymes can recognise and act at specific sites on the viral DNA.

The next step in the discovery of restriction enzymes was taken by Smith *et al* (1970), when they isolated a restriction enzyme from the bacterium *Haemophilus inflenzae.* Smith showed that the restriction enzyme can cut the viral DNA at specific sites. All the DNA termini were pGpTpPy-3' and 5' pPupApCp, showing that these were cut from the double stranded DNA having the following sequence

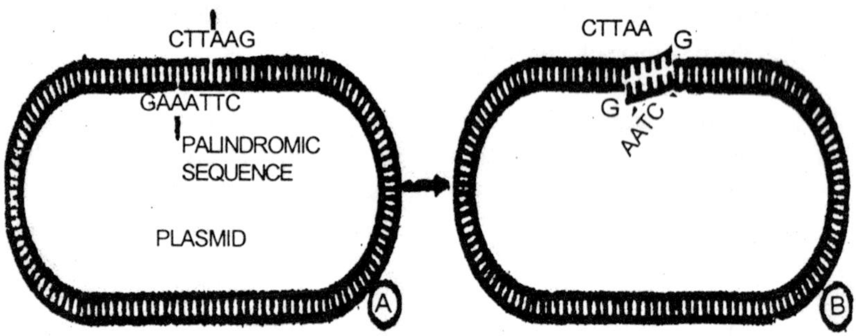

Fig. 11.1 Genetic Engineering
Action of restriction enzymes

5'pGpTpy	|	pPupApCp-3'
3'pGpApPu	|	pPypTpGp-5'

The above example shows that the restriction enzyme from *Haemophilus influenzae* can cut the DNA molecule which exhibit a two fold symmetry i.e., with a palindromic sequence (same sequence running in opposite directions), some restriction endonucleases, their source, recognition sequence and sites of cleavage (indicated by arrow).

Enzyme	**Source**	**Recognition sequence and clearing site**	
		↓	
Eco R1	*Escherichia coli* Ry13	G.AA	TTC
		GTT	AAG
			↑
Hind II	*Haemophilus influenzae*	GTPy ↓	PuAC
		CAPu ↑	PyTG
Hind III	*Haemophilus influenzae* Rd	A AG	CTT
		TTC ↓	GA A
			↑
Hpa I	*Haemophilus parainfluenzae*	GTT ↑	AAC
		CAA ↓	TTG
Hpa II	*Haemophilus parainfluenzae*	CC	GG
		GG	CC
		↑	
Bam HI	*Bacillus amyloliquefaciens*	↓ GGA	TCC
		CCT	AGG
		↓	↑
Bgl III	*Bacillus globigi*	A GA	TCT
		TCT	AG A
			↑
Hae II	*Haemophilus aegyptius*	Pu GG	GCPy
		Py CG ↑	CGPu
Hae III	*Haemophilus aegyptius*	GG ↓	CC
		CC ↑	GG
			↓
Hha I	*Haemophilus haemolyticus*	GC	GC
		C G	CG
		↑	
Pst I	*Providencia stuartii*	CTG	CA G
			↓
		GAC	GTC
		↑ ↓	

Sma I	*Serratia marcescens*	CCC		GGG
		GGG		CCC
		↓	↑	
Taq I	*Thermus aquaticus*	TC		GA
		AG		CT
				↑
Bal I	*Brevibacterium albidum*	TCG	↓	CCA
		ACC	↑	GGT
		↓		
Sal I	*Streptomyces albus* G	G TC		GAC
		CAG		CTG
				↓↑
Xor II	*Xanthomonas oryzae*	CGA		TCG
		G CT		AGC
		↑		
Alu I	*Arthrobacter luteus*	A G		CT
		TC		GA
			↓	
Msp I	*Moraxella* sp.	CC		GG
		GG	↑	CC
		↓		
Mbo I	*Moraxella bovis*	GA		TC
Sau 3 Al	*Staphylococcus aureus* 3A	GA	↓	TC
		CT	↑	AG

Subsequent work by Daniel Nathans (1971), showed that the restriction enzyme discovered by Smith can cut the monkey virus DNA into eleven well defined pieces. By using specific restriction enzymes it is also possible to map the DNA which they cut as their specificity is known. Arber, Nathans and Smith were awarded the 1978 Noble prize in physiology and medicine for their work on restriction enzymes.

In many microorganisms, restriction enzymes are a means of protection against intrudent viral DNA. In the working of the restriction enzyme, the nucleotide sequence of the host DNA is identified and methylated to avoid degradation. As a result of this the DNA gets the protection while the foreign DNA with the same sequence will not be methylated and coonsequently subjects itself for degradation. The whole mechanism is called the RM system (restriction modification), and is wide spread in bacteria.

Classification of restriction enzymes

Restriction enzymes are able to recognise specific nucleotide sequence in the double stranded DNA and cleave both the strands. To date, more than 150 restriction enzymes have been discovered which are known to recognise 40 different nucleotide sequence. Some restriction enzymes, however, may not cut DNA at specific sites. Based on this, restriction enzymes are classified into two, categories, type 1 and type 2. Type 1 restriction enzymes recognise a specific sequence of DNA molecule but the cut is made elsewhere whereas type 2 enzymes make a cut only within the restriction site and produce two single strand breaks, one break in each strand. Type 2 enzymes are most important because they produce broken chain of nucleotides with a palindromic sequence. The following table gives a summary of the difference between type 1 and type 2 eyzymes.

	Type 1	Type 2
1.	Non-specific in cleavage	Cleavage is site-specific
2.	M.W 300,000	Lower MW (20,000 - 1,00,000)
3.	Has non-identical subunits	Has two identical subunits
4.	Require ATP, Mg ++ and generally S-adenosyl	Require only Mg ++ as co-factor
5.	Examples : EcoB, EcoK, EcoPi	Example : EcoRI

Nomenclature of restriction enzymes

The names are designated according to the following principles.

(1) Name of the organism is identified by the first letter of the gene's name and the first two letters of the species name to form a three letter abbreviation in italics. For example *E.coli* = *Eco* and *H.influenzae* = *Hin* etc.

(2) A strain or type identified is written as a subscript. For example, *E.coli* strain k is written as Eco_k.

(3) In cases where the restriction modification systems are genetically specified by a virus or plasmids the extra chromosomal element is identified by a subscript as *Eco RI, Eco PI* etc.

When a particular strain has several restrictions and modification systems, these are identified by Roman numerals. For example Hin dI, Hin dII, Hin dIII etc.

Cleavage brought about by restriction enzymes

Restriction enzymes have an ability to recognise a particular sequence and cleave DNA molecule at that site. Most of the sequences recognised by these enzymes are tetranucleotide or hexanucleotide pairs which show a characteristic two fold symmetry i.e., palindromic sequence. The Eco RII site is very unique as it contains 5 base pairs with the axis of symmetry passing through the middle AT pair as shown below.

CCTGG
CCACC

Some times restriction enzymes from unrelated species can recognise the same nucleotide sequence. For instance, GGCC sequence specific enzymes have been discovered in *Bacillus subtilis, Haemophilus aegypticus* etc. Such enzymes are called isochizomers.

Cleavage sites may be even or staggered. In even cleavage the breaks on the two strands are exactly opposite to each other. Whereas, in staggered breaks, the breaks on the two strands are a little away from each other. Even breaks do not produce unpaired terminal nucleotides, whereas staggered breaks may produce sticky ends.

Use of restriction enzymes

The use of restriction enzymes has proved to be a highly useful tool for the sequence analysis of DNA. It also plays a significant role for gene cloning and amplification in genetic engineering.

The DNA molecule from bacteriophages and many other viruses consists of 5,000 to 50,000 base pairs. It is of fundamental importance to know the sequence of these bases to understand the gene structure and regulation and also for decoding the information stored in these genes. By the use of several restriction enzymes it is possible to cut fragments of DNA as small as 300 bp. These fragments can be used for sequence analysis and mapping.

The DNA segments produced by restriction enzymes can be covalently linked to the plasmid DNA and such recombined DNA can be inserted into bacteria such as *E.coli* for transformation. The transformed cells can now be cultured as separate colonies.

The ability of restriction enzymes to locate and cut at a particular sequence is of great help in identifying and isolating a particular gene.

Identification of a required DNA piece (gene)

Theoretically, identification of a gene involves, the reading of the message back in protein synthesis. During protein synthesis DNA produces mRNA and this in turn identified and allows the alignment of amino acids to synthesize a protein. In the identification of a gene responsible for the production of a particular protein, the process has to be followed backwards. Let us take an instance of the human insulin which is a protein.

The steps involved in the identification of the gene or segment of DNA responsible for insuline are as follows.

(a) Isolation and purification of the protein.

(b) Sequential degradation of the protein : In sequential degradation, not only the qualitative composition of the amino acids are known, but even their sequence of arrangement in the polypeptide chain will be known. After ascertaining the sequence of amino acids, the next step would be :

(c) Identification of codes for each amino acid : This can be done easily as we know the triplet codes for each amino acid. This would give us a length of the mRNA segment equivalent to the length of the polypeptide chain. (For example, if the protein consists of 25 amino acids, the length of the mRNA should be 75 nucleotides).

The nucleotide sequence in mRNA can be easily substituted for the complementary DNA segment. We now have a gene or a segment of DNA which governs the production of insulin.

ISOLATION OF SPECIFIC DNA SEGMENT (GENE)

The DNA of an organism, whose specific gene is required for the purpose of expression in another organism contains thousands of genes. The first problem therefore, is to separate the required gene from the DNA of the organism. There are several techniques to isolate the required gene from the DNA molecule.

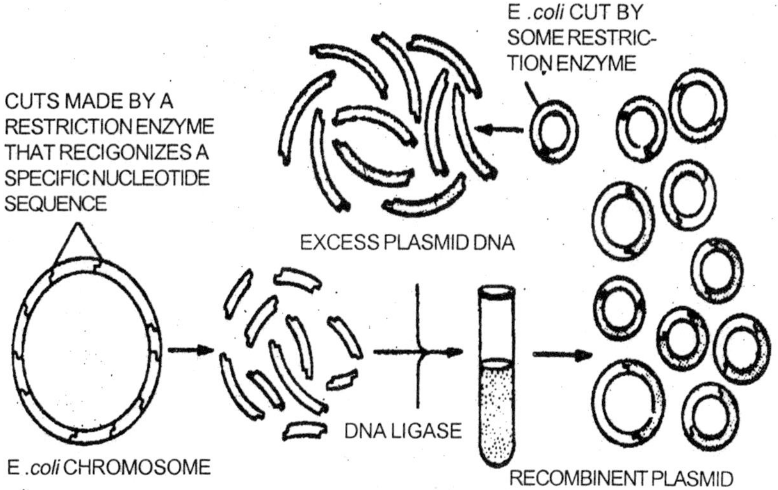

Fig. 11.2 Genetic Engineering

Isolation and recombination of genes using shot gun method

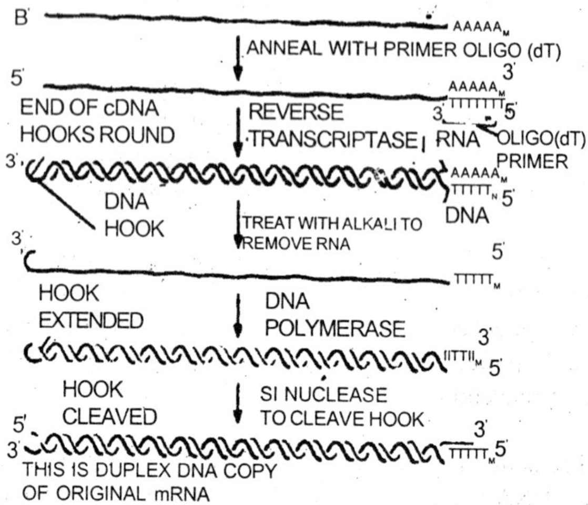

Fig. 11.3 Genetic Engineering
Obtaining cDNA from mRNA

Shotgun method

In this, the entire DNA of the organism is subjected to the action of restriction enzymes and is broken down into a number of fragments. Each of these fragments is then combined with the DNA from a suitable vector to produce recombinant DNA molecule. Subsequently, these recombined DNA molecules are inserted randomly into the host cells such as *E.coli*. These host cells are cultured separately and grown into colonies. The following are the steps of the shotgun method.

1. Using specific restriction enzyme, the isolated DNA from a particular organism is cleaved into many bits. As the restriction enzymes may bring about even or staggered cuts many sticky ends are left. The resultant DNA fragments are of different length.
2. Host cells of *E.coli* are selected, and the outer membrane is first dissolved by a detergent and its DNA is released. The plasmids of *E.coli* are then separated from the main genome by ultracentrifugation. The plasmids are then kept in a solution containing the same restriction enzyme which will cut the plasmid circle into a linear form. Linear plasmid DNA also has overlapping sticky ends.

3. In this step, isolated DNA and linear plasmid DNA are mixed. As a result, random association occurs between the foreign DNA and the plasmid DNA. Subsequently, the foreign DNA gets integrated into the plasmid

DNA to form a long linear DNA molecule, which then becomes a circle due to covalent joining. The resultant DNA is a recombinant or chimeric DNA.

4. The chimeric DNA molecules are now placed in a solution of calcium chloride which also has *E.coli* cells. On heating suddenly upto 42°C, the membranes of bacteria become permeable to chimeric DNA molecules. Thus, the *E.coli* cells incorporate the chimeric DNA plasmids.

5. The bacterial cells containing in the chimeric plasmids is then cultured on a medium. The cells express not only their genes but the genes of the chimeric plasmids also, thus producing a desired product. All these cells are produced by the division of the original bacterial cell having a single chimeric plasmid. Hence, they are called clones. Insertng a chimeric plasmid into a bacterial cell is of a very rare occurrence; it has been estimated that only one is 10^5 of 10^6 cells are capable of undergoing transformation.

In shotgun method, several genes can be cloned at a time and it is an omnibus method because the genes to be cloned are not picked and identified before.

Obtaining DNA from mRNA

In shotgun method, gene cloning is a general one and not specific as the genes to be cloned are not identified and isolated before. Hence, the chances of inserting a particular gene by this method are very rare. Another specific method of isolating gene is to first obtain the messenger RNA transcribed by that particular gene and then obtain DNA from mRNA. DNA from mRNA can be obtained by the following two methods.

Hybridization method

When DNA transcribes and forms mRNA, a complex is formed. The hybridization method depends on this mechanism. At the first step, the total DNA from an organism is isolated and treated with heat or alkali to convert it into a single stranded form. These single strands are then mixed with the specific mRNA obtained from the gene. Obviously, the mRNA chains pair with that section of DNA which has complimentary nucleotides to it, thus forming aDNA-RNA hybrid. This complex can be isolated and the DNA can be separated from the RNA. The single strand DNA can be converted into a double strand DNA by enzymes such as polymerase I. This method is highly useful for isolating ribosomal genes.

Reverse transcriptase method

In the normal transcription, the flow of information is from DNA → RNA → proteins. Transcription of RNA from DNA is catalysed by the enzyme transcriptase also known as DNA dependent RNA polymerase. Temin and Baltimore independently discovered that occasionally the genetic information may flow in the reverse direction also (from RNA to DNA), in some RNA tumour viruses. They named the enzyme *Reverse transcriptase* (RNA directed DNA polymerase). This enzyme catalyses the synthesis of a single strand DNA using an RNA template. This discovery has proved to be highly useful for isolating particular genes.

In reverse transcriptase method, purified mRNA segments of the required gene are used as a template to produce a fragment of single stranded DNA under the influence of the enzyme reverse transcriptase. As this single strand DNA is complementary to mRNA, it is called complimentary DNA or cDNA. Hydrolysis of RNA template will separate a single strand DNA. This single strand DNA is converted into double strand using the enzyme DNA polymerase. Another enzyme S1 nuclease breaks the covalent linkage between two DNA strands.

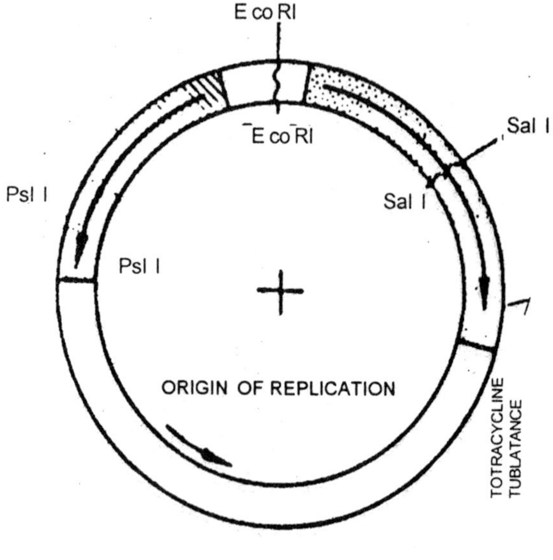

Fig. 11.4 Genetic Engineering
Genetic map of plasmid pBR 322

VECTORS (Carriers or Vehicles)

Vectors are defined as those DNA molecules that can carry a foreign DNA fragment inserted into them without any disturbance in their expression. Vectors are also called vehicle DNAs. Vectors are obtained from a number of sources. These are grouped into categories such as plasmids, bacteriophages, cosmids and phasmids.

Plasmids

These are extra chromosomal, self- replicating and double stranded covalentaly closed circular DNA molecules present in the bacterial cell. A number of plasmids are present in bacterial cells that have sufficient genetic information for their replication. Eventhough they are not necessary for the survival of the bacteria, they carry such necessary information as drug resistance, metal resistance, toxin production, nitrogen fixation etc. Naturally occuring plasmids can also be modified by *in vitro* techniques such as code transformation. Use of plasmids as a vector for DNA cloning was reported for the first time by Cohen et al (1973).

Plasmids are used as cloning vehicles when they possess the following features.

1. It can be readily isolated from the cells.
2. It possesses a single restriction site for one or more restriction enzyme(s).
3. Insertion of a linear molecule at one of these sites does not alter its replication properties.
4. It can be reintroduced into a bacterial cell and cells carrying the plasmid with or without the insert can be selected or identified (Bernard and Helinski, 1980).
5. They do not occur free in nature but are found in bacterial cells.

In many instances, the main objective of coloning is the insertion of a particular fragment of DNA into a suitable plasmid vector and its amplication. Amplication is a process of increasing the number of plasmids in a bacterial cell. In this process, a cell containing a relaxed plasmid is treated with a drug to inhibit protein synthesis. Consequently, replication stops and the relaxed plasmid (eg pBR 322), continues to replicate despite the drug treatment. Addition of chloramphenicol causes the plasmid BR 322 to increase its number to 3,000 per cell. In order to replicate and clone a piece of foreign DNA, it is necessary to possess a sequence of nucleotides which is recognised by the host bacterium as origin of replication. The origin of replication occurs naturally in plasmids. In naturally occurring plasmids, due to exchange of genetic information between plasmids and bacterial chromosome new genes originate in the plasmids. Therefore, plasmids have been artificially developed with specific sequence of genes so that they would integrate foreign DNA without any problem.

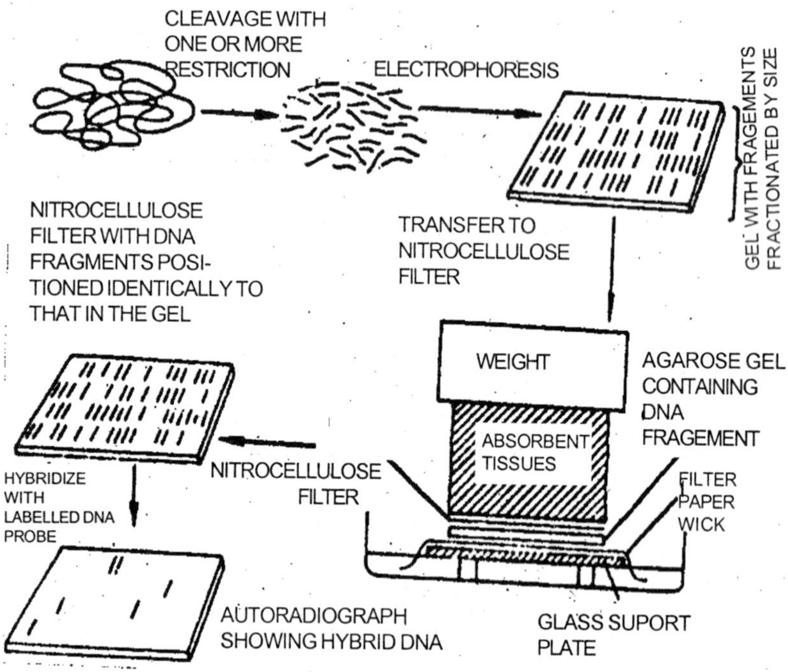

Fig. 11.5 Genetic Engineering
The Southern blotting technique

Some cloning vectors

Cloning vectors	Natural occurrence	Size (Kb)*	Selective marker**
Plasmids			
pACYC 177	*E.coli*	3.7	Amp^r, Kan^r
pBR 322	*E.coli*	4.0	Amp^r, Tet^r
pBR 324	*E.coli*	8.3	Amp^r,, tet^r. El imm
pMB9	*R.col*	5.8	Tet^r
pRK 646	*E.coli*	3.4	Amp^r

Cloning vectors marker**	Natural occurrence	Size (Kb)*	Selective

Plasmids			
pACYC 177	*E.coli*	3.7	Amp^r, Kan^r
pBR 322	*E.coli*	4.0	Amp^r, Tet^r
pBR 324	*E.coli*	8.3	Amp^r,, tet^r. El imm
pMB9	*R.coli*	5.8	Tet^r
pRK 646	*E.coli*	3.4	Amp^r
pC194	Staphylococcus aureus	3.6	Ery$_r$
pSA 0501	S.aureus	4.2	Str^r
pBS 161-1	Bacillus subtilis	3.65	Tet^r
pWWO	Pseudomonas putida	117	Kan^r
Cosmids			
pJC 74	Derived plasmid from ColEL	16	Amp^r Ei imm
pJC 720	do	24	Amp^r, Tet^r
pHC 79	Derivative of PBR 322	6	Amp^r, Tet^r
Virusus			
SV40	Mammalian cells	5.2	-
Phage M13$^+$	*E.coli*	6.4	-
Phage *E.coli*		4.9	-

* 1Kb (Kilobase pairs) = 1,000 base pairs = 0.66 mega dalton;
**. Resistance of ampicillin (Ampr), tetracycline (Tetr), erythromycine (Eryr), streptomycine
(Sttr), Kanamycin (Kanr), rifampicin (Rifr), and colicin production (EL imm).

The plasmid BR 322 is such an artificially constructed plasmid using two plasmids of *E.coli*-- PBR 318 and PBR 322. This plasmid contains the origin of replication *(ori)* derived from a plasmid related to naturally occurring plasmid Col E1. As a result, the BR 322 plasmid replicates at a rate faster thean bacterial genome. It also possesses genes confering resistance to antibiotics such as ampicillin and tetracyclin. It also has a unique recognition site for 20 restriction enzymes. Certain restriction sites, for example Bam HI in the *tet* gene of the plasmid is present within the gene in such a way that the insertion of foreign DNA will inactivate the *tet* gene. As a result, the recombinant plasmid will allow the cells to grow only in the presence of ampicillin, but will not protect them against tetracyclin. This antibiotic resistance will allow the selective isolation of cells which can be cultured on a medium supplemented with ampicillin.

In addition to the *E.coli* plasmids, a plasmid of gram negative bacteria called RK2 plasmid has been developed which has a very broad host range and has a single restriction site which can be acted upon by several endonucleases.

Isolation of bacterial plasmid

The following steps are ncessary for isolating the bacterial plasmids.

1. Treat the prokaryotic cells with detergents which will release plasmid DNA and chromosome from the bacterial cell along with other molecules. This mixture is known as lysate.
2. Treat the lysate with potassium acetate or acitic acid solution. This precipitates chromosomal DNA and some protein molecule.
3. Remove the precipitate from the lysate by igh speed centrifugation. The supernatant contains plasmid DNA along with RNA, protein and some DNA debris of the chromosome.
4. Treat the lysate with RNAase, this will digest the RNA.
5. The lysate is treated with phenol and as a result two layers are formed – phenol layer and water layer. The water layer contains plasmid DNA, and the other contaminants are found in the phenol layer.

Bacteriophages

A virus which attacks the bacteria is called a bacteriophage. A number of bacteriophages with different genetic materials have been reported so far from a variety of bacteria. Some of these phages are ϕ x 174, phages λ, M13, Fd11, R209 etc. Phages also can be used as cloning vectors like plasmids. Of all the phages, the λ phage perhaps is the one that has been used very frequently as a cloning vector. The λ phage contains a proteinaceous head and a long tail. The head portion has 50 genes in its 49 kb genome of which about half of the genes are essential. On attachment to the host (*E.coli*), the linear DNA molecule is injected into the cell. The linear double stranded molecule of DNA becomes a circle by joining through the single strand of 12 nucleotides commonly known as *cos* site. This *cos* site is a very important feature of λ DNA.

The λ phage has two kinds of life cycles – the lytic phase and the lysogenic phase in the bacterial cell. In the lytic pathway, early in the infection the circular DNA replicates by a rolling circle mechanism and produces long molecules of DNA. At the same time, the phage DNA directs the synthesis of proteins to produce the head. In order to pack the DNA into the head, it undergoes cleavage to yield the fragments of such sizes as to fit in the head. Eventually, a tail is attached to the head. In the lysogenic pathway, the phage genome integrates with the bacterial genome at an attachment site (*att*) where it replicates along with the bacterial genome.

The following are the advantage of using phage as cloning vectors – 1. DNA can be packed *in vitro* into phage particles and transduced into *E.coli* with high efficiency, 2. Foreign DNA upto 25 kb in length can be inserted into phage vector, and 3. Screening and storage of recombinant DNA is easier (Dahl et al., 1981). Before using the λ phage vector, it will be essential to remove the genome from the restriction sites for the enzymes commonly used for cloning. Two types of phage cloning vectors have been constructed – these are the *insertion vector* and the *replacement vector.*

Insertion vector. These have unique cleavage sites into which a relatively small piece of foreign DNA can be inserted. These fragments do not affect the functioning of the phage. The upper and lower limits of DNA that may be packed into the phage particles is between 35–53 kb. The maximum size of a foreign DNA that can be inserted is about 18kb.

Replacement vector. These vectors have cleavage sites present on either side of the length of non-essential DNA. As a result of cleavage, left and right arms are formed and each arm has a terminal *cos* site and a longer stuffer region (the non-essential region), which can be substituted by foreign DNA. The maximum size of foreign DNA that can be inserted into this vector depends on how much the phage DNA is non-essential. It has been estimated that about 25–35kb of genome can be replaced with foreign DNA segments. Example of substituted vectors are–gt, WES and λ 1059.

Cosmids

These may be defined as the hybrid vectors derived from the plasmids which contain *cos* site of λ phage. Reported for the first time by Colins and Hohn (1975), cosmids lack genes for encoding viral proteins. Therefore, neither viral particles are formed, in the host cell nor cell lysis occurs. Special features of cosmids similar to plasmids are origin of replication, a marker gene coding for antibiotic resistance, a special cleavage site for the insertion of foreign DNA, and the small cells. Characters of cosmids dissimilar to plasmids are presence of extraphage DNA, and the *cos* site.

Phasmids

When plasmids are inserted into the genome of λ phage, they are said to be phasmids. The process of insertion of the plasmid DNA into the phage genome occurs in the same way as a phage genome is inserted into the bacterial chromosome. Insertion of plasmid into λ phage is done with a view to have a specific site responsible for recombinational insertion of the phage into bacterial chromosome.

Phasmids may be used in many ways. They can be used as a phage cloning vector from which a recombinant vector may be released.

CLONING ORGANISMS

The vectors that are chosen as vehicles to carry the foreign DNA must be cloned in an organism that multiplies rapidly, thereby producing the gene product in a large quantity. Such organisms into which the vectors are inserted for the purpose of gene expression are known as cloning organisms. A number of microorganisms have been used for the purpose of cloning. The most important among them are the various strains of the colon bacterium, *E.coli.* The strains of *E.coli* used as cloning organisms are HB101, h303 and RR1. The cloning organisms should normally be deficient in DNA recombination as it will reduce undesirable hybridization of DNA. Another essential feature of the cloning organism is it should lack restriction enzymes that might degrade foreign DNA. Strains of bacteria that can bud off mini cells are highly useful if the gene expression is the purpose of experiment (mini cells are those which contain all the elements of gene expression except chromosome).

For a long time, *E.coli* has been the most favourite cloning organism in recombinant DNA studies. Of late, yeasts are also being used as cloning vehicles and they have some advantages over bacteria. They do not live in human beings and are non-pathogenic. Fink et al (1974), for the first time introduced a bacterial gene for the production of leucine into baker's yeast. A snail enzyme was used to induce the yeast to take up bacterial DNA.

Techniques used for identification of clones

Several techniques are employed to identify the isolated clones using a specific molecular probe. Some of these techniques are described below :

1. Polyacrylamide gel electrophoresis

When the genomic DNA extracted from a particular source is digested by a specific restriction enzyme it is broken into several fragments. These fragments of different sizes can be separated through a technique called gel electrophoresis. This separation is necessary to identify a specific gene that is required for use. After these are separated, a specific molecular probe can be used to detect the segments which have sequences similar to those in the probe. For instance, if a gene with a specific nucleotide sequence is required to be separated from a mixture, a probe is prepared with nucleotide sequence complementary to the gene. These (molecular probes) are then allowed to hybridize and the particular gene is identified.

In gel electrophoresis, the DNA molecules migrate from end of the gel to the other. The gel is usually a cylindrical structure, or occasionally even may be a slab. The rate of movement of fragments of DNA on the gel is inversely proportional to the size of the molecules. Heavier fragments will remain at the beginning while lighter fragments will move away. Fragments of different sizes appear as discreet bands on the gel and can be isolated for further study.

Separation of DNA molecules using PFGE (pulsed field gel electrophoresis)
Gel electrophoresis is suitable for only small fragments of DNA which can easily get separated. Large size molecules of DNA, however, cannot be separated using this technique. A new technique called PFGE is used for the separation of large sized DNA fragments, some times representing even whole chromosomes. Using the PFGE technique, separation of DNA molecules belonging to each of the sixteen chromosomes of yeast has been made possible. In this technique, short pulses of electric current are used in two different directions and the DNA is embedded in Agarose plugs to avoid fragmentation. Using this technique, genomes of several fungi have been resolved into chromosomal bands.

Blotting techniques
Gel electrophoresis helps in the separation of fragments of DNA or RNA. The separated bands can be stained and absorbed on the gel. This technique, however, will only help in the separation of fragments of DNA but will not tell as to which fragment contains the desired gene. For this purpose, a molecular probe (which is a DNA fragment) consisting of nucleotides complimentary to the gene are used for hybridization and the required gene is identified. For easy identification and detection, the probe is radioactively labelled. To facilitate hybridization, the bands of the genome are usually transferred to a nitrocellulose membrane in a technique called blotting. Blotting technique was first invented by E.M. Southern, and is known as the southern blotting technique. Two other blotting techniques discovered subsequently are named northern blotting technique and western blotting technique. These are described briefly below.

Southern blotting technique. The DNA sample is first cleaved with a restriction enzyme and the sample is loaded on to gel a electrophoresis for separation. The DNA bands that appear in the gel are denatured into single strands using and placed on a glass plate with its two edges immersed in the buffer. A sheet of nitrocellulose membrane is placed on top of the gel and a stack of many paper towels along with a weight are placed on this membrane (see figure). The buffer solution is drawn up by the filter paper folds and passes through the gel and to the nitrocellulose membrane. Finally, buffer moves up to the paper towels. While passing through the gel the buffer carries with it the fragments of single stranded DNA and binds them on to the nitrocellulose membrane as it passes through it (membrane) to the paper towels above. For the complete migration of the buffer it may take a few hours. Subsequently, the paper towels are discarded, the nitrocellulose membrane with the single stranded fragments of DNA attached to it in different positions is baked at 80°C to fix the strands permanently on it. This membrane now has the fragments of single stranded DNA fixed on to it in the same position in which they were on the gel. In otherwords, the nitrocellulose membrane is the replica of the gel with its attached DNA fragments. The nitrocellulose membrane is now dipped in a solution containing labelled gene that is present on the membrane. The radio labelled probe hybridizes with the complimentary fragment of DNA on the membrane. The filter subsequently is removed from the solution and washed thoroughly to remove any unbound fragments of DNA. Later, the filter is placed against a photographic film. Wherever the probe has hybridized with the DNA fragments, there will be markings like black dots. This can now be compared with the original gel and the corresponding DNA band can be identified as the required gene.

Northern blotting technique. This is used to identify the required RNA bands. The technique of southern blotting used for DNA transfer from the gel to the nitrocellulose membrane could not be used for transfer of RNA because RNA would not bind with the nitrocellulose membrane. Alwin *et al* (1979), devised a technique in which the mRNA bands from the gel were blot transferred to a chemically reactive paper prepared by diazotization of aminobenzyloxymethyl paper. RNA fragments can bind themselves on to this paper and the paper can be used for hybridization with radio labelled DNA probes. Subsequently, by autoradiography the hybridized bands can be identified. Some of the researches of Thomas (1980), however have shown that mRNA bands can also be blotted directly on to nitrocellulose membrane using special techniques.

Western blotting technique. This technique is used to detect proteins of a particular speciality. Discovered by Towbin *et al* (1979), in this technique newly coded proteins by a transformed cell can be identified. In this method, however, nucleic acid probes are not used. This technique involves the following steps :

(a) Electrophoretic separation of the required protein on a polyacrylamide gel.
(b) Transfering of the separated peptide chains on to a nitrocellulose membrane.
(c) Using radio labelled antibodies as probes the required peptide chain may be hybridized. (The probe consists of antibodies which bind to the antigenic protein).
(d) Autoradiographic identification of the required hybridized protein.

CONSTRUCTION OF CHIMERIC DNA (insertion of DNA into vector)
Chimeric DNA molecules are produced by the insertion of a foreign DNA segment into a DNA molecule of the vector. These chimeric DNA molecules are the means of the cloning genes or other small DNA sequences. Basically, production of chimeric DNA molecules involves breaking open the vector DNA and insertion of foreign DNA followed by joining of the ends. There are three techniques available for the construction of chimeric DNA. These are – pallindromic sequences (cohesive ends) and staggered cleavage, dA-dT tailing and blunt end ligation.

1. Pallindromes and staggered cleavage (cohesive ends). It has already been explained that a pallindromic sequence has complimentary sequences at the two ends of a single strand. Restriction enzymes cutting the pallindromic sequences may have an even cut or a staggered cut. Result of the staggered cut is, DNA fragments with sticky ends are formed. When another DNA molecule (a foreign DNA molecule) is cut similarly, and when the two DNAs are mixed, both DNAs will combine forming a chimeric DNA.

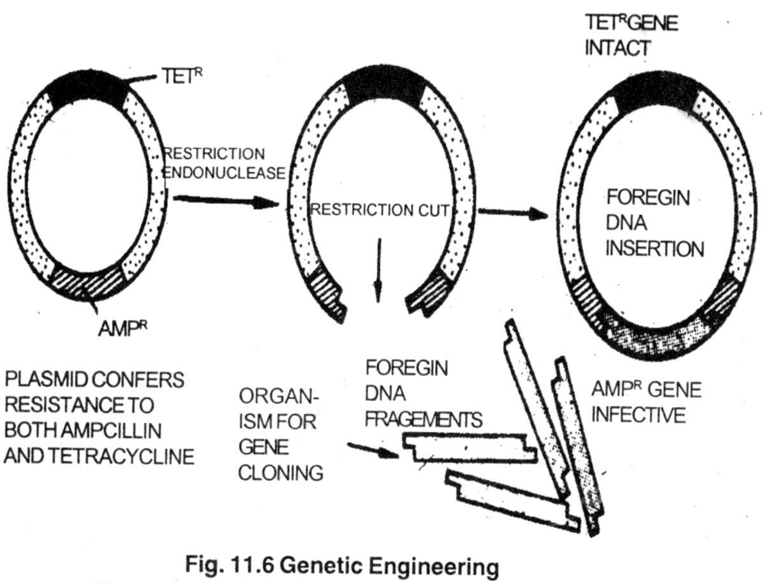

Fig. 11.6 Genetic Engineering
The concept of cloning a foreign DNA in a vector

Discovered for the first time by Boyer and Cohen, this mechanism has an advantage of easily combining two different DNA molecules. Further, it also produces two restriction sites in the chimeric DNA. There are some disadvantages in this method. For instance, the two sticky ends of a vector DNA or of a foreign DNA may themselves join without allowing for the insertion of one into the other.

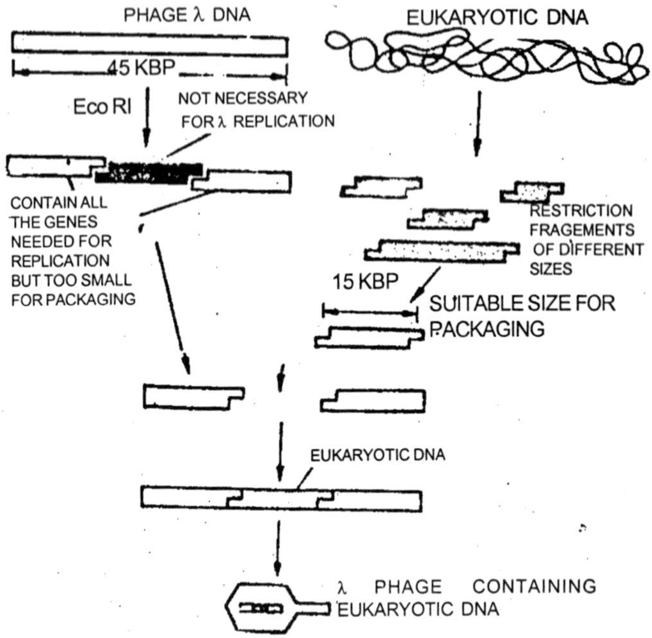

Fig. 11.7 Genetic Engineering
Use of λ phage of cloning

dA - dT tailing (the terminal transferase method)

In this method, DNA can be cut at the desired position both in the vector and in the clone without staggered cleavage. Using the precursor dATP, polydA is added at both the 3' cut ends of the vector with the help of the enzyme terminal transferase. Similarly, using the precursor dTTP poly dT is added at both the 3' cut ends of the DNA sequence to be cloned. The vector and clone can then be joined by anniealing the polydA with polydT tails and then lygating them using the DNA lygase enzyme. In this technique, there is no possibility of the two cut ends of the same DNA molecule joining together because they have the same nucleotides. There is one disadvantage in this technique because it will not be easy to retrieve the cloned DNA as the recognition site of the enzyme is lost due to the addition of poly dA and poly dT. It has been found out that the use of poly dG and poly dC will help in the restoration of the recognition site.

Blunt end ligation

In this technique, a restriction enzyme is used to cut a duplex DNA at the same site in both the strands. The broken ends are then used for joining with the two

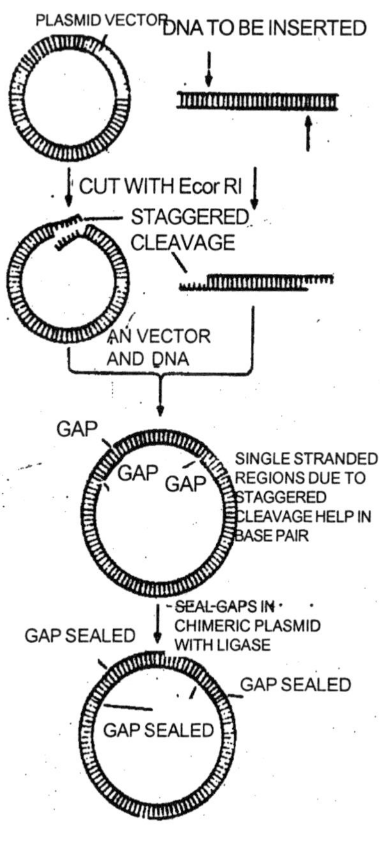

Fig. 11.8 Genetic Engineering
Cloning of a segment of DNA staggered cleavage

ends of foreign DNA molecule irrespective of the sequences present at the broken ends of the two molecules. The blunt end ligaton is made possible by the use of the T4 DNA lygase. This technique also has a disadvantage in that the broken ends of the same DNA molecule may join together.

Construction of genomic and cDNA libraries

In order to isolate required gene or genes from a genom a mixture of clones each carrying DNA fragments derived either from the natural genome or from cDNA is prepared. This mixture contains thousands of clones and is collectively called genomic library. Similarly, when cDNA clones are collected it is known as cDNA library.

Construction of genomic library

Cloning an entire genome in the form of fragments mixed randomly is known as genomic library. This may be done by a method known as shotgun method (see earlier in the same chapter for details). With a genome library, clones can be perpetuated indefinitely in a plasmid vector and retrieved when ever necessary.

Genomic libraries can also be prepared by using a number of restriction endonucleases one at a time so that one will have fragments of different sizes with cuts at different places. This has a disadvantage in that many a time the incision site many be within a gene so that complete genes may be difficult to obtain. This may be overcome by using restriction enzymes which have very short recognition sequences so that the fragments that are obtained will be long and the chances of slicing within a gene are avoided.

cDNA library

It has already been explained above as to how to obtain a cDNA using its

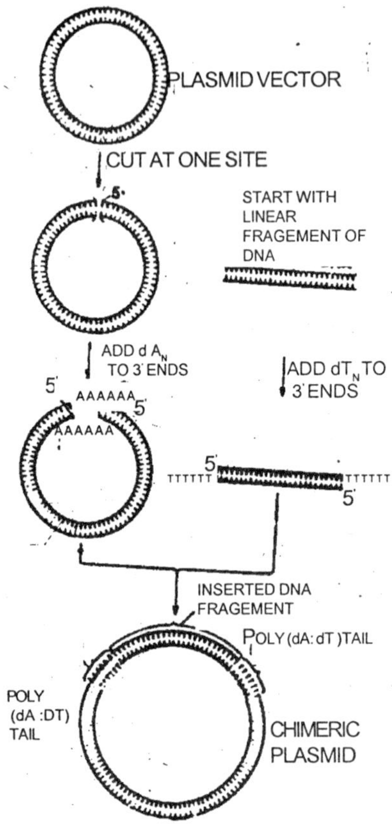

Fig. 11.9 Genetic Engineering Cloning of DNA segment by dA-dT tailing

complimentary mRNA under the influence of the enzyme reverse transcriptase. When once a cDNA molecule is obtained, it can be made double stranded by the use of DNA polymerase. A collection of such cDNA molecules will constitute cDNA library.

Screening of the library

When a genomic or cDNA library is available it can be used for the isolation of a specific gene sequence. For this, the DNA library has to be screened. This is done with the hlep of a technique known as Colony hybridization. In this technique, bacteria carrying chimeric plasmids are cultured into colonies and are then lysed on nitrocellulose filters. Their DNA is denatured *in situ* and fixed on the filter which is then hybridized with a radio actively labelled probe carrying a sequence related to the gene to be isolated. In otherwords, the probe will have nucleotides complimentary to the ones present in the gene to be isolated. Colonies that carry the sequence will be identified by dark spots after radioautography so that the colonies carrying the vector with the desired gene sequence can be recovered from the master plage and used for further experiment.

Transfer of the recombinant DNA into a cloning organism

When once a mixture of recombinant DNA is obtained it has to be transferred to a suitable cloning organism for the purpose of expression. Several techniques are used to transfer the recombinant DNA into suitable cloning cells. Transforming the cloning cells with recombinant DNA is a difficult process, as the cells of bacteria *(E.coli)* have restriction enzymes which will immediately degrade the foreign DNA. To escape from degradation, exponentially growing culture of *E.coli* is pretreated with $CaCl_2$ at low temperature and thereafter the DNA is mixed up. This treatment is known to sufficiently prevent the action of restriction enzymes. When the recombinant plasmids get established in the cells of *E.coli,* the cells are said to be transformed.

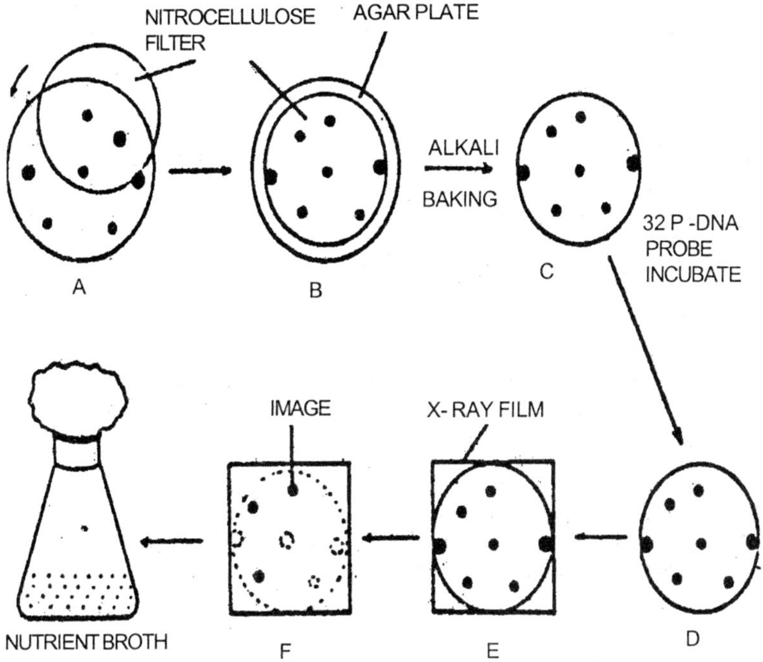

Fig. 11.10 Genetic Engineering
Colony hybridization technique

Instead of plasmids even, λ phages could be used as transforming elements. The entry of the phage into *E.coli* is usually termed transfection (a combination of transformation and information).

Not all the cells will accept recombinant DNA and get transformed. It has been estimated that about 10^5 transformats per microgram of cloned circular plasmid can be generated. The cells which have been transformed will grow into colonies along with other non-transformed ones. Colonies with transformed cells can be identified with suitable techniques.

Expression of cloned DNA

The purpose of genetic engineering is not just to separate and clone the genes in a suitable organism. Isolation and multiplication of genes is only a means ultimately to obtain the gene product namely the required protein. Hence, the recombinant DNA molecules should be allowed to express in the new environment (*E.coli* cells), and produce what ever protein they were producing. For instance, if an insulin gene from human beings is transfered into *E.coli,* the gene should be able to produce insulin molecules using the protein synthesizing machinery of the bacterium. Expression of a cloned gene is a very difficult process. This is more so when a eukaryotic gene is inserted into a prokaryotic organism. As the protein synthesizing genetic machinery of prokaryote is different from eukaryotes, the eukaryotic gene should be provided with the necessary ingredients of a prokaryote to produce the protein. The factors that are necessary for eukaryotic gene to express in the prokaryotic environment are as follows :

(i) Supply of prokaryotic promoter necessary for expression of eukaryotic gene.
(ii) Supply of ribosomal binding sites for the cloning vector.
(iii) Removal of nitrons from mRNA obtained from the eukaryotic genes.
(iv) Inhibition by a bacterial gene responsible for degradation of the protein produced by the eukaryotic gene.

Recovery of the gene product

This is the final step in genetic engineering techniques. After all the hurdles are passed and the eukaryotic gene has successfully expressed itself in the prokaryotic cell, the required product, namely the eukaryotic protein is produced. This can be identified by antigen antibody reaction (for details see genetically engineered insulin later in the same chapter). The protein produced can be purified by standard chemical processes.

APPLICATIONS OF GENETIC ENGINEERING

Scope of genetic engineering (Biotechnology)

The possibilities are immense and the sky seems to be the limit for future biotechnologists. The following are some of the applications of genetic engineering.

I. Gene therapy

A new branch of curative medicine is opened up due to genetic engineering. Many genetic disorders like haemophilia, alkaptoneuria, galactosemia etc., which are due to defective genes can be cured by replacing them (defective genes) with correct genes. Plasmid or virus mediated genes may be incorporated into human systems.

II. Cheap production of drugss

Many of the drugs like vitamins, antibiotics, harmones are now being produced chemically or produced from higher organisms which is a slow and costly process. If the genes for the production of these are introduced into rapidly multiplying bacterial cells, we have virtually myriads of tiny living factories turning out the medicines at a fraction of their original production cost.

III. Gene for new diet

Some of the plants or plant products are presently non-edible for want of enzymes in the human system to digest them. For ex, wood, grass, etc. If the genes that code for the digestive enzymes (for materials like wood etc) could be introduced into the human system, wherein there would virtually be no famine on earth. Another possibility, (which sounds unbelivable at present) could be to introduce genes that can produce the chlorophyll pigment in the animal or human system, wherein we would be solving the food problem once for all !.

IV. Role of genetic engineering in Agriculture

The following are some of the advantages of gene recombination in crop improvement.

(a) Production of new varieties

New varieties of plants that are not present in nature can be produced by the introduction of new genes.

(b) Somatic hybridization

Two crop plants with important characters, which could not be hybridized sexually due to genome incompatibility can be hybridized through protoplast fusion. Production of *Pomato* (Tomato x Potato) is one such example.

(c) Increasing photosynthetic efficiency

Ribulose diphosphate carboxylase (Ribisco) is a key enzyme in the dark fixation of CO_2. Modification of *Rubisco* to reduce photorespiration (Osmond 1981) by elimination of oxygenase activity would increase-the CO_2 fixation rate. The enzyme has two sub units and is known to be coded by seven genes. If these genes could be altered, resulting in the changed *Rubisco,* it will increase the photosynthetic efficiency.

(d) Nitrogen fixation

The roots of leguminous plants normally possess nitrogen fixing nodules, induced by various strains of *Rhizobium.*m As a result, leguminous plants require little or no nitrogen fertilizer. Aerable crops require about 100 kg of nitrogen per hectare. If this can be replaced by N_2 fixing genes, it would lead to a lot of saving in terms of fertilizer cost. Nitrogen fixation is catalyzed by the ATP dependent electron reduction of dinitrogen to ammonia by nitrogenase. If it is possible to introduce the gene to higher plants, it would bring about a revolution in agriculture. This may be achieved by one of the following ways.

1. Transfer of nitrogen fixing gene (*nif*) from the living bacteria like *Klebsiella, Azatabacter* etc., to higher plants. This would enable the plants to fix the N_2 directly. This is still in the experimental stage, as the *nitrogenase,* produced by the *nif* gene is highly oxygen liable and gets degraded in higher organisms.

2. Expansion of host range of symbiosis to crops other than legumes. At present non leguminous plants do not have symbiosis with *Rhizobium.* If it is possible to expand the host range for the bacteria, a lot more plants will have the capacity of N_2 fixation.

3. Increasing the efficiency of symbiotic bacteria to N_2 fixation. By gene manipulation, it may be possible to increase the N_2 fixing capacity of the symbiotic bacteria, which will benefit the plants.

(e) **Herbicide resistance, disease resistance etc.**

By gene manipulation, it will be possible to introduce this character into plants.

In conclusion, it may be said, that genetic engineering is a highly potential tool by which man can alter the macromolecule of inheritance, and thereby change the biological environment to his choice. But the new technique has to be used judiciously, otherwise it can also create untold harm to human society, if unscrupulous scientists use it to create destructive genes. A few of the specific applications of genetic engineering are discussed below in some detail.

Nitrogen fixing genes (*nif* genes)

Nodule formation and nitrogen fixation are two of the important biological processes which are genetically controlled in diazotrophic bacteria. From the genome of *Rhizobium,* several genes coding for nodulation (*nod* genes), and nitrogen fixation (*nif* genes) have been identified and isolated. In free living and symbiotic nitrogen fixers, nodule forming and non-nodule forming *nif* genes are present either on the genome or on the plasmids in their cells.

Nod genes

Nodule formign species of *Rhizobium* contain in their cells an extremely large plasmid called *mega plasmid,* which consists of a number of *nod* genes and *nif* genes (Rosenberg et al, 1981). Both the genes are closely located.

Kondorosi *et al* (1982), were able to transfer the megaplasmids from one species of *Rhizobium* to the other simultaneously transfering the ability of nodule formation.

The nod gene region is of 8.5 kb in nucleotide sequence length, and proteins encoded by the 8.5 kb fragment when transferred by *E.coli* has been determined. The *nod* gene is a cluster made up of 4 genes designated as *nod a, b, c* and *d.* A comparison of *nod* genes between different species of *Rhibozium* shows that about 70% of the nucleotides are homologous.

Nif genes

The *nif* genes are present in several species of Rhibozium and are located on a megaplasmid adjacent to the *nod* genes. *Nif* genes have also been identified in many cyanobacteria such as *Anabena,* where the *nif* genes are located on the main genome.

Early studies on *nif* genes were carried out in *Klebsiella pneumoniae,* and the function of these genes was confirmed by transfering into *E.coli* cells. Researches have also been conducted to identify the product of *nif* gene cluster. The *nif* gene cluster consists of 7 operons.

In cyanobacteria, *nif* genes have been located both in vegetative cells as well as in heterocysts. But there is some difference in the nature of the *nif* gene between the vegetative cells and the heterocysts.

Cloning of nif genes

Researches are being conducted in many countries to transfer the *nif* genes to higher plants especially to monocots. If this is successful it will be indeed a great boon to agriculture as there will be plenty of saving in nitrogen fertilizers, for the plants themselves will be able to fix atmospheric nitrogen. Gene expression of *nif* genes in eukaryotic cells is very difficult because the nitrogenous enzyme required for fixing the nitrogen and controlled by the *nif* gene would get denatured in the presence of oxygen. If this can be overcome, *nif* genes can be easily transferred to eukaryotes. *Nif* genes, however, have been cloned in *E.coli.* One of the methods of transfering *nif* genes is to transfer them into the chloroplast of eukaryotic cells as the translational and transcriptional processes of chloroplasts resemble prokaryotic features. These attempts might succeed because the chloroplasts are able to produce the ATP and reducing power both of which are required for nitrogen fixation (Merrick and Dixon, 1984). But there are major problems for such a transfer of *nif* gene to chloroplasts.
These are:

1. Lack of chloroplast transfering technique.
2. Protection of nitrogenase from oxygen evolved during photosynthesis.

Genetically engineered Insulin

Insulin is an important hormone that plays a vital role in the metabolism of carbohydrates. In patients suffering from diabetes, insulin has to be injected subcutaneously to regulate the blood sugar level. Insulin producing genes can be transferred to bacteria by techniques of genetic engineering and large quantities of insulin may be extracted. This process is cheaper and convenient than the conventional one. We will briefly discuss below the method of procuring insulin by genetic engineering techniques.

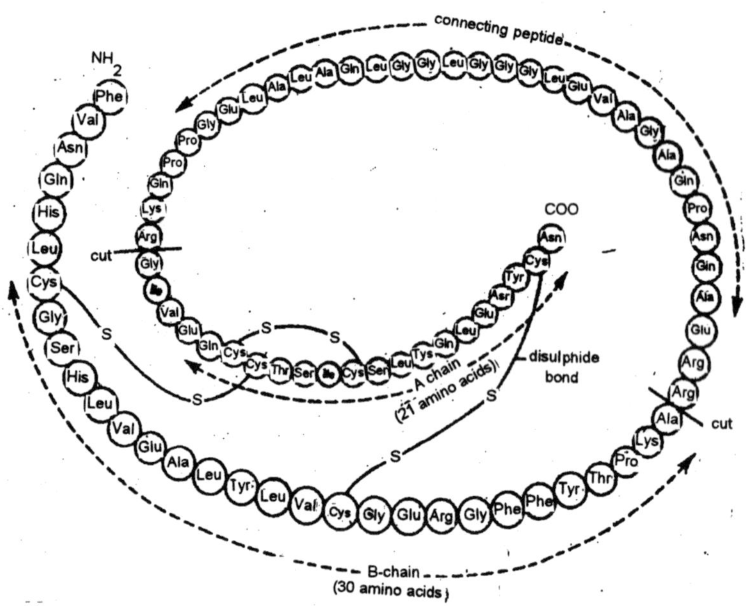

Fig. 11.11 Genetic Engineering
Structure of Insulin molecule

Structure of human insulin

Insulin molecule is composed of two chains – α and β chains of aminoacids. Initially, it is produced as a single long chain called *Pre-proinsulin* which is later cut into a shorter chain called *proinsulin* (composed of 84 aminoacid units). Proinsulin then breaks into α and β chains when a piece from the middle is cut off. Chain α is composed of 21 aminoacids and chain β has 30 amino acids. The remaining 33 aminoacids act as a connecting peptide chain.

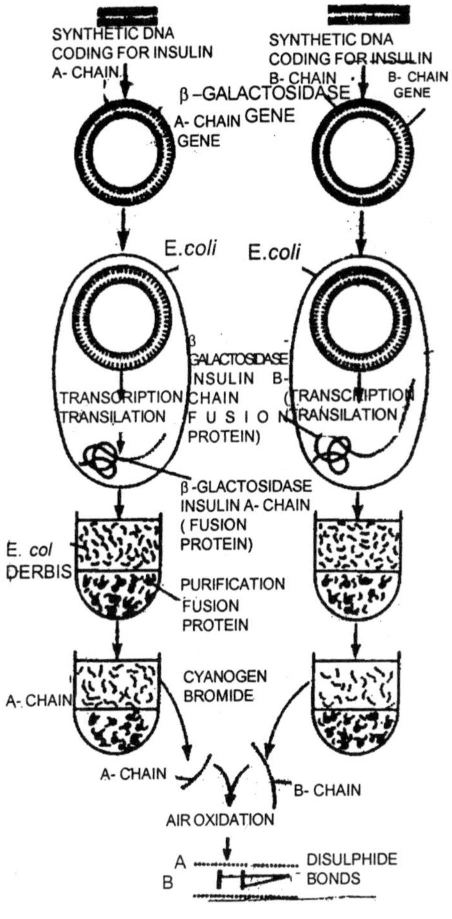

Fig. 11.12 Genetic Engineering
Technique of producing genetically engineered insulin in E.coli

Production of Insulin in E.coli

The genes for insulin chain α and β are identified and located. This may be done first by identifying the codons and then the corresponding mRNAs can be reconstructed. Using these mRNA molecules, and with the help of the enzyme *reverse transcriptase,* the corresponding segments of double stranded DNA molecules are produced. Then with the help of the enzyme *terminal transferase,* the ends of the insulin genes (DNA strands) are extended with short sequence of identical bases.

Plasmids obtained from *E.coli* are opened up at specific portions by the use of restriction enzyme. The insulin genes (DNA strands) of α and β chains are added to the plasmids separately forming recombinant DNA plasmids. Gaps if any in the circular plasmids of α and β chains are filled by lac operon gene for switching on the synthetic process. The plasmids with recombinant DNA for α and β chains are inserted into *E.coli* finally.

The recombinant DNA plasmids multiply in *E.coli* and trigger the synthesis of α and β chains of insulin as a genetic expression. Following large scale production cultures, the insulin chains are obtained following specific purification processes. Finally the α and β chains are chemically bonded by disulphide bonds to produce the complete molecule of insulin.

Gilbert and Villakomaroff (1980), isolated mRNA for insulin from β cells of the pancrease of rat and inserted it into PBR 322 plasmid in the middle of a gene normally coding for the enzyme penicillinase and incorporated it into the cloning organism (*E.coli*), *E.coli* cells produced the hybrid protein consisting of penicillinase + proinsulin. Proinsulin was then separated from penicillinase.

Immunological tests to identify insulin producing colonies

On a culture medium, there will be colonies which are transformed as well as non transformed. In otherwords, colonies with a recombinant DNA and those without recombinant DNA will be there. Colonies with recombinant DNA and successfully expressing the gene will produce the insulin protein. These colonies have to be identified so that they can be picked up for further multiplication. This is done by two immunological techniques – radioactive antibody test and immunoprecipitation test.

Radio active antibody test

in this test a plastic disk to which antibodies of the gene product (insulin) are bound is pressed on to the colonies. If the protein (insulin) is present in any colony, it will bind to the antibody and to the disk. The disk is then placed in a solution containing radio active antibody (to insulin). If any required proteins (insulin) are present on the disk, the radioactive antibody sticks to the proteins on to the disk. The disk is then washed and the autoradiograph will identify colonies if any producing insulin.

Immunoprecipitation test

This test is also used to identify a protein (insulin) producing colony. In this technique, insulin antibodies are added to the agar medium. When the bacterial colonies grow, if any of the colonies contain insulin it gets percipitated surrounding the colonies, due to the action of the antibodies.

Human insulin Humulin produced by the genetic engineering technique was made available to Diabetic patients in September 1982. It was the first therapeutic product to be produced by genetic engineering.

Human growth hormone (HGH)

Also known as Somatotrophin, this hormone is secreted by the anterior of the pituitary and is made up of 190 amino acid residues. The deficiency of this hormone results in dwarfism. For the conventional treatment of dwarfism, HGH is injected into the patients. But the extraction of hormone from the pituitary glands cannot meet the ever increasing demand for this hormone. Genetic engineering procedures have successfully produced somatotrophin.

Double stranded cDNAs were obtained from the mRNA precursors of HGH and then incorporated into bacterial cells where they expressed in non-precursor form. The bacteria were unable to convert the peptides of HGH into biological active forms. A recombinant plasmid containing a full length of HGH cDNA is cleaved with restriction enzymes that release fragments containing the complete HGH coding sequence after codon 24. Several overlapping complimentary oligo nucleotides were joined to form a synthetic strand of a small DNA fragment which contains the coding sequence for the first 24 aminoacids of mature HGH. The synthetic DNA and cDNA molecules are joined together to produce a new fragment which has the complete coding sequence of HGH. This is then ligated to a restriction site a little below the lack promoter operator gene cloned on a plasmid. These plasmids are cloned in *E.coli*. Synthesis of HGH is induced by an inducer (IPTG) of lack operon. The HGH thus obtained is later purified by chemical methods.

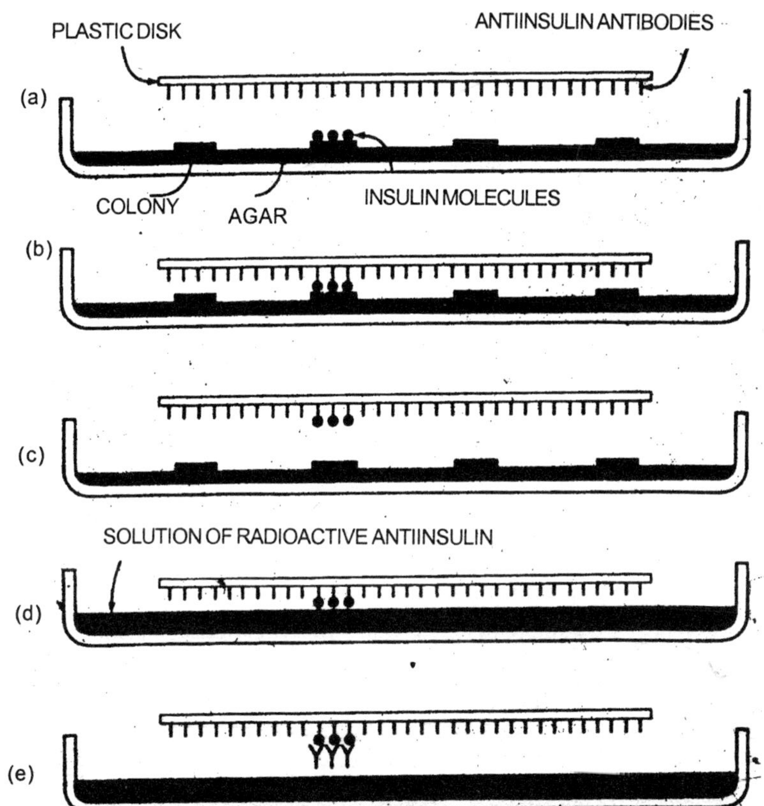

Fig. 11.13 Genetic Engineering
Radioactive antibody test

About 1 lakh molecules of the hormone per cell of *E.coli* have been produced by genetic engineering techniques. One of the problems in this is the attachment of an extra methionine to the molecule. This has to be removed to make the molecule biologicaly active.

Vaccine for Rabies virus (RV)

Hydrophobia caused by mad dog bite used to take a heavy toll of the human life. The credit of discovering a method to cure hydrophobia or rabies goes to Louis Pasteur. Even now however, due to ignorance and not taking the vaccine in time, is causing a number of death of humans and animals in many Asian and African countries. The conventional antivirus vaccine is difficult to obtain and also a tedious process. Researches are being conducted to produce genetically engineered rabies vaccine using *E.coli.* Attempts are made to isolate the mRNA coding for the viral protein from the RV infected cells. The genes coding for the production of a virus, glycoprotein coat has been successfully transferred to the cell of *E.coli.* This seems to the first step in the production of antirabies vaccine because this glycoprotein stimulates the production of antibodies in affected individuals. The Eister Institute in Philadelphia, USA has developed a new genetically engineered vaccinia virus by inserting a small fragment of foreign DNA. This genetically engineered virus could synthesise antirabies vaccine. The virus, however, does not cause rabies in those animals who received the rabies genome but instead activates the immune system against infection. The vaccine has been certified that it can be administered orally to animals.

BIOHAZARDS OF GENETIC ENGINEERING

Recombinant DNA technology popularly called genetic engineering is one of the most potent tools in the arsenal of human knowledge that can be used either for good or for bad. Ever since human civilisation evolved, two developments in the field of science have had the effect of changing the human destiny. One development that took place in the field of physical sciences is that of atomic fission, which gave man a vast amount of energy, which be produced can be used for improvement or for destruction. A development similar to this or much more potent than this has taken place in the biological sciences in the form of recombinant DNA technology. With the power and ability to synthesise the gene, to order the alignment of gene and to regulate its expression, man indeed, has been handed over the very basis of origin and differentiation of life. With this enormously potential tool he can create different types of organisms, vastly modify the existing organisms for his betterment or he can create organisms which are more harmful than the existing ones, and can even spell the deathknell of the very nature which has created him.

The brief discussion above will clearly point out to the fact that research in genetic engineering has to be carefully controlled, manipulated, and should be encouraged only when it is absolutely necessary, and that too with plenty of safeguards. In the paragraphs below, we will briefly outline the biohazards of uncontrolled genetic engineering research and what safeguards have to be followed while following the techniques of recombinant DNA.

Harmful aspects

Genetic engineering involves introducing a foreign DNA into another organism so that it gets transformed. Many a time, when a gene segment is transferred one may be aware of its expression in the original organism. But one is not too sure as to what kind of expression it may have in the new environment i.e., in the new organism. This kind of recombination many a time creates a dangerous new form either deliberately or accidentally. A harmless microorganism might turn out to be a very harmful one after it acquires new genes from another organism. Eventhough it is true that the genes introduced by themselves may not be harmful in their original combination, but when they are introduced into a new set up they might bring about the potentially dangerous expression. If, by chance, a transformed organism comes to carry a deadly disease due to genetic manipulation and if that organism escapes out of laboratory, one can imagine the uncontrolled horror that the organism might bring about. The genetically manipulated organism might be so potent in its pathogenicity that any of the existing drugs may not be able to combat it. For instance, *Streptococcus,* a pathogenic organism was never resistant to penicilein originally in nature. For the purpose of genetic engineering research, a plasmid with a desired gene is introduced into *Streptococcus* and at the same time to identify the transformed organisms, a marker gene such as penicillin resistant gene is itnroduced, when the organism becomes penicillin resistant the antibiotic now would be useless. If such an organism even though created in the laboratory were to escape one can imagine the dangers caused by it.

Escape of transformed microbes into environment

Inspite of precautions being taken, their always exists a possibility that a recombnant orgnanism might escape from laboratory and get permanently entrenched in the ecological system. If the organism in question were to be a parasite, it would cause immense havoc to all those organisms it parasitizes. Microorganisms that live in special ecological niche like hot spring or salt water might not pose such a great problem as they cannot spread widely.

The escape of microorganisms from the laboratory may be by design or by accident. If it is by design, then one has to understand that microbes are allowed to escape deliberately to cause harm to the environment outside. Accidental escape of chimeric microbes from the experimental laboratories may take any one of the following routes.

(1) Laboratory personnel such as scientists, attendants, visitors, sweepers etc. might inadvertently carry the microbes and spread them outside.

(2) Microbes might also find their way out through laboratory facilities such as glassware, drainage of chemicals, spillage of media, and such other inanimate objects.

In addition to the above, non-scientific personnel in the laboratory many a time may not be aware of the stringent precautions that have to be mandatorily taken while dealing with pathogenic microbes. Many a time, even if they are aware, they might just not follow the precautions as they are quite tedious and and cumbersome. Laboratory personnel while constantly dealing with pathogenic microbes many a time might acquire immunity and may not suffer from the consequences of contact with pathogens, but they act as carriers and others who come in contact with them will have to suffer the consequences.

Unintended escape of microbes and possible creation of new microbes possesses perhaps the greatest danger to the environment. In this context, it is worthwhile mentioning the opinion of three international experts on AIDS – Dr.John Seale from London, Dr.Robert Strecker from California, and Dr.Jacob Segal from Berlin. These scientists claimed in 1987 that the AIDS virus was artificially created by American scientists during laboratory experiments when they were trying to discover a cure for cancer by combining parts of Maedy - visna virus found in the sheep and human T-cell Leukaemia virus type 1. They argued that the scientists or other personnel in the laboratory could have been infected with the new virus through a minute cut or even by inhaling (Ignacimuthu, 1995).

Plus and Minus points of genetic engineering research

A great controversy is going on currently in the International Community involving biologists, social scientists, politicians and religious people regarding the desirability or otherwise of conducting genetic engineering research. While many people argue that genetic engineering research would be a harbinger of prosperity, others are equally vehement in opposing that research in genetic engineering would cause untold harm to human life as it is trying to tinker with the very basis of life and evolutionary process. The arguments for and against recombinant DNA technology are summarised below.

Arguments favouring genetic engineering research

1. The transfer of genetic material from eukaryotes to prokaryotes, and *vice versa* have been going on even in nature perhaps on a very small scale. If nature were to intend to prevent the transfer of a genetic material, evolution

would have surely brought in a perfect genetic barrier preventing such a transfer. In fact, ever since the origin of life, evolution could proceed only because nature took to genetic engineering.

2. The dangers of genetic engineering research perhaps will recede with time as people and scientists become experienced and accumulate knowledge.

3. Enough safe guards are already available in research laboratories dealing with pathogenic organisms. They can be made much more stringent. In any case, the beneficial aspects of combating the disease perhaps far more outweigh the possible dangers.

4. Genetically transformed microbes have been designed with a special purpose. Hence, the argument that they would compete with the other normal microbes and perhaps might be picked by natural selection appears to be imaginary. Further, with attenuated characters, the transformed organisms will not be able to survive unless special conditions such as the laboratory environment are provided with.

5. The idea that by acquicuring a new gene the microorganisms become pathogenic is not true, because in any genetic transformation the host organism will always reject a virulent gene combination and the unwanted genes are generally lysed by restriction enzymes and even if the genes express themselves, the gene product, namely the foreign protein gets degraded by proteolytic enzymes.
6. All types of genetic engineering research need not be dubbed as harmful and dangerous. For instance, research in the field of agriculture, dairy and others will not pose any problem. Researches in the medical field, however, can be continued with stringent regulations and control. At the same time however, one cannot simply push aside the vast benefits that have accrued to mankind because of improved health care products from genetic engineering. Hence, total stoppage of genetic engineering research would weaken the ability of man to understand and control nature and ultimately it is not in the interest of man himself.

Arguments against genetic engineering research

1. Unintended or accidental escape of microbes from the laboratory is possible through many routes. This escape of the transformed organism into the environment will pose great danger.

2. Genetic engineering research actually tinkers with the very basis of the evolutionary process. Nature indeed has been a very efficient genetic

engineer and it has been doing this work very slowly and steadily over a period, of millions of years. Transfer of genetic material and production of new traits in nature takes millions of years to be achieved. If this is sought to be achieved in a very short period it amounts to short circuiting the normal process and the consequences could be really hazardous. When an individual is fashioned out of natural process, it will be a package of both weakness and strength. If we were to tinker with this process it would naturally disturb the fine tuned balance and relationship that is existing between the organisms since millions of years.

3. The intended benefits of genetic engineering research can be actually achieved through other research methodologists. To achieve these goals one need not resort to such dangerous techniques like genetic engineering. The benefit of genetic engineering research does not justify the cost (not just in terms of money) involved.

4. With the kind of diverse political set up prevailing in different countries of the world where dictators, fundamentalists still have a very powerful say, this technique can definitely be misused and there is every possibility that these people might unleash a biological warfare on countries whom they consider inimical to them. For instance, water bodies of a country could be infected with new versions of pathogens or their environment might be sprayed with pathogenic microbes. All these possibilities should make one think whether it is worthwhile indulging in this kind of research.

Safe guards in genetic engineering research

The scientific committee at large and biologists in general, have been thinking for a very long time as to how to introduce some sense into genetic engineering research. While it may not be prudent to indulge in genetic engineering for problems which can be solved otherwise, it is equally true that genetic engineering cannot be totally given up as it indeed has the potential of solving some of the serious problems confronting the human race, which other research methodologies cannot hope to solve. The only answer to this being, research in genetic engineering should be properly regulated, monitored and if necessary with stringent international control.

A committee headed by Paulberg (1973-74), an eminent scientist, held discussions with other interested persons and published a small bulletin outlining the risks involved in genetic engineering research. In 1975, an International group of scientists who were very much concerned with genetic engineering research convened a meeting in California, USA and considered the ethical and safety implications of genetic engineering research. An outcome of this meeting was the formulation of a series of guidelines that sought to regulate and in some cases even prohibit research in recombinant DNA. In the USA, the National Institute of Health (NIH) issued guidelines in the year 1976 for US scientists involved in genetic engineering research. The guidelines required among other things that any research proposal involving recombinant DNA should be submitted to a special committee of NIH for approval. NIH also graded the laboratories into four levels (P1 to P4) where genetic engineering research of different levels were recommended. For instance, experiments involving DNA from primates and animal viruses involving harmful genes could be carried out only in extra secure P4 laboratories. In 1982, an advisory panel of NIH - the recombinant DNA advisory committee (RAC) published revised guidelines. According to these new regulations, recombinant DNA experiments were graded into 5 classes, and of these it is essential that three levels of research must obtain NIH approval even before initiation. These include researches involving drug resistance of microorganisms, toxin synthesizing genes and operations involving release of genetically engineered organisms into environment (as those used for pest control).

On the 26th January, 1982, a meeting was held of the parlimentary assembly of the council of Europe in Strasbury and recommended certain guidelines aimed at protecting the individual from the inheritant dangers of genetic manipulation. Three areas were mentioned in particular – genome map of human cells, gene therapy of somatic cells and gene therapy of germ cells. The report of the council concluded that inspite of the possible advantages of genetic engineering research, man made decisions should not substitute the free play of nature and everyone's right not to be genetically manipulated should be affirmed (Ignacimuthu, 1995).

Outside USA, the regulations on genetic engineering research are actually as stringent as they ought to be. France and Germany are considering whether to restrict research in genetic engineering because of the ethical questions involved. Recombinant DNA guidelines of France are ambiguous about the deliberate release of genetically altered organisms. In Germany, however, release into environment of genetically altered organisms are strictly prohibited. Many social activist groups are opposing gene therapy and such other genetic engineering parctices in terms of ethical and moral parameters.

In India, genetic engineering research is still in infacy, yet considering the importance of it, Government of India has constituted four committees for this purpose.

1. Recombinant DNA Advisory Committee (RDAC). To evolve guidelines.
2. Institutional Biosafety Committee (IBSC). For implementation of guidelines.

3. Review Committee of Genetic manipulation (RCGM). To recommended special suitable conditions for experiments and trial.
4. Genetic Engineering approval committee (GEAC). For large scale manufacturing, environmental release of genetically altered organisms/ products and import/export of genetically altered organisms.

The followng may be considered as some of the safeguards that should involve any research in genetic engineering.

1. Harmful genes, such as those coding for toxins and antibiotics should not be cloned without extreme precautions.

2. Animal genes, genes of tumour viruses should not ordinarily be cloned in bacteria or others microbes.

3. Laboratory facilities should be such that there should be virtually no possibility of the escape of genetically altered microbe. In addition to scientists, even the other ordinarily personnel of the laboratory should be thoroughly briefed about the dangers of carelessness before they are employed. Workshops on precautionary measures should be conducted periodically so that no one violates them. There should be mandatory check of laboratory equipments of cabinets, hoods and other glassware to see that they are microfree when they are taken out of the laboratory. Laboratory should also possess special equipments like negative pressure rooms, safety cabinets, special traps and drainage etc., to prevent the possible escape of the microbe.

4. Use of microbes which can live only in special environmental conditions should be picked up for genetic engineering research, because even if these organisms escape into the environment after genetic alteration, they will not survive in the normal environmental conditions so that the chances of their rapid spread are very less.

5. As far as possible, cloning vectors should be non-conjugative plasmids because even if they are uncontrolled they will not be able to promote their own transfer by conjugation.

6. The cloned vector should be sufficiently modified are altered in such a way that it will not be able to express itself in any environment, but can do so only under special circumstances like a particular host strain that are normally not available The genetic alteration must be such that in case it is transferred to another microbe, it should prove lethal to it.
7. The cloning organisms should be modified in such a way that after cloning,

they depend upon very special conditions of existence which can be provided only in a laboratory and are not available in the environment.

8. Attempts are being made to produce "crippled microbes" for use in genetic engineering research. The genome of these microbes is altered and disabled in such a way that if they escape from the special laboratory conditions they will not survive. The following are some of the instances of crippled microbes for use in genetic engineering research.

Diaminopimelic acid (DAP), is an important component of the cell membrane in *E.coli*. Genetically altered *E.coli* were produced which were DAP deficient. These cannot survive unless supplemented by DAP. Hence, these organisms even if they escape outside cannot survive. Such organisms are ideal tools for genetic manipulation. Precautions are, however, necessary because many a time microbes are known to undergo mutation very fast. In fact in the above example, the DAP deficient *E.coli* was very soon mutated back and was able to produce DAP. Scientists however, discovered that there was more then one gene involved in the production of DAP and deleted that also. But the bacteria would not yield so easily to the manipulation of the microbiologists. The DAP deficient bacteria again survived by producing a sticky substance called Calanic acid around their protoplast. But then, the calanic producing gene was again removed. This will clearly reveal that our knowledge on the genome of microbes would always remain incomplete because we will be able to identify a gene only when it expresses itself after natural selection, for instance calanic acid producing gene was not selected to express when DAP producing gene is there. In the absence of the latter, the former was selected. Who knows in the absence of calanic acid producing gene which other bacterial gene which is hybernating may get selected and express itself. After all life is a struggle and no individual will give up its existence so easily.

Genetic engineering research and ethics

Man is a product of evolution and, however much he may think that he is different from other organisms because of his intelligence, the stark truth is that he is definitely a part of biological spectrum whether he likes it or not. Man is neither the first step nor the last step in the evolution of life. The process of evolution which has produced so many diverse types of organisms has also produced man and all these organisms that form the biological spectrum were not produced within a day. Even the evolution of human beings took several millions of years. It is quite true that the millions of organisms in the present day biological spectrum are produced out of nature genetic engineering. But one must realise that these have been crafted painstakingly and patiently over the course of millions of years. Man's attempt of fast genetic engineering seeks to destabilise the delicate balance that has been inbuilt into the relationship of all organisms. Further, the diversity of life and its glorious uncertanity is the spice that makes life meaningful. Present day world consists of beneficial organisms, harmful organisms, bad individuals, good individuals etc. Variety should be there, only then one can understand the difference. If every thing is sought to be made uniform then the very essence of life is lost. Man should seriously think whether it is in the interest of nature and his own interest to pursue with unbridled genetic engineering research. Because, as per the process of natural selection, the genes that are not expressing in a given set of environmental conditions might express in a new environment and who knows what they might contain.

12

CULTURE OF MICROBES

Microbial sutdy involves various methodologies. Microbes present in natural conditions grow, mutliply and die. In the process, they affect our life in various ways both for the good as well as bad. Naturally, then we have to understand their life processes (microbes), as to how they are affected if certain requirements are not available. In natural conditions, microbes are rarely present in pure condition; there will always be mixed populations. Under such circumstances, it will be difficult to study them. In order to study the various life processes of microbes, they must be cultured on artificial media under laboratory conditions; isolated in pure form, and only then a meaningful study is possible.

A culture may be defined as growth of microbes under contamination free condition in the laboratory on a suitable medium.

The term culture is usually employed for an induced growth of microbes under *in vitro* (Latin; *vitro* - glass) conditions. Microorganisms as the very name indicates are extremely minute in size. It is difficult to obtain any information by the study of a single individual. Hence, microbes are always cultured to obtain populations of cells. Microbial populations are usually referred to as a **colony** (pl. Colonies). A colony develops from the multiplication of a single cell, and hence the population of cells are always homogenous. When a colony consists of only one type of cell it is called pure culture. In a mixed culture there will be cells of different types.

CULTURE MEDIUM

Microrganisms can be cultured in the laboratory under aseptic conditions using sterilized glass ware and other appliances. In order to cultivate the microbes, they should be provided nutrition on a medium in nearly as natural a way as possible. Preparation of suitable nutrient substances is the single most important factor in the cultivation of microbes in the laboratory. Nutrient substance prepared in a suitable manner (so that the microbes can grow on them), is referred to as the *culture medium* (Pl. media). Different microorganisms require different nutrient materials. Thus there will be different types of media. Basically media may be classified into the following four types –

1. Natural (empirical) media
2. Semisynthetic media
3. Synthetic media and
4. Living media.

1. Natural (empirical) media

Natural media do not have any addition of specific nutrients. Natural food substances such as milk, vegetable juices, diluted blood, meat extracts etc. have been used since a long time by microbiologists for culturing microbes. This is also called an empirical medium because the exact chemical composition and the specific percentage of the chemical components is not known. Now a days peptone (digested protein rich in amino acids that provides nitrogen and carbon) is a major ingredient in natural media. The sources of peptone may be animals or plants, and in composition they (peptones) are highly variable. The chemical composition of peptones are partially known (both in concentration and identity). They are known to possess peptides, polypeptides, aminoacids, carbohydrates etc. in addition to many organic and inorganic nutrients. Peptones seem to provide the microbes in any easily digestible form P, S, N and C.

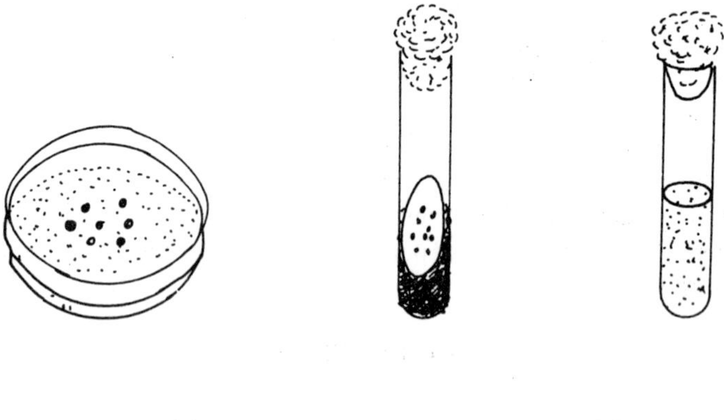

Fig. 12.1 Culture of Microbes
Method of culture, Petridish (left); Agar slant (centre); Broth (right)

Another important ingredient that is frequently used in natural media are the meat extracts and infusions. Beef extract and meat infusion (prepared by soaking fresh ground meat in water), are very commonly used in natural media. Beef extract is a rich source of minerals, organic micronutrients, proteins and carbohydrates. Yeast extract is often added to the beef extract to supplement vitamins. Beef extract may be used in liquid form when it is called *infusion broth* or *nutrient* or *nutrient solution.*

Many vegetable and fruit extracts are also used in natural media. These include extract of tomato, orange, apple etc. The choice of the extract depends on the type of microbe to be cultured. The natural medium with the addition of these extracts, then becomes a *special medium.*

Semisynthetic media

A semisynthetic medium is one in which the chemical composition (of the medium) is partially known. In other words, a semisynthetic medium has a natural medium (unknown chemical composition) to which certain specific nutrients (known chemical composition) are added. A medium which has agar becomes a semisynthetic medium. Some of the semisynthetic media are–porato dextrose agar (PDA), czapek dox agar, oat meal agar, beef peptone and nutrient agar etc.

Synthetic media

A synthetic medium is one in which all the chemical ingredients are mixed in definite proportions. They are usually prepared using chemically pure known inorganic and organic compounds. These media are also called **chemically defined media** as they have properly identified (defined) ingredients in specific proportions. Formation of a synthetic medium requires a prior understanding of proportions. Formation of a synthetic medium also requires a prior understanding of the nutritional requirement of the microbes to be cultured. Some of the synthetic media may have only inorganic components. For instance, *Thiobacillus Thiooxidans,* a soil inhabiting chemilithotroph can be cultured on a medium containing – $(NH_4)\,2SO_4$ 0.2G; Mg SO_4 $7H_2O$-0.5G; KH_2PO_4-3.0g; $CaCl_2$-0.25g; S (powdered) - 10.00g; distilled water in required quantity to make the volume to one litre.

In modern microbiological laboratories, dehydrated media (readily available) are used. These are available in ready to use mixtures which need to be weighed and added to a specific quantity of water.

A synthetic medium may be classified into two categories – General purpose medium and selective medium. A general purpose medium also known as enriched medium has standard ingredients known to help microbial growth, and can be used to culture a variety of microorganisms. A selective medium which has particularly selected ingredients is used for the culture of particular microbes only. Another kind of medium is differential medium used for biological assay of vitamins, antibiotics etc. This is called the assay medium. Some of the commonly used synthetic, media, are – Czapeck solution, Richards solutions, Ranlin's solution, Martins rose Bengal medium etc.

Living media

A living medium consists of living cells or tissues (even callus or organs), which are used for the culture of strictly parasitic organisms like viruses or rickettsiae which can not be cultured on a non living medium. For the cultivation of viruses normally chicks are used.

Other classifications of culture media

Media may also be divided into four categories on the basis of consistency ie., their physical state. These are–liquid media, semisolid media, liquefiable solid media and solid media.

Another classification is based on the application or function of the medium. Based on this, media are classified as follows -

1. Cultivation media
2. Storage media
3. Enrichment media
4. Differential media
5. Assay media and
6. Maintenance media

Basic ingredients of culture media

Media are of various categories based on the purpose for which they are used, and as such they contain a wide array of organic and inorganic ingredients in varying proportions. Often a microbiologist devises his own medium to suit a specific culture of microbes. There are some basic ingredients which, however, find a place in most of the media. Some of these are described below.

1. Protein hydrolysate

Protein extractions may be obtained from a wide variety of sources including both plants and animals. The following substances are used as sources of protein hydrolysates – meat, fish (fresh), caesin (milk protein), gelatin, keratin, yeast extract, soya meal, cotton seed, sunflower seed, groundnut meal etc.

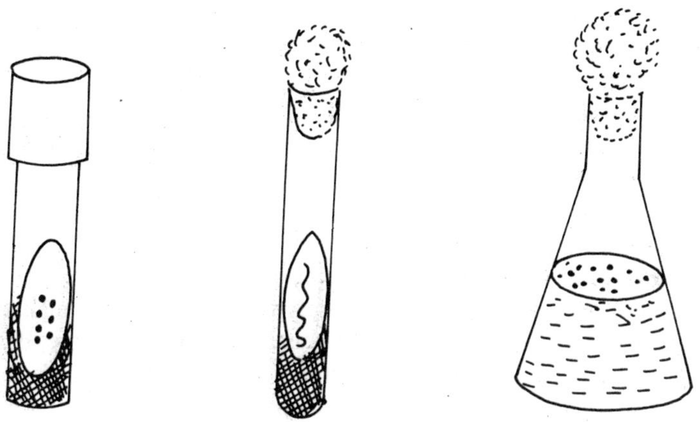

Fig. 12.2 Culture of Microbes
Methods of culture (Contd), Macartys slant (left); Streaking (centre); Flask (right)

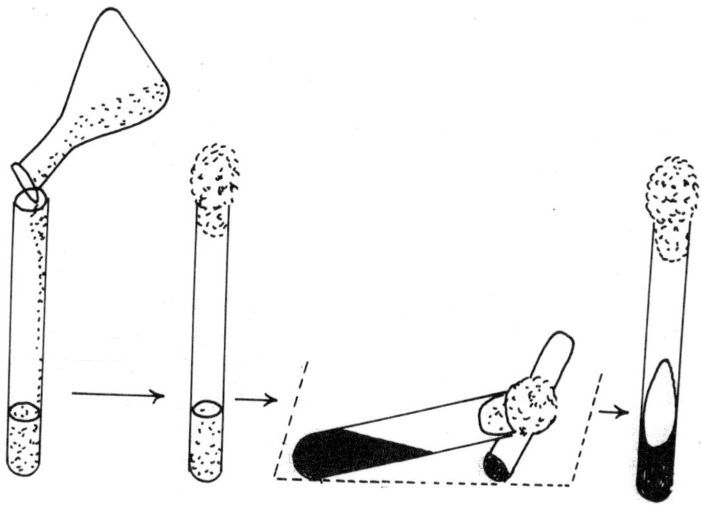

Fig. 12.3 Culture of Microbes
Preparation of Agar slant

2. Meat extracts

These should be pure, clean and must not have any toxic substances. Metal content and fermentable carbohydrates should be low.

3. Yeast extract

This is available in powder form and consists of dehydrated cells of *Saccharomyces*. Nutritionally yeast extract is rich in vitamins, carbohydrates, minerals, aminoacids, peptides etc.

4. Malt extract

This is prepared from the germinated seeds of cereals such as, barley. Malted barley seeds are ground and a syrup is prepared. Nutritionally malt extract consists of a variety of sugars such as – maltose, hexose, sucrose, dextrose etc. Other nutrients present in malt extract are– lipids, organic acids, phosphorus and sulphur compounds, vitamins (particularly B complex) etc.

5. Agar

Agar is a polysaccharide which can change from solid to liquid depending on the temperature. It has no nutrients but it is used for gelling the medium. It gets solidified below 40°C. Agar (also called china grass) is obtained from a number of sea weeds belonging to the class *Rhodophyta*. Agar yielding red algae are – *Gelidium, Gracilaria, Hypnea, Chondrus, Pterocladia, Gelidiella, Gigartina, Ceramium, Campleophora* etc. These algae which yield agar are called *Agarophytes*.

Chemically agar is a mixture of two polysaccharides – aggarose and agaropectin in the ratio of 70:30. In addition to polysaccharides, agar also contains – calcium (o.8% - 18%), chloride (0.06-0.24%), sulphate (0.7-20%), magnesium (0.07% to 0.2%), nitrogen (0.09-0.24%), iron (21 - 324 ppm) etc.

6. Gelatin

This is a protein derived by the degradation of collagen, which is a fibrous protein present in the connective tissues of animals. Gelatin may also be obtained from defatted and demineralized bone matrix ash. The ash is processed for the extraction of gelatin. Many phaeophycean and rhodophycean algae also yield gelatin.

7. Bile salts

These are obtained from freshly collected bile of oxen. The components of bile liquid are – bile pigments, bile salts, mucin, proteins, lipids, enzymes and inorganic ions. Bile containing media are used in the culture of bile tolerant microorganisms.

8. Carbohydrates

A variety of simple sugars, disaccharides and complex carbohydrates are used in the culture media. The following are used in the media (Bhattacharya - 1986).

1. Pentose – arabinose, xylose, rhamnose etc.
2. Hexose - glucose, fructose, galactose etc.
3. Disaccharides - sucrose, maltose, lactose, trehalose etc.
4. Trisaccharides - raffianose etc.
5. Polysacharides - starch, insulin, dextrin, glycogen etc.
6. Alcohol - glycerol, crythritol, adonitol, mannitol, dulcitol, sorbitol etc.
7. Glucosides – salicin, aesculin.

Carbohydrates are prone to hydrolysis on heating and may also interact with salts which may retard the growth of microbes.

The following precautions will prevent the undesirable effects of carbohydrate decomposing (Bhattacharya, 1986).

1. Before autoclaving ,sugars should be completely dissolved.
2. Dextrose – Peptone media of high concentration should not be autoclaved.
3. The prescribed duration, pressure and temperature of autoclaving should be carefully adhered to.
4. pH should be maintained below 8.0.

9. Minerals

Some mineral elements have to be added for the culture of some microbes. These are K, Mg., Mn and SO_4. Some trace elements such as Cu, Co, Zn, Fe, Mg, molecules and Ca are also necessary.

10. Chelating agents

These are organic compounds which combine with heavy metals and keep them in solution in the media. Many substances such as – citrate, succinate, acelate etc. can complex with capsid, mg, fe etc. to prevent them from forming insoluble compounds. Certain aminoacids such as cysteine, histidine and glycine are also used as chelating agents. Presently, however, the best chelating agent used is EDTA (ethylene diaminotetraccetic acid).

TYPES OF CULTURES

Microorganisms can be cultured in media of various types depending on the requirement. The media used for culturing may be categorized into two depending on the consistency ie., solid or liquid. Solid media are prepared by adding either agar or gelatin to the nutrients in liquid, while liquid media do not have agar. The choice of the medium depends on the type of microbe to the cultured. The two types of cultures – liquid and solid are discussed below in some detail.

Liquid culture (liquid medium)

This is also known as the broth culture. A broth culture may be defined as a sterile liquid nutrient solution which does not have any solidifying medium (agar).

Broths are of different types

The choice depends on the requirement of microbe to be cultured. In most of the cultures two types of broth are used – **nutrient broth** and **glucose broth.** Broth cultures are ideal where growth performance of microbes are to be studied by techniques such as, turbidementric method. Turbidity or cloudiness in the nutrient solution indicates concentrated growth of microbes which are evenly dispersed in the medium. Turbidity may be measured in a spectrophotometer or colourimeter, where optical density indicates the measure of growth rate..

Broth cultures are also suitable for studying pellicle formation wherein, fungi cultured in liquid medium form large mats floating on the surface. Fungal nutritional studies also can be carried out using a broth culture. Study of the rate of sedimentation of bacterial cells is also possible using a broth culture.

Nutrient broth and glucose broth are two of the general liquid media used for the cultivation of a variety of bacteria. The following is the method used for the preparation of broth culture.

Requirement

1. *Nutrient broth*
   a. Peptone 5.0g
   b. Beef extract 3.0g (pH 7.0)
   c. Distilled water 1000ml
2. *Glucose broth*
   a. Peptone 10.0g
   b. Glucose 5.0g
   c. Sodium chloride 5.0g (pH 7.3)
   d. Distilled water 1000ml

In addition to the above, the following are also required –
1NHC1, 1Naoh, pH meter, distilled water, hot plate, autoclave, culture tubes, flasks etc.

Procedure

Nutrient broth. Put the weighed amount of peptone (5g), and beef extract (3g), in 500ml of distilled water. Heat the mixture and agitate with glass rods to dissolve the constituents. Add distilled water to make the volume to one litre. Adjust the pH of the medium to 7.0 either by adding acid or alkali depending on the necessity. Take test tubes and pour 10ml each. Conical flasks also may be used instead of test tubes, in which case pour 25ml to each flask. Apply cotton plugs. Autoclave at 121^{0}C under 151b pressure for 15 minutes. Allow the autoclave to cool. Remove the broth tubes (flasks), and store at room temperature for use.

Glucose broth may be prepared in the same fashion as the nutrient broth.

Shake cultures

These are nothing but liquid cultures. The culture flasks contain a large quantity of medium. Many a time the nutrients settle down to the bottom of the flask. As a result nutrients are not uniformly distributed all over the medium. This will affect the growth of the microbes and they do not develop properly. In order to overcome this, the culture flasks have to be agitated or shaken frequently to allow uniform distribution of nutrients. Instead of shaking the flask manually, this is a platform on which are placed the flasks. The flask shaker has an electric motor which shakes the platform and the flasks kept on it.

Solid cultures

In liquid cultures, the microbes are not fixed to the place of their origin and colonies cannot be studied or separated as they (microbes) are floating. This can be overcome by the use of solid media. Solid media can be prepared by the addition of agar to the medium which solidifies at room temperature. The solid surface of the medium is ideal for the spores or cells of microbes as they get trapped and grow very well. A colony originates from a single cell and it cannot shift to any other place.

Solid medium can be used either in test tubes or in petriplates. Accordingly, the solid media may be classified into agar plates (in petri plates), and agar slants (in test tubes).

Preparation of agar plates

1. Clean petriplates and wipe them dry.
2. Prepare the culture medium (including agar which melts on heating to 96^{0}C and hardens into a jelly on cooling to 40-45^{0}C.)
3. Carefully open the petriplate and pour the medium into the petriplate and close the lid.
4. Conduct the operations near a flame.

5. After pouring the medium, tilt the plate back and forth to allow for even distribution.
6. Allow the medium to settle and solidify.
7. After the medium gets solidified, keep them in an inverted fashion in an incubator to prevent the condensation of vapour into water drops.

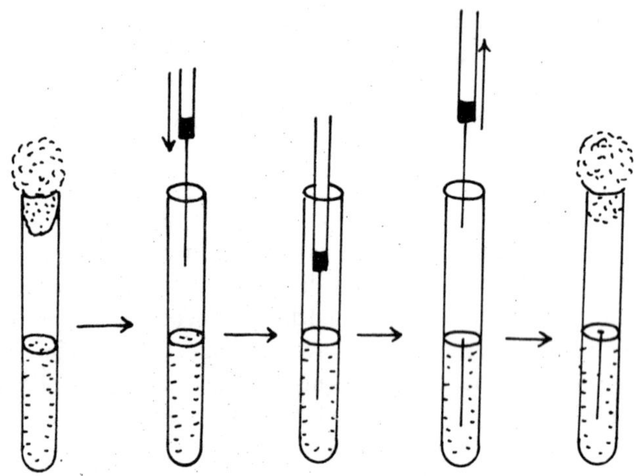

Fig. 12.4 Culture of Microbes
Agar stab culture

Preparation of agar slants

Agar slants are usually prepared in tubes. After the medium (about 10-20 ml) is poured into tubes, they are autoclaved. After removing from the autoclave the tubes are kept in a slanting position and the medium is allowed to solidify. After hardening, the surface of the medium will be slanting. The purpose of slanting is to increase the superficial area for the growth of the microbes, otherwise the surface would be as wide as the diameter of the tube.

Agar slants are ideally suited to study the cultural characters of microorganisms. Slants are also used for maintenance and transfer of cultures.

Agar stabs

Here also, the media is poured into the tubes, but instead of keeping the tubes in a slanting position, they are kept straight. As a result the surface of the solidified medium is even. Stab cultures are used to study the O_2 requirement of microbes, motility of microbes and to preserve the culture for a longer period.

Specific media for particular microbes

The description of the media given above is a general purpose one routinely used for the culture of microbes. However, not all microorganisms can be cultured on these media. Hence, special types of media are to be prepared, and particular nutrients needed have to be added to facilitate the growth of microbes. Some of these media are described below.

Nutrient agar and broth

This medium is used for the isolation of bacteria and actinomycetes. The components of the media are as follows.

a. Peptone 5.0g
b. Beef extract 3.0g
c. Nacl 5.0g
d. Distilled water 1000ml
e. Agar 15g

(Note : for nutrient broth, omit agar)

In case of nutrient broth the final volume should be made up to 1 litre. pH to be adjusted to 7.0.

Potato Dextrose Agar (PDA) medium

This medium is a typical solid medium used for the culture of a number of bacteria and fungi. The medium may be used in petriplates, test tube slants or test tube stabs. PDA is highly useful for the isolation of plant pathogenic fungi. Acidifying the medium with a few drops (for every 100ml melted agar) of 25% lactic acid is highly suitable for growth plant pathogens. The following is a detailed procedure for the prepartion of PDA.

Requirements

Peeled potato tubers (200g), dextrose (20g), agar (20g), corning beakers, potato peeler, knife etc.

Procedure

1. Take 500ml of water in a 5 litre beaker.
2. Peel the skin of potato; cut into small pieces and boil in water for about 30 minutes to make a thick paste.
3. Add some water to the paste and filter through a cheese cloth; squeeze out all the liquid.
4. Take 500ml of distilled water in a beaker and heat it.

5. Dissolve agar in water by gently stirring.
6. Mix the potato extract with agar solution.
7. Add water to make up the total volume to1 litre.
8. Distribute the nutrient liquid to either conical flasks or test tubes (quantity), nutrient liquid in each flask/tube depends on the requirement.
9. Cover the flask/tubes with cotton plugs.
10. Autoclave at 121°C or on 15 minutes under 151b pressure.
11. After removing from the autoclave, the tubes may be kept in slanting position or upright (for deep agar tubes).
12. After solidifying, the medium may be used for inoculation or may be stored under refrigeration.
13. If agar plates are required, the medium may be poured into petriplates under strict aseptic conditions and allowed to solidify before using for cultures.

Yeast extract mannitol agar

The composition of this medium is as follows

Mannital	10.0g
Dipotassium Hydrogen phosphate	0.5g
Magnesium sulphate	$7H_2O$ 0.2g
Sodium chloride	1.0g
Yeast extract	1.0g
Congo red (1% solution)	2.5ml
Agar	20.0g
Distilled water	100ml

Martin's medium

It is also called Martin's Rose bengal streptomycin agar. The composition of the medium is as follows.

Dextrose	10.0g
Peptone	5.0g
Potassium dihydrogen phosphatge	1.0g
Magnesium sulphate	1.0g
Rose bengal (one part in 30,000 parts of the medium)	
Agar	20.0g
Streptomycin	30.0g
Distilled water	100ml

Streptomycin vial has to be opened aseptically only at the time of preparation of medium. To one gram of streptomycin and 100 ml of double distilled water. Add 0.31 of this solution to each 100 ml of the basal rose bengal medium after it is cooled to 45°C.

Selective and enrichment media

These are the media that are specially meant to culture specific microbes, and are not generally used for routine cultures.

A selective medium is one which permits and promotes the growth of some specific group or type of organisms while preventing or retarding the growth of other microbes.

An enrich medium helps in the increased growth of only one kind of a microbe. In other words, an enrichment medium 'enriches' (increases the concentration) the microbial content in the medium.

Mannital salt agar medium

This is a typical example of a selective medium. As has been pointed out already a selective medium not only promotes the growth of specific microbes but also retards the growth of others and thus it helps in the isolation of bacteria. Selective action of the medium is due to the addition of certain chemicals to the medium. For instance, addition of crystal violet dye selectively inhibits the growth of gram positive bacteria and is added to a medium to culture enteric bacteria (gram negative bacteria). In the same way, addition of sodium azide at specific concentration into the medium will help in isolating lactic acid bacteria. There are many dyes used for selectively promoting the growth of *Pseudomonas* (cetrimide), halophytic bacteria (tetrazolium) etc.

Mannital salt agar is used for the culture of pathogenic stayphylococci, as most other bacteria are inhibited by the high concentration of sodium chloride in the medium. The fermenting ability of the staphylococci colonies induced by a yellow halo can be detected by mannitol present in the medium. Mannitol is metabolised and fermented by staphylococci and the acid produced can be detected by phenol red, a - pH indicator. The following is the composition and procedure for the preparation of the medium.

Mannital	10.0g
Beef extract	1.0g
Peptone	10.0g
Sodium chloride	75.0g
Phenol red	0.025g
Agar	15.0g
Distilled water	100ml

Procedure

Dissolve the components (beef extract, peptone, sodium chloride, mannitol, phenol red) in 500ml of distilled water and add agar solution (dissolve 15g, agar in 500ml of water). Make the total volume to one litre. Pour the medium into suitable culture tubes/ conical flasks and autoclave for 15 minutes under 151b pressure.

After the medium is sterilized and sufficiently cooled, inoculate the medium with *Staphylococcus aureus*. If necessary pour the medium into petridishes for the purpose of culturing. Incubate the inoculated plates/tubes at 37°C for 48hrs and observe the growth of bacteria.

Enrichment medium

Enrichment or enriched medium is one in which the nutritional components are balanced in such a way as to selectively enhance the growth of a certain microbe in a mixed population of the inoculum. This method is suitable for those microbes which are in extremely small numbers in the sample, and whose growth is slower than the others. The basis behind an enrichment medium is perfect regulation of nutrients, and culture conditions such as temperature, oxygen, light, pH etc. in such a way that it provides optimal environment and promotes the growth of the required microbe. For example, if a medium containing salt solution with $NaNO_2$ (pH 8.5) is inoculated with a sample from the garden soil and is incubated in the dark (in the presence of air) at a temperature of 25-30°C, pure colonies of *Nitrobacter* develop. Obviously even though garden soil sample consists of other microbes, given the culture conditions, they will not grow. Acterotrophic bacteria çan be enriched by the addition of plant or animal extracts into nutrient broth or agar.

Enrichment medium is ideal for the isolation and purification of specific microbes.

Differential medium

A differential medium is one which will cause certain colonies to develop differentially from others by producing a characteristic change in the microbial growth or in the medium surrounding the colonies. Differential medium is employed to determine differential reactions which will permit presumptive medium. It can be used to differentiate hemolytic properties of streptococci by their reactions on the medium. Thus streptococci may be distinguished in a mixed culture. Red blood cells get hemolyzed around the colonies of streptococci and form a clear zone. Such a clear zone is not formed around non hemoloytic bacterial colonies. Hemolytic zones are classified into alpha, beta and gamma corresponding to a greenish halo, a clear zone and no reaction respectively. The constituents of blood agar medium are as follows.

Infusion from beef heart 500.0g

Tryptose	5.0g
Sodium chloride	5.0g
Agar	15.0g
Distilled water	100ml

Dissolve the ingredients given above except agar in 500ml of distilled water, and add it to agar solution (15 gms of agar dissolved in 500ml of distilled water). Autoclave the medium for about 15 minutes under 151b pressure. Cool the sterile defibrinated blood and mix gently but thoroughly. Pour the medium into suitable culture plates. Inoculate with *Streptococcus faecallis* and incubate at 37°C for 48hrs. Observe the hemolytic reaction of bacterial colonies.

The following zones are seen –

(a) Greenish discolouration of the medium surrounding the bacterial colonies (α hemolysis) due to the breakdown of RBCs with the reduction of hemoglobin to methemoglobin.

(b) Clearzone surrounding the bacterial colonies (ß hemolysi) with complete destruction of RBCs and use of hemoglobin by bacteria.

(c) No reaction zone on the medium (γ hemolysis neither discolouration nor a clear zone) indicating absence of hemolysis.

INOCULATION

Artificial induction of microbes into the culture medium is called inoculation. Inoculation is the most important technique for facilitating a proper growth of microorganisms under aseptic conditions.

Microorganisms from the original stock culture are to be introduced into either petriplates or agar slants prepared in culture tubes (for preparation of agar slants see earlier in the same chapter). The procedure of inoculation into an agar slant culture is as fol'ows.

1. Hold the tube containing the stock culture and the tube containing the culture medium in the left hand. Hold the inoculating needle or plantium loop in the right hand.
2. Hold the needle/loop on to a flame of bunsen burner until it becomes red.
3. Remove the cotton plugs of both the tubes with the finger of the right hand, and simultaneously expose the mouth of the tubes to the flame device.

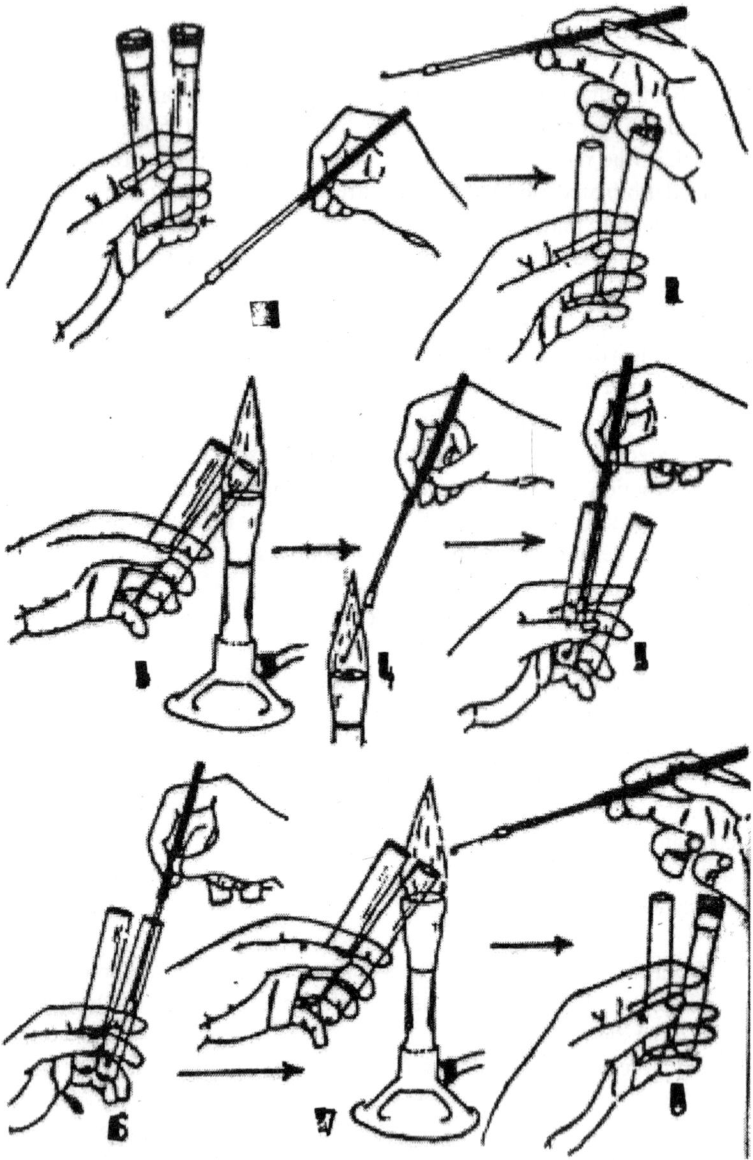

Fig. 12.5 Culture of Microbes
Procedure for inoculation

4. Immediately after the removal of the plug and flaming the mouth, flame the loop/needle again and dip it into the tube containing the stock culture.
5. Take out a little quantity of the inoculum (from the stock culture), and immediately introduce it into the other tube containing the culture medium. The inoculum should be rubbed on to the surface of the agar medium.
6. Flame the needle and the mouths of both the tubes again once or twice.
7. Plug the tubes immediately and transfer the tubes for incubation under suitable conditions.

Precautions

(a) All steps particularly after removing the cotton plugs are to be carried out as quickly as possible to prevent contamination.

(b) After removing from the tube, till replacement, cotton plugs should be held in the hand only and *must not be kept on a table or anywhere.*

PURE CULTURE TECHNIQUES

Microbes, in nature, are never found in pure populations. In the source – water, air or soil, microbes are always found as a mixed population. Even in the infected plants or animals, the affected parts harbour a variety of microorganisms.

In the laboratory study of microorganisms, a mixed population is of a disadvantage. For instance, a specific microbe on which some experiments have to be carried out may be in very small numbers. In such an instance if one has to chemically analyse a metabolic product produced by that specific microbe it will be practically impossible as the quantity is very less. Further, in a mixed population one group of microbes may adversely influence the function of another. Hence, any meaningful work on any microbe is well nigh impossible in a mixed population. The solution for this is to obtain pure population of required microbes. The advantages of pure populations (colonies) of microbesin, a laboratory study are –

1. Growth conditions/requirements may be clearly studied.
2. Metabolic products may be analysed.
3. Large populations of pure colonies are available for genetic studies.

Isolation of a single microbe for the purpose of pure culture is easier said than done. A great variety of laboratory techniques are to be involved to isolate required microbes from natural habitats. Culturing of microbes from natural populations with diverse microbes will often lead to overlapping of colonies on the medium making it difficult to separate them. Hence, it becomes necessary to follow methods which allow colonies to develop distinctly and separated from each other to permit easy manipulation. Colonies which are distinct tend to be pure and may be picked up for further culturing.

Several laboratory techniques have been devised to obtain a pure culture from mixed culture developed from natural populations. It must be pointed out however, that liquid media by their very nature are unsuitable for pure cultures. Colonies grown in liquid media can never be fixed and localized. Solid agar media are most suitable because, they can trap and fix a cell or a spore of microbe, and lead to the development of a colony in the place of original location.

The techniques to obtain pure cultures may be classified as follows.

1. **Common methods**
   a. Pure plate method
   b. Streak plate method
   c. Spread plate method
   d. Serial dilution method

2. **Special methods**
   a. Single cell isolation
      i. Capillary pippette method
      ii. Micromanipulator method
   b. Enrichment method
   c. Selective media method
   d. Differential media method

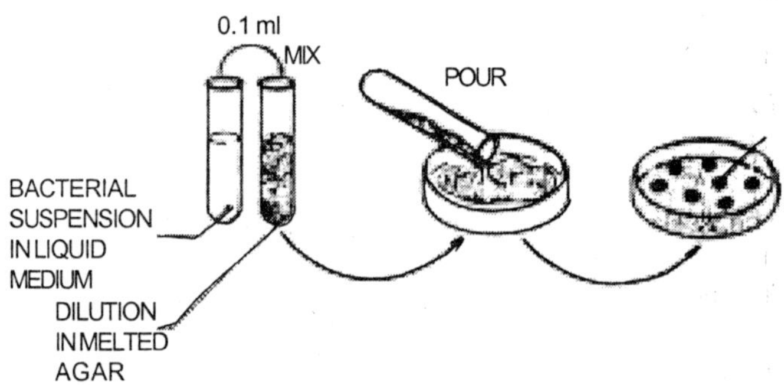

Fig. 12.6 Culture of Microbes
Pour plate technique

Pure plate method : The principle involved in this method is to dilute the sample inoculum mixed with the melted agar medium. This allows for a thorough and uniform distribution of cells or spores of microbes within the medium. The medium is then poured into peptridishes, allowed to solidify and incubated under suitable conditions. As a result of this dilution discreet and well separated colonies develop in the culture media. As the inoculum is distributed evenly in the medium both surface as well as subsurface (within the agar medium), colonies develop in the medium.

Procedure of pure plate method

(a) Take four petridishes; sterilize them properly in a hot air oven. Remove them and mark the numbers 1 to 4 with the help of a glaze marker.
(b) Prepare four tubes of nutrient agar with each tube having 12-15ml of medium. Cool the tubes to 45°C and number them (1 to 4) (see earlier in the same chapter for the details of medium).
(c) Prepare a suspension of the source material such as soil sample, sewage or bacterial inoculum etc. in 20ml of sterile distilled water.
(d) Transfer the microbial suspension into the four tubes (marked 1-4), containing nutrient agar medium as follows.

I. Tube

1. Transfer one loopful of the medium, stir the loop in the medium. Gently roll the tube between the palms to facilitate even dispersion of the inoculum in the medium. Keep the tube in water bath for incubation at 45°C.

II. Tube

2. Collect two loopful of inoculum (mixed in the medium) from tube number one and transfer it to tube no two. Follow the same procedure to evenly distribute the inoculum (as in tube one). Keep the tube for incubation in water bath.

III. Tube

3. Transfer two loopful of inoculum from tube no 2 to tube no 3. Follow the same procedure and proceed for incubation.

IV. Tube

4. This will be a control with no microbial inoculum.

Transfer of cultures to petriplates

1. Arrange the petriplates serially 1 to 4.
2. Take out the incubated tube no. 1 from water broth. Wipe the outersurface perfectly dry. Carefully remove the cotton plug, flame the mouth of the tube and carefully pour the contents into petriplate no. 1 by raising the lid

slightly. Immediately close the lid to prevent any contamination. Gently rotate the petriplate to allow even distribution of the medium.

3. Similarly pour the contents of tubes 2,3 and 4 to petriplated 2,3 and 4 respectively.
4. After the medium solidifes, place the petriplates in an incubator at 30ºC in an inverted position for 48hrs.

Note : The petriplates are to be kept in inverted position (ie., the lid should be below), in order to prevent the condensed water drops from falling back into the medium.

Observation

Colonies appearing in plate no.1 are not distinct, small in extent and overlap with one another. In contrast to this, in plate 2 the colonies are large, discreet and generally do not mix with other coloneis. Plate number 3 will have much larger and sparsely distributed (no crowding) colonies with distinct shape, colour, texture, size and form. This is due to the fact that from plate 1 to 3, there is progressive dilution of the inoculum and only a less number of microbes are allowed to grow on plate 3. Plate no 4 will have no microbial growth as there was no inoculum in tube no 4. In plate 2, and more particularly in plate 3, colonies with diverse characters will develop both on the surface and subsurface of the medium.

Streak plate method

In this method, a small amount of mixed (concentrated) inoculum is placed on the solid agar medium (in petriplate) with the help of a needle/loop. From the point of origin, thin and gentle lines of the needle/loop are drawn over the surface of the medium.

The purpose of streaking is to thin out the inoculum successively so that microbes get separated. A second plate also may be streaked from the same loop/needle without reinoculation. In the streak, microbes are crowded and colonies develop closely, but as the streaking proceeds, cells get separated as the needle contains fewer and fewer cells. Hence, at the last stretch of the streak few, well developed and clearly separated colonies will develop. It will be thus possible to obtain pure cultures. Various designs of streaking can be made on the medium as shown in figures. The following is a generalized procedure used in streak plate method.

1. Melt the autoclaved nutrient agar medium in a boiling water bath and cool to 45ºC.

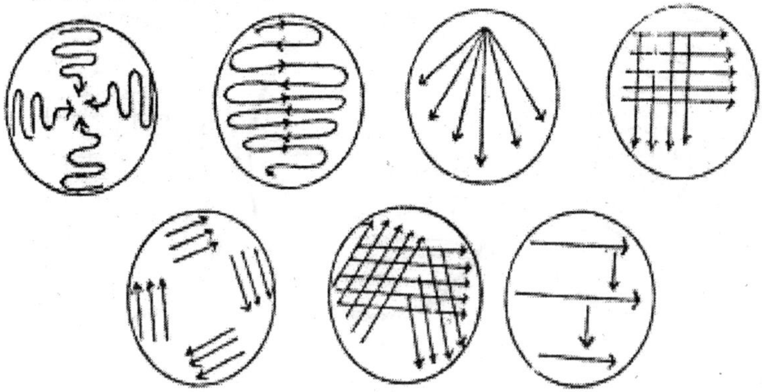

Fig. 12.7 Culture of Microbes
Methods of streaking

2. Pour the medium (about 15ml), into several petriplates carefully. Rotate the petriplates to allow for uniform distribution of the medium. Allow the medium to solidify. Strict aseptic conditions are to be followed to prevent contamination.
3. Take a small amount of bacterial suspension on to the needle or loop. Carefully lift slightly the lid and streak on the medium in any fashion as shown in the figure. Different designs of streaking mayh be done on different petriplates using the same sample source. Do not take a suspension a second time while streaking a plate.
4. Incubate the petriplates according the procedure at 30°C for 48hrs.

Observation

Observe the different petriplates. It will be noticed that well developed and discrete colonies have developed at the end of streaks. The effect of different designs of streaking on the growth and separation of colonies may also be observed. Colonies that are clearly may further be transfered to agar slants/ petriplates to grow as pure cultures.

Sub culturing

Transfer of microbial colonies from one culture to another for the purpose of isolation and purification is called subculturing. Great care has to be taken during transfer to prevent contamination. Bacterial cultures kept in the same medium for a long time will not survive. Ideally, cultures are to be transfered to fresh media at an interval of 15 - 20 days.

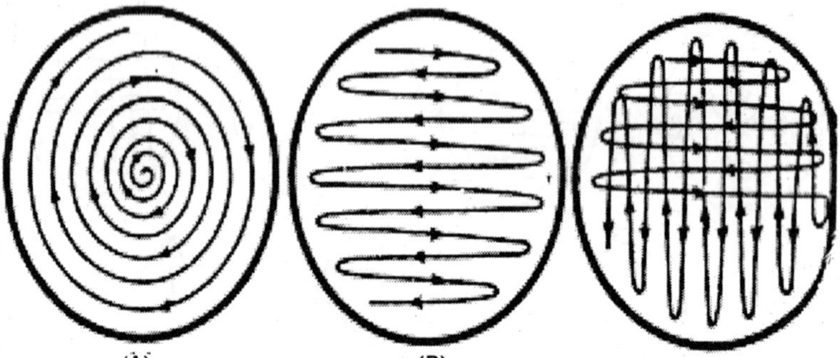

(A) (B) (C)

Fig. 12.8 Culture of Microbes
Methods of streaking (contd)

Spread plate method

In this method, an aliquot of the inoculum is placed on the medium surface and is spread uniformly with a bent sterile glass rod. This will help in diluting the cell concentration in the inoculum. By this method cell aggregates get separated. When colonies develop, they may be picked up and further spread on a fresh medium. Several such spreads may be needed before obtaining the pure cultures of diplococci, streptococci and staphylocci.

Serial dilution method

Among the methods used for purifying cultures, this is by far the simplest and most widely used method. In this method the original inoculum is subjected to serial successive dilutions, so that the concentration of the microbes (in a fixed quantity of the liquid), gradually becomes less and less. When these serial dilutions are plated, colonies will appear discrete and far removed from one another.

The principle of serial dilution may be explained as follows. If we prepare a 10ml sample of soil suspension and assume that it contains 1000 microbes, then each ml of this solution consists of 100 microbes. Take one ml of this and add 9ml of sterile distilled water to make the volume to 10ml. This dilution will have 1:10 or 100 microbes per 10ml, or 10 microbes per one ml. This may be called as 10^{1} concentration of the original sample. From this dilution if we take one ml and add 9ml of sterile distilled water to make it to 10ml, then each ml will have one microbe. This concentration is 1:100 or 10^{-2} of the original concentration. This dilution of one microbe per ml is the starting point of pure culture. Further culturing will yield microbes of one type otherwise known as pure culture.

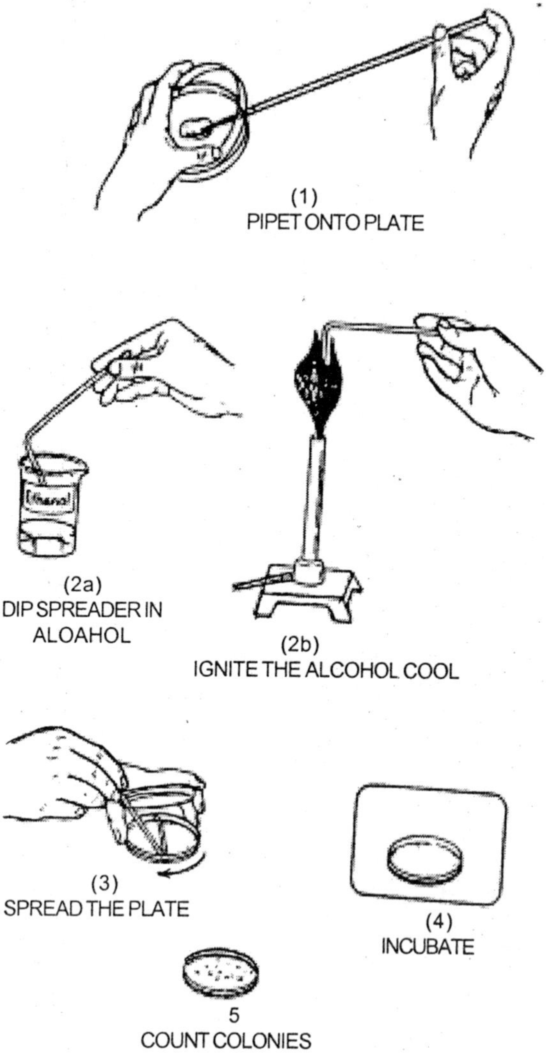

Fig. 12.9 Culture of Microbes
Spread plate technique

Procedure for serial dilution (A sample experiment)

1. Take 6 graduated sterile pipettes of 2ml each and also 12 sterilized petridishes.
2. Take 6 test tubes each having 9ml of sterile distilled water. Arrange them serially after marking the numbers 1 to 6.
3. Prepare nutrient agar medium, melt it; cool to 45°C and keep in water bath at 45°C.
4. With the help of a sterile pipetee take one ml or one gm of soil or sewage or any source inoculum and transfer to tube no 1. The tube now contains 10ml of liquid i.e., 9ml of water and one ml of inoculum. In otherwords, the inoculum is diluted 10times (10^{-1}). Mark the dilution factor on the tube.
5. Using another sterile pipettee take 2ml of the diluted inoculum (10^{-1}) from tube No.1 and trasfer one ml each to two petridishes. All transfers are to be

done following strict aseptic conditions.

6. Using the same pipettee collect one ml of sample from tube no 1 and transfer it to tube no.2. Mix thoroughly by vigorous shaking. This tube now will have 10ml liquid (9ml water +1ml 10^{-}) with the diiution of the inoculum being 1/1000 or 10^{-2}.
7. Collect 2ml of sample from tube No.2 and transfer one ml each to two petridishes for culturing.
8. In the same way as explained above, continue serial dilutions upto 10^{-5}, and accordingly transfer one ml each of diluted samples to petriplates for further growth.
9. Add 7.5ml of the nutrient agar medium to petriplates containing different dilutions of the sample. Gently shake the petriplates to allow even distribution of the medium.
10. After the medium solidifies, incubate the petriplates at 30°C in an inverted position.

Observation

Afteer 48hrs the developed microbial colonies can be observed. As it can be seen from the growth of microbes, the colonies become large, less crowded and finally completely discrete as the dilution factor increases. Coionies clearly separated are pure cultures and they can be picked up for further culture on agar slants. This will give the pure culture of an organism.

Advangate of serial dilution technique

This can be used for qunatitative estimation of microbes in the original sample. Suppose we use one ml of 10^{-3} dilution, and 15 cells are present per ml in this dilution, the total number of cells per ml of the original sample is 15 x 1000 (dilution factor) or 15000 per ml.

Disadvantages of serial dilution technique

Serial dilution technique has some limitations also. For instance, microbes which have different nutritional requirements will not develop on the same medium. Another problem here is that species of microbes which are in large number in the original sample, tend to occupy place on the medium in all dilutions.

Special methods

These include single cell method, enrichment culture method etc.

Single cell method. In this, a single microbial cell is picked up from the culture medium and is allowed to grow. Two methods are employed to isolate single cells. These are:

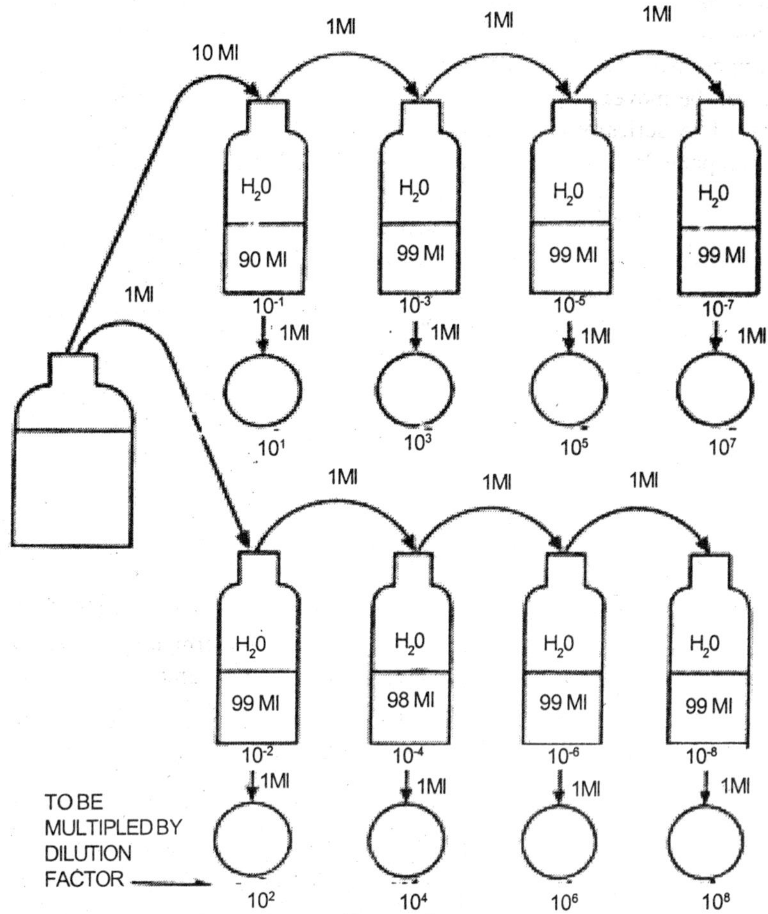

Fig. 12.10 Culture of Microbes
Serial dilution technique

1. Capillary pipette method

Several droplets of a suitably diluted culture medium are placed on a sterile glass cover slip using a sterile capillary pipette. Each drop of the medium is observed under microscope. A drop which contains only one microbe is removed through a capillary pipette and inoculated on to a fresh medium. The individual microbial cell multiples and produces a colony of pure culture.

2. Micromanipulator method

Micromanipulators are devices that can pick up a single microbial cell from a colony of mixed culture. Micromanipulators are used in conjunction with microscopes to pick up a single bacterial cell from hanging drop preparations. The micromanipulator consists of a micropipette with micrometer adjustments and can be moved in any direction, left, right, up, down or backward and forward. A series of hanging drops are placed on a sterile cover slip by micropipette. With the help of the microscope, a hanging drop with a single microbe is located. The single microbial cell is gently sucked into the micropipette and transfered on to a large drop of sterile medium on another coverslip. After the cell multiplies here, the microbes are transfered to a suitable culture medium. Micromanipulator method makes one reasonably sure of pure culture coming from a single cell. The device, however, has to be used with skill and precision.

Othermethods of pure culture

Enrichment culture, selective media culture, and differential media culture also can be used for obtaining pure cultures. These have already been explained earlier in the chapter.

CULTURAL CHARACTERS OF MICROBES

Identification of microorganisms is based on several criteria like morphology, staining reaction, genetic systems etc. Many a time the form, texture, growth pattern etc. which constitute the culture characters also greatly help in the identification of microbes. This is particularly true in the identification of bacterial colonies. For instance, the two strains of *Diplococcus pneumoniae,* viz the virulent and the non virulent can be identified based on their colony appearance. The virulent strain forms smooth shiny colonies, whereas non virulent strain form rough roll colonies.

Microorganisms show diverse culture characters and the diversity also depends on the type of medium used for culturing. Growth parameters such as size, colour, texture, margin, elevation, consistency etc., are of value in the identification of colonies. However, not all the above characters can be studied on all the media. While the above mentioned characters are studied in solid media, the amount of growth and distribution can be studied in liquid cultures such as nutrient broth. We will study below in detail the various cultural characters when grown in different types of media.

1. Colonies on agar plates

Following characters of colonies can be studied on agar plates.

A. Size

The size of the colonies varies from very small (pinpoint), measuring only a mm or less to large colonies ranging between 5–10mm in diameter. While using size as a criterion one should take into consideration only those colonies which are situtated far apart. Colonies situated close by tend to be small due to competition for nutrition. Further, the colonies which have grown for sufficient period only should be measured, as young colonies tend to be smaller in size. Unusually large colonies are encountered in bacteria like *Protes* and *Bacillus* where they spread on the entire medium on petriplates.

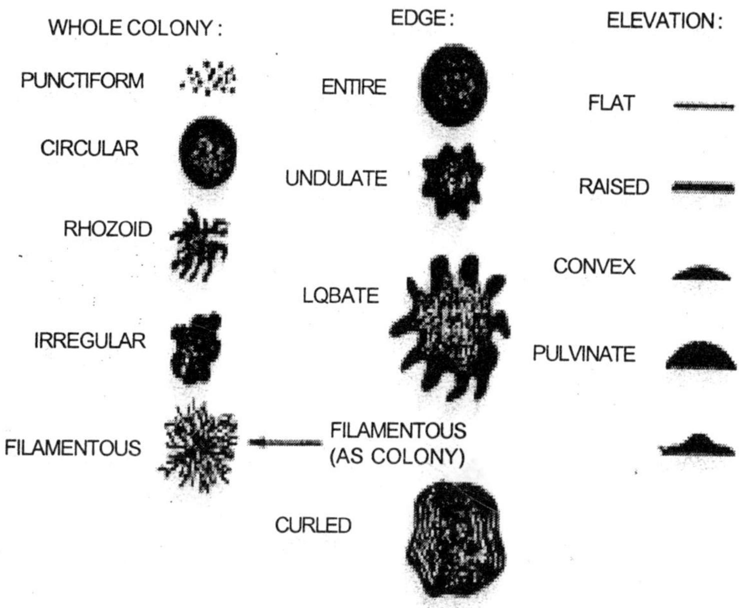

Fig. 12.11 Culture of Microbes
Colony Characters

B. Margin or edge

The contour or the periphery of the colonies may assume various shapes or patterns depending on the microbial species. The margin may have any of the following shapes.

1. *Entire* - Smooth margin
2. *Undulate* - Wavy margin
3. *Lobat* - Lobed margin
4. *Erose*
5. *Filamentous* - Margin extended into thread like growth
6. *Curled* - Folded margin.

C. Elevation

The surface view of the colony is called elevation. The following types of elevations are seen in bacterial colonies.

1. *Flat* - On the same level as the medium
2. *Raised* - Slightly above the level of the medium
3. *Convex* - Raised above in a convex shape
4. *Pulvinate* - Appearing
5. *Umbonat* - Raised above, but with a hood

D. Optical features

Colonies may be opaque, translucent or opalscent.

E. Texture

Depending on the microbial species, the colonies may have varied surfaces like

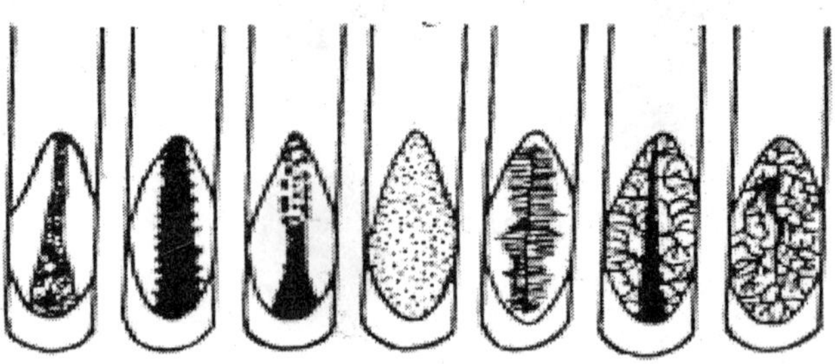

Fig. 12.12 Culture of Microbes
Colony characters (contd) various growth types in streak cultures (for explanation see text)

1. Smooth, shiny or glistening
2. Rough, dull granular or mattle
3. Mucoid (slimy or gummy)
4. Wrinkled or folded surface

It has been found out that for many pathogenic bacteria, the shiny or glistening surface is associated with virulence or toxicity. Colonies which are shiny have capsulated cells, while rough colonies have cells without capsules.

F. Pigmentation or chromogenesis

Many microbes produce intracellular pigments as a result of metabolism and these pigments impart various colouration to the colonies. The following are some of the pigmented colonies (Pelczar, 1986).

Flectobacillus major	Pink
Servatia marcescens	Red
Chromabacterium violaceum	Violet
Staphylococcus aureus	Gold
Micrococcus luteus	Yellow
Derxia gummosa	Brown
Bacteroides melaninogenicus	Black

The pigments produced by bacteria may be *water insoluble and water soluble.* If the pigments are insoluble, only the colonies are coloured, while the medium remains colourless.On the other hand, if the pigments are water soluble, the colour diffuses into the medium and stain it also. For example, in *Pseudomonas aeruginosa,* a blue water soluble pigment (Pyocyanin) stains the medium. Some pigments are partially water soluble (as in *pseudomonas chlororaphis),* and form green crystals (chlororaphin) around the colonies.

Certain types of water soluble pigments are fluorescent i.e., the medium around the colonies show white or green brilliance on exposure to uv light.

Characteristic pigmentation, however, will only develop when all cultural conditions are ideal.

Form : The total appearance of the colony is called form. This is of the following types.

1. Punctiform - Like dots
2. Circular - Like circles
3. Filamentous - With thread like structures constituting the colony
4. Irregular - No definite shape
5. Rhizoid - Thin threads

6. Spindle - Colonies appearing like spindles

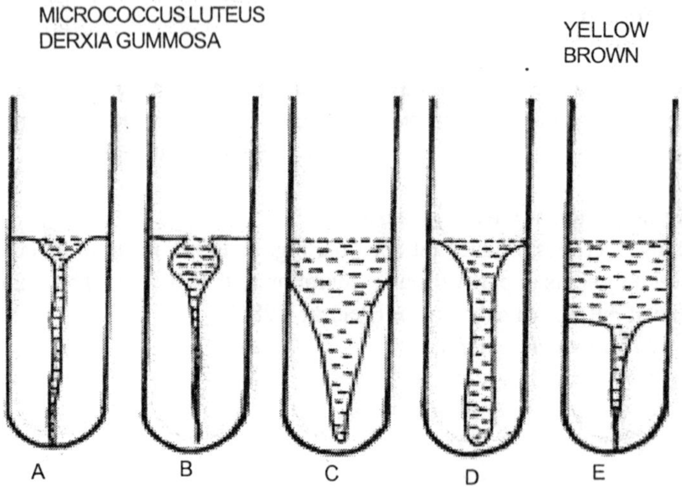

Fig. 12.13 Culture of Microbes
Colony characters (contd) - gelatin stab
Cultures – **A.** Crateriform; **B.** Napiform, **C.** Infundibuliform; **D.** Saccate; **E.** Stratiform

AGAR SLANTS

Cultural characters of bacteria in agar slants are of the following types.

1. Amount - Scanty, moderate or abundant
2. Margin - Entire or irregular
3. Texture - Same as for
4. Form - This is of the following types
   a. Filiform - Like a small tube
   b. Echinulate - With pointed outgrowths
   c. Beaded - Like beads
   d. Effuse - Spread over the slant surface
   e. Arborescent - Branched like a tree
   f. Rhiozoid - Thin rihzoid like growth

3. GELATIN STABS

This is basically of two types -

1. **Growth along the line of stab** (puncture) **:** Growth may be confined along the line of inoculation or it may diffuse a little away from it. This is further classified into the following types -

a. Filiform – Thin tubes

b. Beaded – Small beads
c. Papillate – Thin finger like projections
d. Villous – Very thin projections
e. Arborescent – Branched like a tree

2. Growth by liquefaction : Here, growth will proceed in various funnel like designs by liquifying the medium along the line of growth. This is of the following types -

a. Crateriform
b. Napiform - Broad above and suddenly tapering below
c. Infundibuliform
d. Saccate - Like a small bag
e. Stratiform

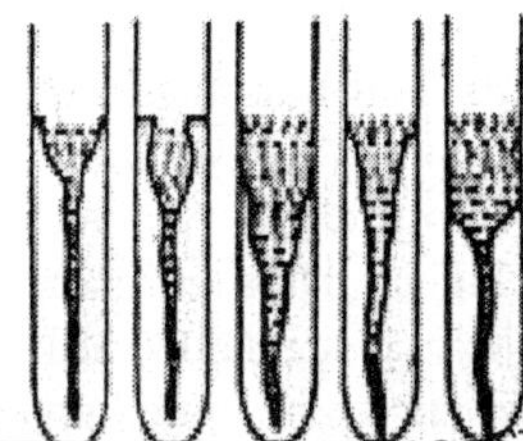

Fig. 12.14 Culture of Microbes
Colony characters (contd) gelatin stab cultures showing line of puncture

3. Nutrient broth : In liquid medium such as nutrient broth, features like shape, form, texture etc. cannot be observed. The growth characters observed are –

a. Amount - Scanty, moderate or abundant
b. Distribution and Type - Cells may be distributed uniformly throughout the medium imparting all round turbidity or it may be confined to the surface or bottom of the broth. The following types of accumulations are seen –

1. Flocculant
2. Ring
3. Pellicle
4. Membraneous
5. Sediment

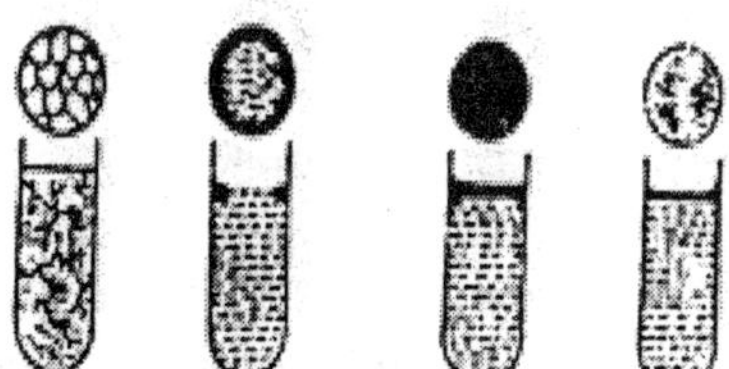

Fig. 12.15 Culture of Microbes
Colony characters (contd) surface growth types in nutrient broth

Somephysical conditionsrequiredforbacterial growth in a culture media

Apart from the type of media for the bacterial growth, many physical conditions of the environment are very much essential for the optimum growth. Unless these conditions are provided, bacterial colonies do not develop properly. It must, however be pointed out that there is no uniformity for these conditions applicable to all kinds of microbes or all kinds of media. These physical conditions have to be studied for every type of microbe that is to be cultured and the optimum of these conditions has to be fixed for a particular experiment. Some of these conditions are temperature, pH, oxygen, relative humidity etc. These conditions (as they influence microbial growth) are discussed below.

1. Effect of temperature

Temperature is one of the most important physical factors that has a great influence on the growth of microorganisms. Microorganisms like bacteria, are unlike higher plants and animals, in that they lack a homeostatic mechanism which helps in regulating the heat generated by metabolism. Hence, any minor alteration in temperature has a great effect on the growth and physiology of bacteria. Initially, however an increase in temperature range will increase the metabolic rate. But this is, however, subjected to a certain limit. Initially (within a limited temperature range), there is a two fold increase in the rate of enzyme catalysis for every 10°C rise in temperature.

Based on the temperature tolerance and its influence on growth, bacteria may be classified into the following categories.

1. **Psychrophiles** : Optimum temperature ranges between 0°C and 20°C.
2. **Mesophiles** : Optimum temperature between 20°C and 40°C.
3. **Thermophiles** : Optimum temperature between 40°C and 80°C.

Thermophiles are of two categories. These are facultative thermophiles with the optimum temperature range between 45°C to 60°C, and obligate thermophiles with optimum temperature above 60°C. Normally, however, temperature in the range between 50°C and 100°C are lethal for bacterial cells and spores. Some endospores, however can withstand it. A very high temperature preferred by bacteria, is a genetic trait, which results in enzyme production with varying temperature requirements.

For the growth of bacteria there are three cardinal temperature points. These are a minimum, an optimum and a maximum. The minimum growth temperature may be defined as the lowest at which a species can grow and below this, growth comes to a standsstill. An optimum temperature range is one, where the growth of a bacterial species is maximum. Starting from the minimum temperature, an increase to reach the optimum temperature will have a beneficial effect on the growth of bacteria. From the minimum temperature to optimum there is a logarithmic where bacterial growth is still possible but the growth will not be maximum. Any increase beyond maximum will be deleterious. From this discussion it becomes obvious that from the minimum to optimum, bacterial growth reaches the maximum. At the maximum temperature there will be a stand still in the growth rate of bacteria.

For the in vitro growth of bacteria, the duration of exposure at a particular temperature range is very essential. For instance, to decide the lethal effects of high temperature on bacterial growth there are two methods. These are the *thermal death point* (TDP) and *thermal death time* (TDT). The TDP is a temperature at which an organism is killed in 10 minutes of exposure and TDT is the time required to kill cells/ spores suspension at a given temperature.

Effect of pH

The pH is the logermathic symbol of the reciprocal of hyrdrogen ion concentration in gram atmos/per litre. For instance, a pH range of four indicates a concentration of 10^{-4} or 0.0001gm atoms of hydrogen ions in 1 litre of solution (for a detailed discussion of pH and the methods used for its determination – please see College Microbiology Vol. 1, by the same author). In simple terms, a pH may be defined as the relative acidity or alkalinity of a medium or a solution. pH is expressed in a range from 0–14; in this 0 – 7 is acidic and 7–14 is alkaline and 7 itself is called neutral.

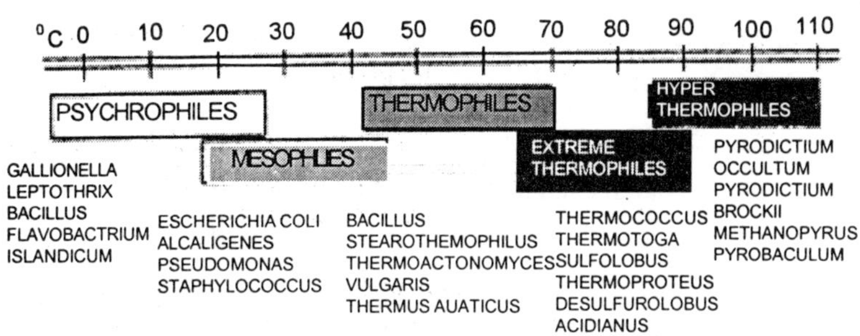

Fig. 12.16 Culture of Microbes
Growth pattern of microbes with reference to temperature

Changes in pH have a profound influence on the growth and metabolism of all organisms including microorganisms. Enzyme systems on which depends all biological reactions are known to be highly sensitive to pH changes. Any change in pH range would render the enzyme non functional, thereby affecting the growth of microbes like the cardinal temperature ranges, the organisms and the maximum. Above the maximum pH organism fails to grow, and at the optimum pH the growth will be maximum and in the minimum pH growth is just possible. A pH range below the minimum would have a retarding influence on the growth of microbes.

It has to be understood, however, that when the bacteria fail to grow it is primarily due to the non functioning of the enzyme in a pH range that is unsuitable to them.

There is no general, minimum, optimum and maximum pH range that can be fixed as standard for the growth of microorganisms. Different microbes have different pH ranges suitable for them. And each species has a preference for a specific pH range that may be broad or limited. Most of the times optimum growth of a microbe occurs within a narrow optimum pH range. Experiments have shown that for many bacteria, the specific pH range is between 4 and 9. With the optimum ranging between 6.5 - 7.5 pH. There are, however, a good number of acidic pH. For instance, *Mycoderma aceti* and *Thiobacillus thioxidans* prefer an acidic pH below 4. *T. thioxidans,* infact prefer a full acidic pH ie., 0. This is the reason why organic acids are used to preserve foods like fermented dairy products.

Fungi, in general, are more tolerant to higher acidic conditions than bacteria. Their optimum activities are between a pH of 4 and 6.

The specific pH requirements reflect an organism's ability and adaptation to their original natural conditions. The pH preference of specific microbes are taken into consideration while preparing selective media.

In culture conditions the metabolic products of bacteria like organic acids, from the degradation of carbohydrates and alkalies from protein breakdown, may actually inhibit the growth due to their influence on pH. This is the reason why buffers are added to the media to take care of acidic or alkalinity influence and to neutralise them. Generally amphoteric compounds such as, peptones and aminoacids found in the complex media act as natural buffers. In chemically defined media phosphates salts are often used to function as buffers.

Effect of relative humidity

Water is one of the basic ingredients necessary for the growth of all microorganisms. Some microorganisms however, are less sensitive to relative humidity (this may be defined as the amount of water vapour present in the air expressed as a percentage of maximum that the air could hold at a given temperature) and can survive in the absence of required quantity of moisture.

Such microbes, however, recommence growth from hybernation at the availability of suitable moisture conditions. There are some microbes which are extremely sensitive to moisture and die in the absence of it. The different water requirements of the microorganisms can be studied in the the laboratory under different moisture conditions.

Different relative humidity conditions can be created in the laboratory by using pure water vapours, or water saturated with certain chemicals where in the relative humidity comes down. For instance, pure water vapours will create 100% relative humidity (RH). Chemical solution containing $CaSO_4$ $5H_2O$ gives 95% RH. Growth of microorganisms get generally inhibited when the RH is below 62%.

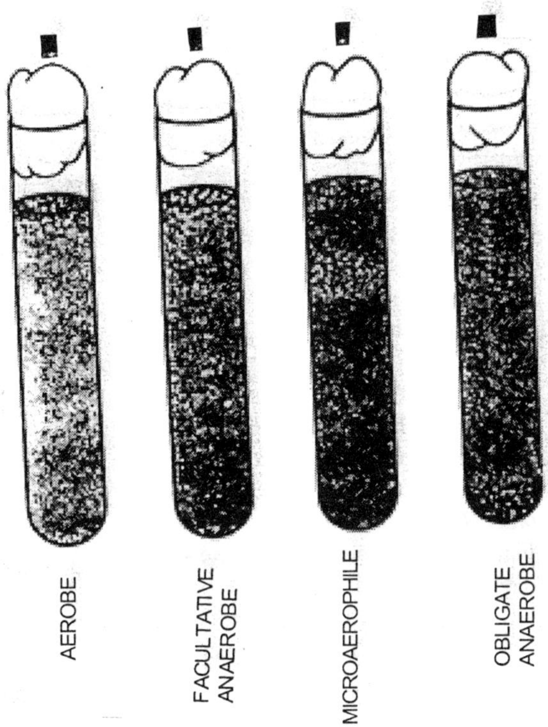

Fig. 12.17 Culture of Microbes

Effect of oxygen

Based on their requirement of oxygen, microorganisms can be classified into three major groups – these are *aerobic, anaerobic* and *microaerophilic.* Microbes which require the presence of oxygen for their growth are called aerobic; those which can grow only in the absence of oxygen are called anaerobic, and those that require limited amount of oxygen and increased amount of carbon dioxide (7-10%) for their growth, are known as microaerophilic. Aerobic and anaerobic microorganisms include obligate and facultative individuals. Obligate aerobes such as, acetic acid bacteria can grow only in the presence of molecular oxygen as they respire aerobically. In contrast with this, there are obligate anaerobics like *Clostridium,* which can grow only in the absence of oxygen because they lack catalase and the resultant accumulation of hydrogen peroxide H_2O_2 is lethal to them.

Some of the anaerobic microorganisms that manage to grow with the presence of oxygen but cannot utilise it are called *aerotolerant.* Some of the facultative microbes can grow both in the presence and absence of oxygen.

The effect of oxygen on the growth of various types of microorganism can be determined by observing them in a tube of solid medium following shake tube inoculation.

Effect of other physical factors

Several other parameters like osmotic pressure, salt, growth patterns of microbes with reference to oxygen concentration, and presence of metals, dyes and different wave lengths of light, including UV rays have a profound influence on the growth of microbes in the culture medium.

Cultivation of anaerobic bacteria

Microbes differ in their ability to use or not to use atmospheric oxygen. While some require molecular oxygen for the respiration (*aerobes*), others do not require it (*anaerobes*). There are various grades between these –

1. **Facultative anaerobes** - Anaerobic but can function in the presence of oxygen.
2. **Microaerophiles** - Require oxygen but too much or too little is detrimental to them.
3. **Obligate anaerobes** - Strictly anaerobic, may even die in the presence of oxygen.
4. **Aerotolerant** - Anaerobic, survive in air, but do not reproduce.

Thus one can see that a wide array of bacteria can be categorized based on their reaction to the presence or absence of oxygen. This is due to the different enzyme systems present in these types of bacteria.

Several methods are employed by microbiologists to selectively grow anaerobic bacteria. Some of these are described below.

1. Pre reduced medium

This involves freeing the culture medium of oxygen prior to inoculation (removal of oxygen is said to be reduction). During preparation, the culture medium is boiled for a few minutes to drive out most of the dissolved oxygen. Reducing agents such as, cysteine, is added to the medium to further decrease the level of oxygen content. Oxygen free N_2 is bubbled through the medium to maintain anaerobic conditions. The medium is then distributed to suitable culture tubes which are also flushed with oxygen free N_2. The tubes are further autoclaved and stored before use. At the time of inoculation, the tubes are flushed with oxygen free CO_2 by means of a cannula; plugged again and incubated.

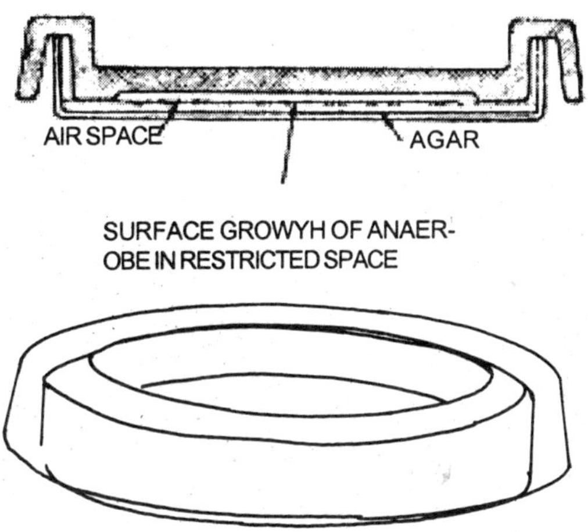

Fig. 12.18 Culture of Microbes

Anaerobic culture plate (air trapping technique)

2. Air trapping method

Air trapping method is one of the most effective method of absorbing oxygen from a layer of trapped air by the use of a special hood or cover. This special cover when kept on a specially prepared agar medium, covers the medium all over except for a thin gap over the medium where air gets trapped. Special chemicals will be added to the medium which will absorb the trapped air (oxygen). Some of these chemicals are – sodium thioglycollate (0.1%) or cysteine, and hydrochloride (0.2%). By this procedure oxygen content of the medium is considerably reduced and only anaerobic microorganisms grow. A special anaerobic medium given below is normally used for culture.

Thioglycollate dextrose medium

Yeast extract	5.0g
Casitone or Trypticase	15.0g
Sodium chloride	2.5g
1-cysteine	0.25g
Sodium thioglycollate	0.5g
Agar	0.75g
Methylene blue 2%	0.1g
Distilled water	1 litre

The medium should be dispersed into culture tubes to form a column about 3-4 inches deep, pH should be maintained at 6.8, if the medium is to be stored for a longer period, it should be heated before use to remove dissolved air.

3. Roll tube method

Also called shake culture method, in this the medium is boiled for 10 minutes to expell the air. The medium is then to be cooled to 45°C, tube is then gently rolled between the palms or shaken gently to distribute the inoculum throughout the medium. The culture tubes are quickly cooled by immersing them in cool water. The tubes are now incubated under suitable conditions. Later, on observation, it will be noticed that different microbes grow at different depths as follows –

1. *Strict anaerobes* – Grow deep in the medium and will not appear anywhere near the surface.
2. *Facultative anaerobes* – Found throughout the medium as well as near the surface.
3. *Microaerophilic* - Grow a little below the surface.
4. *Aerobes* – Grows on the surface of the medium.

The medium used in the above experiment is tryptone yeast glucose agar medium.

4. Deep media method

In this method, aerobic tissues or cells such as germinating seeds, plant cells, fruit or other oxygen using material is added to the medium (broth). After sometime, the medium is boiled and cooled to 45°C and inoculated with a suitable anaerobic microbe. Strict anaerobic microorganisms develop deep in the medium.

5. Phosphorous stick method

Petriplates containing culture media are kept in a closed air tight vessel. Phosphorous sticks are burnt in the vessel to remove oxygen. The sticks continue to burn till all the oxygen gets exhausted and ultimately only the inert gases and partial vacuum remain. A small container of water should be kept to trap P_2O_3, formed as a result of combustion.

6. Pyrogallol method

Certain chemicals such as, a mixture of NaOH and pyrogallol are known to absorb oxygen in a closed vessel where petridishes can be kept. Pyrogallic acid reacts with oxygen and leaves only inert gases, nitrogen and partial vacuum. The procedure is as follows :

(a) Take two agar slants and lable them with the microorganism to be inoculated.
(b) Inoculate the medium under aseptic conditions.
(c) Replace the cotton plug and ignite it with the flame of a bunsen burner.
(d) Using a glass rod, push the burning cotton plug deep into the tube until it nearly touches the surface of the slant.
(e) Put crystals of pyrogallol into the tube on the cotton plug. Add 2ml of sodium hydroxide (4%) into the tube.
(g) Invert the tubes and incubate them at the inverted position at 37°C for 24–48 hours.
(h) Growth of anaerobic bacteria may be observed on the slants.

7. Anaerobic chamber method

Oxygen from the medium may be completely removed by using special anaerobic jars (chambers), such as Brewer's jar, Brown's jar. McIntosh jar, Fields jar or Gas pack system. In these H_2 CO_2 or N_2 can be introduced through a special inlet. The jars have a special air lock where in the culture plates are kept. The air locks are evacuated and refilled with N_2. From the air lock, the plates are kept into the main chamber. Hydrogen is slowly introduced into the medium. This reacts with whatever O_2 is present, forming water. This reaction is catalysed by heating electrically a palladium catalyst. Water, that is formed in this reaction is absorbed by $CaCl_2$. After this, the medium is inoculated with a suitable anaerobic bacteria.

Another device that is simpler than the one explained above is a gas pack system. The gas pack consists of dry chemicals which when hydrated release CO_2 and H_2

A gas pack as the one mentioned above, is introduced into a sealed jar which contains an inoculated medium. Addition of water (10ml) into the gas pack causes the release of CO_2, and CO_2. H_2 reacts with O_2 on the heated palladium catalyst to form water bringing in anaerobic conditions. The CO_2 evolved also aids in the growth of strict anaerobics. An anaerobic indicator ship introduced into the system changes from blue to colourless in the absence of oxygen.

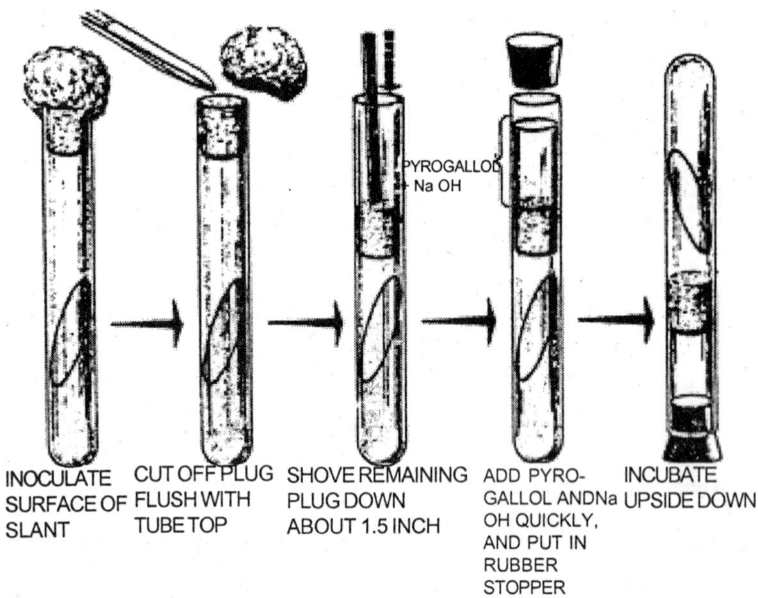

Fig. 12.19 Culture of Microbes
Pyrogallol NaoH method for anaerobic cultivation of bacteria

ISOLATION OF SOIL ALGAE

Thousands of types of algae occur in nature in all possible types of environmental conditions. They abound however, in oceans, salt lakes, freshwater lakes, ponds and rivers. Some algae are also found on damp soils, rocks, tree barks and even in the bodies of plants and animals. In size, algae range from extremely minute to the largest. Some of the sea kelps are more than 90 feet in length. Algae are both prokaryotic (if blue green algae are included), and eukaryotic.

In a microbiological laboratory for the purpose of study, pure cultures of algae are necessary. In general, techniques used for bacteria and fungi are applicable to microalgae. The medium normally used for isolation of soil algae is Beneck's broth. The following are the requirements to prepare Beneck's broth medium.

KNO	0.2g
$CaCO_3$	0.1g
$FeCl_3$	2.0 drops
Glass distilled water	1000.0 ml

Suitable soil sample

Procedure : Required amounts of the chemicals mentioned above are dissolved in 200ml of distilled water, and the volume is made up to a litre by the addition of more distilled water. The above components will give Beneck's liquid medium. Solid medium can also be prepared by the addition of 50 x 20gms of agar per litre of Beneck's solution. The pH of the medium should be adjusted to between 7 and 7.5.

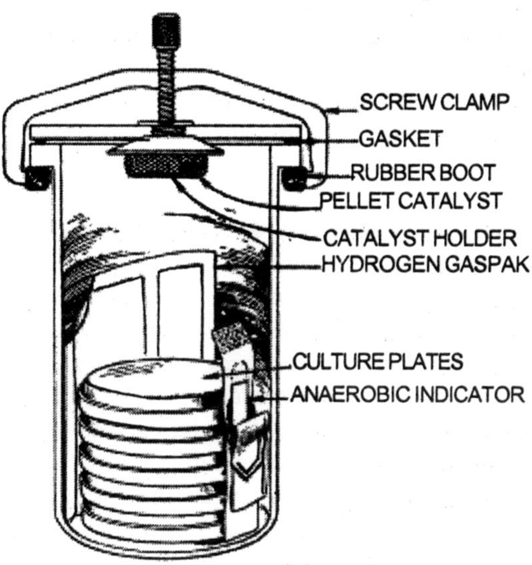

Fig. 12.20 Culture of Microbes

The gas pack system for cultivation of anaerobic bacteria

Algal cultures may be prepared as follows :

1. Pour the broth solution into suitable conical flasks. The conical flasks are filled to a depth of 5cms from the bottom. Plug the mouth of the flasks with cotton wool.
2. Autoclave the medium in conical flasks at a pressure of 20lb for 20 minutes.
3. Allow the medium to cool and remove them from autoclave.
4. Remove the cotton plug from the mouth of conical flasks and add one gram each of soil; immediately replace the cotton plugs. These operations have to be carried out very quickly and under aseptic conditions to prevent contamination.
5. Incubate the culture medium with the inoculum at 30-35^{0}C in an illuminated growth chamber (usually fitted with 60w tungsten lamps).

Observation : After about 2 weeks of incubation, profuse growth of algae can be seen on the medium.

The algal cultures obtained from soil are mixed cultures. They may be further purified by carefully transfering a few cells of filaments under aseptic conditions to a different culture medium. These can be further subcultured.

1. During autoclaving, phosphate should not be mixed with other ingredients. It should be autoclaved separately, and aseptically added to the medium after cooling.
2. The intensity of the light in the growth chamber should ordinarily be 1/3rd of the intensity of normal sunlight.

METHODS OF PRESERVATION OF CULTURES

When once microorganisms are cultured in the laboratory, they are usually subcultured on agar plates or slants at regular intervals to maintain viability. Some of the microbes routinely used for experiments should always be available for any immediate experimentation. In any established microbiological laboratory, ready stocks of properly identified microorganisms will always be available for any kind of experimentation. In order to have a ready availability of cultures, the culture should be maintained properly and in viable condition.

Maintenance of cultures is a very lengthy process and the persons involved should be highly skilled to see that at no time, a given culture should lose vialibility. There are several methods involved in maintaining the cultures. Some of these are described below.

1. Use of refrigator or cold room storage

Live cultures on a culture medium can be successfully stored in refrigators or cold rooms, when the temperature is maintained at 4^{0}C. At this temperature range the metabolic activities of microbes slows down greatly but do not all together stop. As a result, bacterial metabolism will be very slow and only less quantity of nutrients will be utilized. This method, however cannot be used for a very long time for not only the nutrients get utilized but also waste products get accumulated killing the microbes. Refrigerator or cold room storage is of use only for short time preservation of cultures. Subculturing is necessary if the period exceeds two weeks.

2. Paraffin method

This is a very simple and cost effective method of preserving cultures of bacteria and fungi for several years at room temperature. In this method, sterile liquid paraffin is poured over the slant culture of microbes, and stored upright at room temperature. The layer of paraffin prevents dehydration of the medium and ensures anaerobic conditions. As a result, the microbes remain in dormat condition. It has been seen that in some instances bacteria remained viable for even upto 15 or 20 years. Cultures can also be maintained by covering the agar slants with a layer of sterile mineral oil about 1/2inch above the surface of the slant.

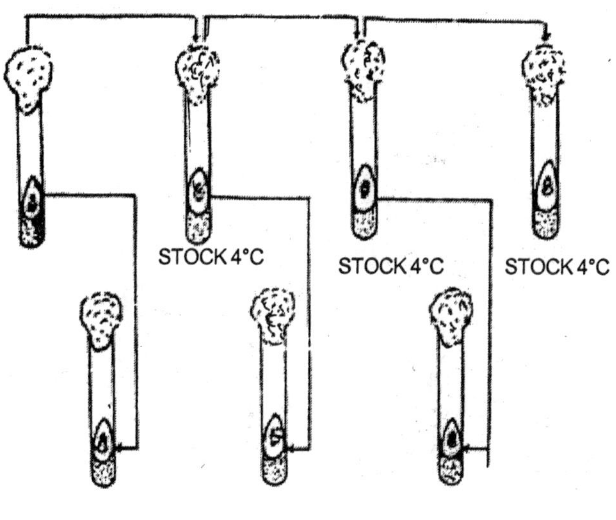

Fig. 12.21 Culture of Microbes
Preservation of culture by periodic transfer

Maintenance by periodic transfer of culture

The usual method of maintaining a culture is to inoculate the microbe in agar slant and incubate the medium for a short period. This gives sufficient time for the microbes to reach the optimum growth phase. Subsequently, the culture can be stored at a low temperature. As and when required, they can be taken out and cultured on a fresh medium and then stored again under cold temperature conditions. Periodic subculturing, however, might induce mutations in the microbes.

Preservation at - 40°C in glycerol

Cultures can be preserved for a number of years in glycerol, at a temperature of - 40°C in a freezer. In this method, about 2ml of glycerol solution is added on to the agar slant culture. The culture can be emulsified by shaking the culture. Emulsion is then transfered to ampoules, with each ampoule having 5ml of the culture. These ampoules are placed in a mixture of industrial mathylated spirit and carbon dioxide and are frozen rapidly to - 70°C. The ampoules are then removed and placed directly in a deep freeze at - 40°C. For utilization of the stock cultures, the ampoules are kept in a water bath at 45°C for about a few seconds and then used for plate cultures.

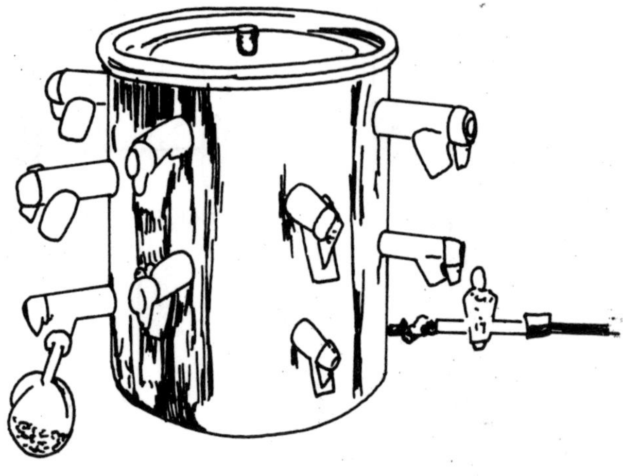

Fig. 12.22 Culture of Microbes
Preservation of cultures by lyophilization

Freeze drying

This method is also called lyophilization. Lyophilization, may be defined as the rapid dehydration of the organisms while they are in a frozen stage. In this method, a thin film of frozen bacteria are dried in a vacuum and subsequently sealed in ampoules. Freeze drying is one of the most frequently used methods of culture preservation employed in many microbiological laboratories all over the world.

Lyophilization protects the damage caused to microbial cells during the process of dehydration. As the organism is freeze dried, the metabolic rate greatly slows down and the microbes remain in a dormant stage for several years.

Generally in the process of drying, elimination of water results in damage to the many chemical compounds and the integrity of the cell is lost. In the freeze drying process, however, the culture is frozen first and water is removed by sublimation from the ice as vapours. As the suspension is not in the liquid condition, the process does not damage the structure. The freeze dried materials can easily be brought back to the normal stage without any damage to the original condition.

In the freeze drying process, the cultures are rapidly frozen at - 70°C, and then dehydrated by vacuum. The tubes are then sealed and stored in the dark in refrigerators at 40°C.

Freezing in liquid nitrogen at temperatures of - 196°C also retards the metabolic rate and the cells remain viable for very long periods.

Freeze drying involves several steps. These are the following –

Predrying requirement for cultures

The type of culture media used is an important criterion in the freeze drying cultures. For some microbes, the predrying culture and maintenance may be the same or different. For example - for *Bacillus macroides,* the predrying culture and maintenance culture are the same, whereas for *Bacillus subtills,* the predrying culture is soil extract agar whereas for the maintenance it is nutrient agar.

The predrying media should have a rich concentration of the microbes. The age of the culture is another important criterion because the cultures that have reached the optimum growth phase survive better than the cultures that are still in the growth phase.

Ampoule preparation

The ampoules used in the preservation of the culture should be made of neutral glass and have an internal diameter of 6mm. Ampoules should first be disinfected and then rinsed in distilled water. After plugging with the cotton wool, they should be sterilized with an autoclave for 20 minutes under 20 pounds pressure.

Harvesting the culture

Cultures grown on agar slants should be harvested for 3-5 days after incubation. 2ml of suspending fluid such as, *Mist Desiccans* containing sterile horse serum, glucose and nutrient broth should be added over each agar slant and the surface should be gently rubbed with a Pasteur's pipette. The suspensions should immediately be transfered to the ampoules.

Primary drying

The filled ampoules should be placed in a stand and the neck of the tube should be flamed. The ampoules should then be centrifuged. Special centrifuges are used for this purpose. These centrifuges have a primary drying chamber and come in a variety of models. Most of these models consist of a drying chamber, where the centrifuge head is located and heated under vacuum. The centrifuge head should be allowed to run at 500rpm. Usually, the primary drying is allowed to proceed between $2^{1/2}$ to 4hrs during which, more than 90% of free water is removed. Air is then allowed to slowly enter into the vacuum chamber. The centrifuge head is then removed from the machine and the ampoules are plugged again with cotton wool.

Secondary drying

The ampoules are then transferred to a secondary dryer which consists of high grade phosphorous pentoxide contained in a tray, to which is attached a vacuum pump. The ampoules are left on this dryer for 18-20hrs upon which the moisture content is reduced to 1%. The ampoules are checked for maintenance of vacuum, and sealed with flame. The culture in the ampoule is now a light powdery substance. The ampoule can be stored at 4°C.

Reconstitution of the culture from ampoules

At the time of utilization, the tip of the ampoule may be cut using a file. The cotton plug should then be carefully removed with the help of forceps. The mouth of the ampoule should be flamed again and a fresh sterilised cotton wool can be plugged.

Sordelli's method of preservation of cultures

This is a method of preservation simpler than the freeze drying method, but as reliable as the latter. This method can be safely used when the samples to be preserved are small in quantity.

In this method, the culture should be preserved or incubated on a solid medium for the required period. The inoculum is emulsified in a loopful of horse serum and is deposited on the inner wall of the small tube (8 x 60mm) which is then put into another larger tube (10 x 150mm). A small quantity of phosphorous pentoxide is placed at the bottom of the outer tube with the help of a glass rod. The inner tube must be placed in such a way that it is held over the bottom of the outer tube but not directly touching the chemical placed at the bottom. The outer tube, is then connected to a vacuum pump, and after the air is removed, the outer tube is sealed. This tube containing the culture can be stored at room temperature away from light.

Preservation of soil cultures

Culture of soil microbes can be maintained in the soil medium. The method of preservation is as follows. A sample of soil is passed through a sieve of 2mm mesh, and collected in sufficiently large test tubes. To these tubes is added 1% solution of dextrose and mixed thoroughly. These tubes are then kept in a boiling water for about 15 minutes and then autoclaved for about one hour at 13 pounds pressure on two successive days.

Pure culture of soil microbes can be kept in this medium and preserved for a number of months under refrigeration.

Preservation of plate cultures

The following are the various methods for the preservation of cultures on petriplates.

1. Put a drop of formaldehyde on the inside of the lid. The formaldehyde solution acts as fixative and will not alter the growth of the already existing colonies.
2. The plates can also be kept under refrigerator at 4^{0}C which will retard further growth.
3. The petriplates temporarily sealed with vaseline which will prevent dehydration of the medium. After the atmosphere inside the media gets exhausted; colonies will not grow further.
4. The clean embedding resin may be poured over the surface of the plate culture this will also help in preservation.

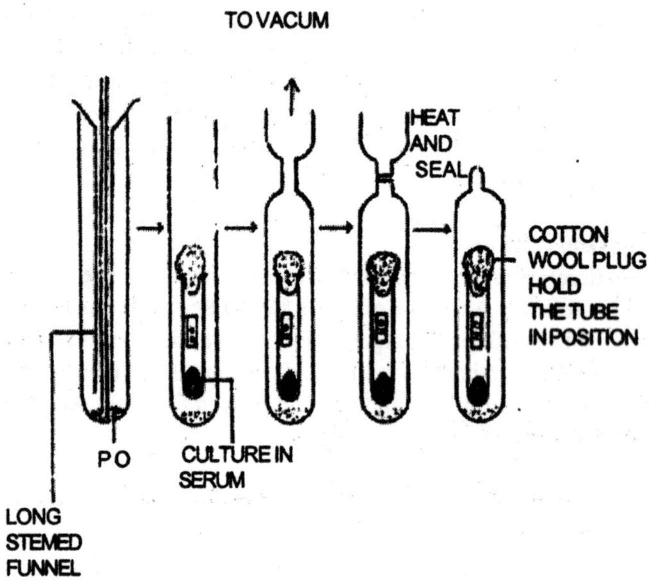

Fig. 12.23 Culture of Microbes
Preservation of cultures by Sordelli's method

Preservation of fungal cultures of fumigation

In fungal cultures infestation by mites is a serious problem of maintenance. Mites get into the culture through various means, such as soil, plant materials etc. The mites that develop on the culture feed on the fungal hyphae and move from one culture to the other through cotton plugs. The development of mites can be prevented by addition of 0.01% lindane powder (an insecticide) to the sabouraud dextrose sugar medium. This kills the mites within 3 minutes. Alternatively the cultures may be dry fumigated with pra-dichloro-benzene

13

MICROBIAL GROWTH AND REPRODUCTION

One of the unique properties of living organisms is their ability to grow and reproduce. It is this unique capacity that has been responsible for the maintenance of life ever since it originated on this planet a few billion years ago.

Microorganism growth is simply an increase in size as most of them are made up of only one cell. In unicellular microorganisms cell division is a means of multiplication (reproduction), and growth. Whenever a cell division occurs, the two daughter cells become two individuals; the cells, however, increase their size (grow) and attain maturity.

Nutritional requirements

Microorganisms, in order to grow must obtain from the environment, all the substances that they require for the synthesis of the cell materials and for the generation of energy. These substances are termed nutrients. In nature, microbes draw from their habitats all the necessary *nutrients*. While culturing bacteria, however, these nutrients must be made available to them in the medium. In addition to nutrients, water plays a key role in the culture of microbes. Although water is not regarded as a nutrient by itself, its presence in the medium is a must, as otherwise biological reactions will never take place. A microbe can absorb nutrients only in the form of a solution.

Microbes are extremely diverse in their nutritional requirements, not only in terms of quality, but in terms of quantity also. A culture medium must, therefore have all the required materials for the growth of a particular microbe. Practically thousands of media have been proposed for the cultivation of bacteria, often with reasons for the presence of a particular nutrient.

The design of a culture medium, therefore should be based on *Principles of nutrition* and must fulfill the minimal nutritional requirements that are essential for the synthesis of cell substances.

Elementary nutrient requirements

The chemical composition of cells, which is broadly constant throughout the living world indicates the materials required for microbial growth. Water accounts for some 80 to 90 per cent of the total weight of cells, and is a major nutrient in quantitative terms. Besides water, the elementary composition of the cells is divided into ten macro elements which are present in all cells. These are carbon, hydrogen, oxygen, nitrogen, sulphur, phosphorous, sodium, potassium, calcium, magnesium and iron (C, H, O, N, S, P, K, Na, Ca, Mg, Fe). Of these, hydrogen, oxygen, carbon, nitrogen, phosphorous and sulphur account for 95 per cent of the cellular dry weight. Trace elements required for microbial growth are – manganese, molybdinum, zinc, copper, cobalt, nickel, vanadium, boron, chlorine, selenium, silicon, tungsten etc. Heavy metals are mostly part of the enzyme systems, involved in the metabolism of inorganic elements and compounds.

All the required metallic elements can be supplied as nutrients in the form of cations of inorganic salts. Potassium, magnesium, calcium and iron are required in relatively large amounts and should be included as salts in the media. Phosphorous, a non metallic element can be provided as phosphate salt.

It should be understood, however, that some biological groups have additional specific mineral requirements; for example, certain marine bacteria require high concentration of sodium.

Carbon requirement

Photosynthetic and chemosynthetic bacteria typically use CO_2 as the principal source of their carbon requirement. The conversion of inorganic carbon into organic cellular constituent is a reductive process requiring the input of energy. In phototrophic and chemotrophic bacteria, the energy obtained from light or by the oxidation of inorganic compounds is used to reduce CO_2 to carbohydrates. Heterotrophic bacteria, however, obtain their carbon requirement directly from organic compounds. In this process, however, there is no need for any further reduction, as all organic substrates are at the same general oxidation level like the cell constituents.

The carbon that enters the cell system has to meet two requirements of the cell.

1. Carbon needed for the biosynthetic pathways of the cell.
2. Provide the energy necessary for the biological activities of the cell. For

this, the carbon compounds directly enter into energy releasing metabolic pathways ultimately releasing CO_2.

In the case of aerobic bacteria, organic carbon gets completely metabolized and CO_2 is released, in the case of fermentive bacteria, however, as the organic carbon is incompletely oxidized, a mixture of CO_2 and organic compounds are excreted.

Many of the microorganisms, such as *Escherichia coli* can use a single carbon source (organic compound) as the starting point, and using their synthetic machinery can produce all their requirements. Some other microbes, however, require more than one carbon source as their primary requirement. This is due to the fact that their metabolic machinery cannot synthesize all their diverse carbon requirement. As a result, some of the required ones have to be added directly to the medium. These additional organic nutrients have a purely biosynthetic functon being required as precursors of certain organic cell constituents. These are often called the growth factors.

Microorganisms are extremely diverse, both with respect to the kind and number of organic compounds which they can use as the main carbon source. Some microorganisms are highly versatile in that they can use any of the carbon compounds as their energy source, while some are highly specialized and require a specific organic compound for their growth. For example, certain cellulose decomposing bacteria can use only cellulose.

Some microbes require gaseous CO_2 itself as a nutrient in very small amounts. This requirement however, is met by the CO_2 naturally released by the organisms themselves during metabolism. It has been known that a few bacteria and fungi require a relatively high concentration of CO_2 (5–10 percent), for growth in the media.

Nitrogen, Sulphur and Phosphorous requirements

Nitrogen and Suphur occur in the organic compounds of cell, mainly as amino group or sulphydryl groups in the aminoacids (proteins). Most of the photosynthetic and many of the non photosynthesizing bacteria take up these elements (Nitrogen and Sulphur) in the oxidized inorganic state as nitrates and sulphates; thus their biosynthetic utilization requires a preliminary reduction. There are certain microbes which cannot bring about a reduction before uptake, and they have to be provided N_2 in a reduced form (ammonium salts). Reduced sulphur requirement can be met by inorganic compounds such as sulphides.

Nitrogen and Sulphur requirements can be met by organic compounds that contain these elements in the reduced state. Amino acids and other protein degradation products (Peptones), are the source of organic nitrogen and sulphur. Some of these compounds can also provide simultaneously the carbon requirements of the bacteria.

There are some unique bacteria which can use gaseous nitrogen of the atmosphere as their nitrogen source (Eg., *Clostridium, Azatobacter, Rhizobium* etc). Such bacteria are termed *Nitrogen fixers* and the process involves a preliminary reduction of N_2 to ammonia. Some of the cyanobacteria (*Nostoc, Anabena* etc) are also capable of biological nitrogen fixation.

Within the cell, phosphorous occurs as phosphate and phosphate esters. Phosphates occurs in the phospholipids of membranes and phosphate diesters from the backbone linkage of micromolecules such as nuclei acids. The role of phosphates needs hardly to be emphasized. They participate in all energy conversion reactions.

Other mineral requirements

In addition to the four major elements (C, N, P and S), microorganisms also need a large number of minerals in trace quantities. Most of these trace elements are required for the functioning of the enzyme systems. Each of these minerals perform a specific function and their presence in the medium in optimum concentration decides the growth of microorganisms. The following table gives a summary of the role of some minerals in microbial nutrition.

Some physiological functions of trace elements

Elements	**Functions**
Potassium	Cofactor of some enzymes
Magnesium	Cofactor in enzymatic reactions, binds enzymes to substrates, constituent of chlorophyll
Iron	Constituent of cytochromes and other heme and non-heme proteins, cofactor of many enzymes
Calcium	Cofactor for enzymes such as proteinases
Magnesium	Cofactor for some enzymes, can replace Mg in some cases
Copper, Zinc, Molybdenum	Inorganic constituents of some enzymes
Cobalt	Inorganic constituents of some enzymes

Role of Oxygen : Aerobic bacteria and other microorganis require molecular oxygen for their respiration (for details see chapter 1).

Growth factors

Organic nutrients, that cannot be synthesized by the microbe, but essential for its growth are collectively called growth factors. Most of the growth factors fall into one of the following three classes (Stainer et al, 1986) –

1. *Amino acids* – Required as constituents of proteins.
2. *Purines and pyrimidines* – they are necessary as constituents of nuclei acids.
3. *Vitamins* – These present a diverse group of organic compounds constituting the prosthetic group of some enzyme systems. The following is a table of vitamins and their role in microbial metabolism.

Table

Relation of some water -soluble vitamins to Coenzymes

Vitamin	Coenzyme	Enzymatic reactions involving the Coenzyme Form
Nicotinic acid (Niacin)	Pyridine nucleotide coenzymes (NAD + and NADP +)	Dehydrogenations
Riboflavin (Vitamin B_2)	Flavin nucleotides (FAD and FMN)	Some dehydrogenations, electron transport
Thiamin (Vitamin B_1)	Thiamin Pyrophiosphate (Cocarboxylase)	Decarboxylations and some group-gransfer reactions
Pyridoxine	Pyridoxal Phosphate	Amino acid metabolism Transmission Deamination Decarboxylation
Pantothenic acid	Coenzyme A	Keto-acid oxidation, fatty acid metabolism
Folic acid	Tetrahydrofolic acid	Transfer of one-carbon units
Biotin	Biotin	Co_2 fixation, carboxyl transfer
Cobalamin (Vitamin B_{12})	Various cobalamin derivative	Molecular rearrangement reactions

Growth factors are required by microorganisms in extremely minute quantities as they satisfy some specific requirements.

Nutritional forms

Microorganisms obtain nutrition by different methods. This constitutes the basis for their categorization into various types depending on the source of obtaining their nutrition. With reference to obtaining inorganic nutrients, all microorganisms obtain them directly from the environment in their finished form. However, with reference to the organic metabolites, there are various patterns by which microbes procure them from the environment. Basically from the point of view of the source of nutrition whether inorganic or organic, microorganisms can be divided two categories – *autotrophs* and *heterotrophs*. Autotrophic microorganisms can survive in an exclusively inorganic environment. Using this as the source of basic energy, their cellular machinery can synthesise all their organic nutritional requirements. Heterotrophic organisms on the other hand require an input of readymade organic food as they do not possess the synthetic machinery to produce organic metabolites by the reduction of carbon.

Autotrophic microorganisms, which can manufacture food from inorganic sources require two basic ingredients – 1. A supply of inorganic raw materials 2. A source of external energy to fabricate the inorganic materials into organic nutrients. Based on the mode of the input of the external energy, autotrophic microorganisms can further be classified into *photosynthesizers* (photoautotrophic), and *chemosynthesizers*. Photosynthesizing microorganisms utilise solar energy to reduce the inorganic carbon into organic carbon. Chemosynthesizing microorganisms on the other hand utilise the oxidation of inorganic rawmaterials as a source of external energy for the reduction of inorganic carbon.

Microorganisms can be classified either based on the method of food procurement (Autotrophic or heterotrophic), or they may be also classified from the point of view of the source of input of external energy (photosynthetic or chemosynthetic). A categorisation of microorganisms is also possible taking both the bases mentioned above as a criteria for classification. On this basis, the following categories of microorganisms can be identified.

AUTOTROPHIC MICROORGANISMS

All types of phototrophic microorganisms use the environmental carbondioxide as source of their inorganic carbon. This is reduced to carbonhydrates by the addition of hydrogen obtained from water or other sources. Autotrophic microorganisms are classified into following categories.

1. Photoautotrophs

Most of the pigmented bacteria and microalgae belong to this group. In all the photosynthesizing microbes, solar radiation is the basic energy source and to trap this radiation, one are more varieties of chlorophyll are present in the organisms.

In all photoautotrophic microbes, the source of hydrogen is water except in photoautotrophic bacteria.

In the case of photoautotrophic bacteria, the source of hydrogen is not water and as such oxygen is not a byproduct of photosynthesis. Bacteria such as these usually live in sulphur rich environments (sulphur springs), where hydrogen sulphide is normally in abundance. This substance serves as the hydrogen source for photoautotrophic bacteria.

Two groups of photoautotrophic bacteria have been identified, based on the type of pigmentation and other features. These are the **purple sulphur bacteria** and **green sulphur bacteria.** The purple sulfur bacteria possess a special type of chlorophyll known as *bacteriochlorophyll,* which is green colour but the colour is masked by yellow carotenoid pigments. Green sulphur bacteria possess the pigment *chlorobian* or *bacteriviridin.* This pigment is different from bacteriochlorophyll. Additionally, the green colour here is not masked by carotenoids.

Light ——— Special chlorophyll ——— Energy

S (byproduct)

H_2S ———

H_2

CO_2 ——— Carbohydrate

In the above mechanism of photosynthesis, elemental sulphur is the byproduct. It gets stored inside the cells in purple sulphur bacteria, and in the case of green sulphur bacteria it is excreted.

2. Chemoautotrophs

Chemoautotrophs microorganisms are entirely made up of bacteria. Generally, they are pigmentless and as such there is no mechanism in them to use the light as a source of energy. In addition to the input of inorganic carbon, they have a variety of inorganic sources whose oxidation provides the necessary energy. These inorganic metabolites, in most of the instances are combined with oxygen in the cells resulting in the release of energy and a variety of inorganic byproducts. In the subsequent food manufacture, water and carbon dioxide act as the inorganic raw materials. The general pattern of food manufacture in chemoautotrophs is as follows :

+0

Inorganic Metabolites ——— Inorganic Byproducts

Energy

O (byproduct)

H_2O ————————

H_2

CO_2 ———————————— Carbohydrate

Chemoautotrophic microorganisms have been classified into the following types, based on the type of inorganic compound whose oxidation serves as a source of energy. These are sulphur bacteria, iron bacteria, nitrifying bacteria, hydrogen bacteria etc.

Sulphur bacteria

These absorb either hydrogen sulphide or molecular sulphur from the environment and combine them with molecular oxygen. The resulting energy is used for reduction of carbon. Elemental sulphur in the form of granules is deposited. Sometimes, sulphate may be the byproduct which may be excreted, or it might become a part of the mineral content of the cell.

Iron bacteria

These bacteria oxidise iron compounds and convert them into insoluble substances, releasing energy during the process.

Nitrifying bacteria

There are two types of nitrifying bacteria–nitrite bacteria and nitrate bacteria. Nitrite bacteria use ammonia as the source whereas nitrate bacteria use nitrite ions as the source.

Hydrogen bacteria

These utilise molecular hydrogen as the source and combine it with oxygen releasing energy during the process.

Note : More details regarding the oxidation of inorganic metoblites by chemoautotrophic bacteria is given in the chapter on aerobic respiration.)

PHOTOHETEROTROPHIC MICROORGANISMS

These are regarded as the intermediate forms between phototrophs and heterotrophs. The characteristic feature of these nutritional forms is that the external energy source is light (hence, basically these are photosynthesizers). But they require the input of organic raw materials unlike in the case of autotrophic organisms. As such, these organisms are refered to as photoheterotrophs. A small group of photosynthetic bacteria namely **purple**

non sulphur bacteria are typically photoheterotrophs. Like the purple sulphur bacteria these bacteria also possess bacteriochlorophyll as well as the red and yellow cartenoids, masking the green colour.

Purple non sulphur bacteria absorb organic rawmaterials from the environment but these cannot be directly used as food. Instead, these organic materials only serve as a source of hydrogen and hydrogen released from this process is then used for reducing the carbon dioxide. During the oxidation of organic metabolites to release hydrogen, carbon dioxide is also released and these two combine resulting in the form of carbohydrates. The scheme of food production is as follows.

Light ———— Bacterio chlorophyll ———— energy

Organic byproducts

Organic rawmaterials ————

H_2

CO_2 ———————— Carbohydrate

Purple nonsulphur bacteria can manufacture food as shown above only if light is present and oxygen is *absent.*

Another kind of categorisation of the autotrophic microorganisms is possible when we emphasise two criteria (source of rawmaterial – inorganic or organic and source of energy – light or by chemical oxidation) simultaneously. On this basis, we can recognise four categories of microorganisms.

Photolithotrophs
These use light as a source of energy and inorganic materials provide carbon necessary for reduction into carbonhydrates.

Chemolithotrophs
Here, chemicals act as both the source of energy as well as the source of carbon. The energy for reduction comes from the oxidation of inorganic compounds.

Photoorganotrophs
In these microbes, light is the source of energy and reduced organic materials provide the source of carbon.

Chemoorganotrophs
These organisms use organic chemicals as a source of energy as well as directly as food.

The following table gives a summary of the classification of nutritional forms;

	Food	
	Manufactured from inorganic source	absorbed from prefabricated organic source
From PHOTOSYNTHESIZERS Light source PRIMARY ENERGY ENERGY	PHOTOAUTOTROPHS (PHOTOLITHOTROPHS) Purple sulphur bacteria Green sulphur bacteria Microalgae (except colourless ones)	PHOTOHETEROTROPHS (PHOTOORGANOTROPHS) Purple nonsulphur bacteria
From Chemical CHEMOSYNTHESIZERS	CHEMOAUTOTROPHS (CHEMOLITHOTROPHS) Sulphur bacteria iron bacteria Hydrogen bacteria nitrifying bacteria	CHEMOHETEROTROPHS (CHEMOORGANOTROPHS) Saprophytic bacteria symbiotic bacteria fungi Protozoa colourlesss micro-algae
	AUTOTROPHS	HETEROTROPHS

HETEROTROPHIC MICROORGANISMS

These are pigmenetless bacteria and eventhough they absorb H_2O and CO_2 from the environment, they cannot be utilised as a source of energy. Hence, they have to absorb readymade food from the environment and use it directly as food. This means the survival of heterotrophs depends on the presence of Autotrophic organisms.

Majority of the heterotrophic microorganisms are chemoheterotrophs in the sense their source of nutrition is readymade organic substances. Chemoheterotrophs are further classified into holotrophs, saptrotrophs and symbionts.

Holotrophs are free living bulk feeders

They ingest the food and get them digested in their body. Saprotrophs or saprophytes, live on dead and decaying organic remains and absorb the readymade food from them. Many of the bacteria and fungi belong to this category. These microorganisms decompose the organic matter and absorb the nutrient molecules directly. Thus saprotrophs are also called decomposers. As a result of their decomposng activity, saprotrophs constitute a vital link in the global nutriate cycles.

THE PHASES OF GROWTH

The growth of bacterial population may be studied by inoculating a nutrient solution with a small number of microorganisms. When all the nutritional and other parameters are ideal, the cells continue to multiply, until one or the other necessary factors reaches exhaustion and growth gets retarded. If during this period, the culture medium is not replenished with fresh nutrients or waste products (of microbial metabolism) removed, the growth in such a closed system is called **batch culture.** In other words, batch culture may be defined as 'growth of microorganisms in a limited volume of liquid medium, where there is no replenishment of nutrients nor is there any elimination of waste products. Such a closed system (culture) is very helpful in studying the growth characters of microorganisms. When bacteria are transferred to a known volume of culture, their population increase undergoes a characteristic sequence in the rate of increase of cell numbers. This increase obeys the same laws as does the growth of multicellular organisms. *Thus a batch culture behaves like a multicellular organism with its genetically determined stages or phases of growth.*

After the initial inoculation into the medium, increase in cell number may be ascertained by counting them in a specified aliquot at periodic intervals (every hour for the first 24 hours). When the logarithm of the counted cells (viable cells only) is plotted against a growth curve is obtained. Basically, this growth curve is sigmoid (as it is no universally), and exhibits various phases or stages of growth.

Growth curve

A growth curve is the graphic representation of the rate of multiplication of cells plotted against time. For all organisms including microbes the growth curve is sigmoid. There is an initial period which appears to have practically no growth, followed by a spurt in growth. This is followed by a stationery or steady state and finally a decline in the number of viable cells. Based on this behaviour, four stages or phases of growth may be identified. These are– **initial slow phase** or **lag phase; log phase** or **the exponential phase; the stationery phase** and **the death phase** or **the declining phase.**

While the above four phases are clearly recognizable stages, it has to be understood however, that there is no sudden shift from one phase to another in the entire microbial population. The change from one phase to another is gradual, as all bacterial cells will not be in the same stage of growth (except in synchronous cultures). As a result, there will be several intermediate phases between the four standard phases of growth. Taking this factor into consideration growth in microbes may be divided into the following phases :

1. Initial phase (lag phase)
2. Phase of accelerated growth
3. The log phase
4. Phase of decrease in growth rate
5. Stationary phase (the study stage phase)
6. Phase of increase in death rate
7. The declining phase
8. The survival phase

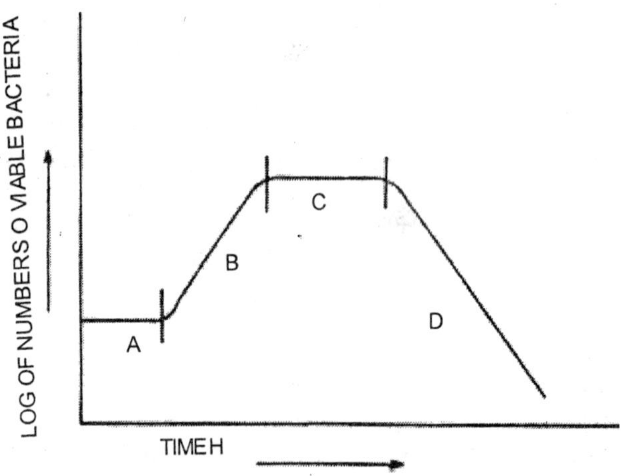

Fig. 13.1 Microbial Growth and Reproduction
The growth curve of microorganisms showing. **A.** Lag phase, **B.** Log phase, **C.** Stationary phase, **D.** Death phase

1. The lag phase

This phase is characterised by the initial period, where there is no increase in the number of cells. The lag phase is said to occupy the time interval between inoculation and establishement of maximum division rate. The actual duration of the lag phase depends on the previous culture history, the age of the culture, and the components of the nutrient medium. During the lag phase, there is an increase in cell size but there will not be any increase in cell number as there is no multiplication. Nevertheless, it is during this period that the cells are very active physiologically absorbing nutrition from the medium and reaching a phase that prepares them for cell division. Studies have indicated that there is maximum synthetic activity during the lag phase resulting in increase of total proteins, RNA, ribosomes, enzymes etc. There is also an increase in the amount of cellular phosphorous.

The inoculum that has been obtained from the previous culture will have cells that would have adjusted themselves to the growth conditions existing in the old medium. After transfer to the new medium, the cells require a little time to adjust themselves to the new physical and nutritional environment.

It has already been pointed out that the duration of the lag phase depends on the stage of growth of the cells in the inoculum in the previous culture. If the cells of the inoculum are obtained from a culture growing in the log phase, the culture displays very little lag phase in the population increase. On the other hand, if the inoculumis obtained from a culture of dormant cells, the lag phase will continue for a longer period.

On the new culture medium, the nutritional environment many a time requires the synthesis of new enzymes to breakdown specific substrate material. A typical example of substrate induced enzyme synthesis is a phenomenon called **diauxie** (biphasic growth). The appearance of biphasic growth or a double growth cycle is found in media which are complex, and have a mixture of several substrates. For instance in a mixture of glucose and sorbitol, *Echerichia coli* will utilise glucose first, because glucose will induce the synthesise of those enzymes required for its digestion and simultaneously suppress the synthesis of enzymes required for sorbitol utilisation. Sorbitol degrading enzymes are only produced after all the glucose has been metabolised. Hence, in these cultures the two lag phases can be explained by this regulatory mechanisms.

The most notable change in the chemical composition of bacteria during lag phase is the 8–12 fold increase of the RNA concentration.

2. Phase of accelerated growth

This is said to be an intermediate phase between the lag phase and the log phase. At the end of the lag phase after the cells are sufficiently physiologically mature, division sets in. All the cells, however, will not start dividing simultaneously because all of them would not have completed the lag phase at the same time. A large number of cells however would have completed the lag phase. Gradually, however, the cells reach physiological maturity and start dividing. On plotting the growth rate, we will observe that the rate of multiplication increase with time. Finally, the rate of multiplication reaches a maximum at the end of this phase. This is said to be the transition phase between the lag and the log phases. It has been observed that during the transition phase the cells are extremely sensitive to alterations in growth conditions like temperature, osmotic pressure and disinfectant chemicals.

3. The log phase (exponential phase)

This phase is characterised by the division of the cells at a constant rate. This constant and steady division results in the doubling of the population. At this stage, the log of the number of cells plotted against time will show a straight line indicating a geometric increase. The cells are said to be in a balanced growth stage, and the mass and volume of the cells increase by the same factor in a manner that the composition of the cells and the concentration of metabolites remain constant for some period of time.

During this phase, cells are actually smaller in size as they are constantly dividing. Physiologically, the cells are said to be very young and actively multiplying. With reference to the physiological activity, the entire population of cells seem to be in a uniform state.

The multiplication of the cells and the doubling time actually depends on the organism in question and the prevailing growth conditions. Enterobacteria can double in a time of 15–30 minutes; *E.coli* growing at a temperature, 37°C can double in 20 minutes. On the otherh and *Nitrosomonas* and *Nitrobacter* have a doubling time of 5–10 hours.

The log phase is a most suitable period for studying biochemical and physiological properties of cells, because in most cells the metabolic activity will be at the peak. Like in the previous phase, the cells are extremely sensitive to the influence of environmental factors such as pH, temperature, aeration, concentration of substrates etc.

4. Phase of decrease in growth rate

During this phase, the rate of cell division comes down and the cells do not multiply with the same rapid sequence as during the log phase. The retardation of the rate of cell division is due to a number of factors. The most important being the gradual decrease in the quantity of available nutrients and the accumulation of toxic waste (product of microbial metabolism). This is a transition phase before the cell division stops completely.

5. Stationary phase (the steady stage phase)

This phase is said to begin when the cells can no longer divide as the amount of available nutrient has almost reached the minimum. The growth rate of the culture gradually gets decreased before all the nutrients get exhausted. In this phase, a high population from the limitation of substrate, other factors such as very high concentration of cells, low partial pressure of oxygen and accumulation of toxic waste products also lead to the initiation of the stationary phase.

At the cellular level during the stationary phase, reserved food materials get consumed, a proportion of ribosomes may be degraded and enzymes may still be synthesised. A viable population count at this stage shows no change.

In fermentation technology involving microbial metabolism (penicilin production), the stationary phase is the real production phase. In biotechnology, this phase is said to be the *idiophase* (production phase). The bacterial mass that is produced at the time of the stationary phase is known as the yield.

6. Phase of increasing death rate

During this phase ,there is a gradual decrease in the number of the viable cells with increase in time. Due to a number of factors, the death of the cells gradually increases and reaches a maximum at the end of this phase. This is a transition phase between the stationary phase and the declining or the death phase.

7. The declining phase

Also known as the logarthemic death phase, during this phase the number of cells decreases exponentially, that is 50% of the surviving cells die in each successive equal time interval. For example, if a population decreases from 10 lakhs to 5 lakhs in the first hour, in the next hour it decreases to 2.5 lakhs, and in the next hour it comes down to 1.25 lakhs, and so on. Thus, here the rate of decline in the number of cells is constant, and it can be calculated using the same formula as for the calculation of logarthemic increase. The calculation indicates the time required to reduce the population by half. This calculation is very useful while considering the germicidal effect of an agent.

The death phase and the cause of the bacterial death in the normal nutrient media have not been thoroughly investigated (Schlegel, 1993). A variety of conditions, however, are known to cause bacterial death. The death rate of bacteria is not uniform just as in the case of cell increase. Some species such as *Neisseria gonorrhoea,* die rapidly as the cells are very susceptible to autolysis. Autolysis is known to be brought about by enzymes called autolysins.

8. The survival phase

This is also known as the decrease in growth rate. During this phase the rate of the death of the cells ultimately reaches an equilibrium, wherein the rate of growth and the rate of death finally balance each other and a very low population of cells is maintained. This stage is called the survival phase. Survival depends on type of organism and other environmental conditions. Some microbes die off within 4–5 days while others might remain viable for months and even years. The formation of endospore at this stage gives an unique opportunity for the cells to provide for very long periods even after the all vegetative cells are eliminated. This is also a stage where, perhaps the mutant forms induced recently find the conditions favourable for rapid growth and get selected.

Growth curve parameters

When the growth of microbes in a batch culture is followed by dry weight determinations, the growth parameters of interest are the yield, the exponentail growth rate and the duration of the lag phase.

Yield may be defined as the difference between the initial and the maximum bacterial mass. It is expressed in terms of grams of dry weight. Of special importance here is the relationship of the yield to substrate consumption. When both these values are expressed in weight units, the ratio is said to be the yield co-efficient.

The exponential growth rate is a measure of the speed of cellular growth in the log phase. It is calculated from the bacterial concentration at initial and final time.

The lag phase like the exponential growth rate is an important parameter for considering the properties of the organism and the suitability of the medium for the bacterial growth.

Factors influencing the bacterial growth (Refer Chapter 1)

Continuous culture of microorganisms

In a batch, which is the ideal medium for the study of growth phases, the cultural conditions undergo continuous change. The bacterial growth curve rises and substrate concentration comes down. As the bacterial growth curve indicates, there are a number of transitional phases between the standard growth phases indicating not all the cells are in the same physiological condition.

There are cells on one extreme and there are dying cells on the other. For many, physiological conditions and the state of the microbes vary. It is ideal, however, for the purpose of metabolic study of an organism, all the cells are in an identical state of growth. In other words, it is desirable to maintain a microbial population in the log phase of the growth in a constant environment. This can be accomplished by a technique called continuous culture technique wherein, there is constant addition of new growth medium and concomitant withdrawl of equal volume of the bacterial culture. By eliminating a quantity of bacterial population and at the same time adding fresh nutrient medium, the growth state of the population is maintained at a particular stage.

A variety of methods have been evolved to maintain a continuous culture of microorganisms for the purpose of study. Two such devices are the *chemostat* and the *turbidostat.*

Chemostat

The chemostat is a mechanism for the continuous cultivation of the microbes. The system consists of a culture vessel and a reservoir. In the culture vessel, the medium is thoroughly mixed to obtain maximum homogeniety. Fresh nutrient medium is supplied to the culture vessel from a reservoir of sterile medium at a constant predetermined rate. The volume of materials in the culture vessel is kept constant by a device which allows for the withdrawal of some amount of culture medium alongwith accumulated waste products and old or dead cells. At the same time, fresh culture medium is added, so that the level is maintained constantly.

The level of growth is maintained by limiting the concentration of a particular nutrient in the medium. The growth limiting nutrient is added to the medium at such a concentration below that required for maximum growth in a batch culture.

The steady rate of bacterial population in a chemostat is maintained by a delicate balance between the growth rate and the dilution rate. Dilution rate may be defined as the ratio of inflowing medium per hour to the volume of culture. In the chemostat, the maximum stability is attained within a range of dilution rates where the density changes get altered very slightly at low dilution. As a result, the change in the bacterial mass is O and the bacterial concentration remains constant. In other words, the exponential increase in cell mass is balanced by a negative exponential process of the same magnitude.

Calculation of growth in a chemostat

As has been explained above, constant growth is maintained in a chemostat through a balance between outflow and inflow. The rate of loss of cells through overflow can be expressed as

$$\frac{dM}{dt} = \frac{F}{V} = DM$$

Where F=flow rate, V=culture volume. Here, flow rate is measured in culture volumes per hour. The expression F/VM is called the dilution rate (D). In a continuous culture,

KM=DM

K=D

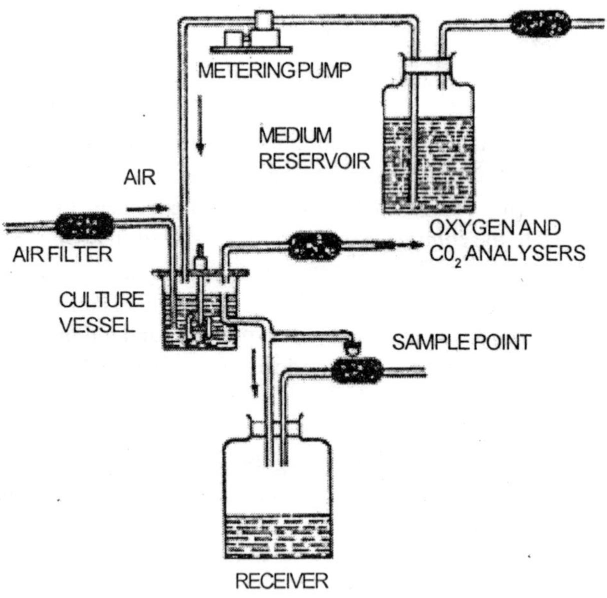

Fig. 13.2 Microbial Growth and Reproduction
Continuous culture of bacteria using a chemostat

In the above, the growth rate (K), equals the rate of dilution (addition of sterile nutrient), in a continuous culture. Under such conditions, the growth of the culture is rather linear than exponential, while the multiplication of individual cells is exponential.

Turbiodostat is another kind of a continuous culture technique. In this, the system has an optical sensor which measures the optical density in the growth vessel. Changes in the density of the culture medium retard the passage of the light through the culture. These changes automatically activate the mechanism that regulates the inflow of the nutrient and the outflow of the waste from the culture vessel. Turbiodostat is a more precise and sophisticated technique than the chemostat as maximum sensitivity and stability are achieved through optical measurements.

Dialysis technique is another device which can maintain the culture in the logarthmic phase for longer periods.

Continuous culture systems prove very beneficial in research studies as they can provide a constant source of cells at a particular stage of growth. Additionally, the system also allows the cells to be grown continuously in limiting concentration of nutrients. Such a kind of growth provides information while studying a particular organism, as to the nature of the limiting substrate. Such a system is very ideal to isolate a specific organism that catabolizes a particular nutrient. Combining this with enrichment cultures, specific microbes may be easily isolated.

Synchronous cultures

These are composed of populations of cells that are in identical stage of growth in their life cycle. In this culture, all the cells in the medium will divide simultaneously; will grow into cells and divide again in the next generation at the same time. Thus, the entire population of microbes is uniform with reference to their growth phases. Synchronous cultures are of great advantage to study the various factors of microbial growth like organisation and differentiation as they provide a large amount of identical material. These analysis are difficult to study in a single microbial cell. Measurements of growth parameters conducted on this population can be simply applied on single individual cells also.

A number of techniques are employed to obtain synchronous cultures. These include physical separation of cells, manipulation of culture, by varying chemical composition etc.

Growth rate and generation time

Bacteria multiply asexually in large numbers mainly by means of binary fission. In binary fission, which is an amitotic division, each cell divides to produce two daughter cells and they in turn divide to produce 4 cells, 8 cells and so on. Thus, the multiplication corresponds to a geometric progression as shown below :

$2^0 - 2^1 - 2^3 - 2^n$ - etc.

In other words, the cells increase as 1—2—4—8—16 etc.

Generation time has been defined as the time required for the cell to divide to produce two cells or for a population to double itself. The generation time or the time required for doubling of the population varies with the type of organism and also with the cultural conditions. The table gives a comparative generation time for some of the microorganisms.

Organism	**Temperature (^{0}C)**	**Generation Time (Min)**
Bacillus stearothermophilus	60	11
Escherichia coli	37	20
Bacillus subtilis	37	27
Bacillus mycoides	37	28
Staphylococcus aureus	37	28
Streptococcus lactis	37	30
Pseudomonas putida	30	45
Lactobacillus acidophilus	37	75
Vibrio marinus	15	80
Mycobacterium tuberculosis	37	360
Bradyrhizobium japonicum	25	400
Nostoc japonicum	25	570
Anabaena cylindrica	25	840
Treponmea pallidum	37	1980

Mathematical expression of generation time

When the number of cells in the initial inoculum as well as the growth conditions in the medium are known, it will be possible to calculate the time required for a population to double itself. The following data are necessary to calculate the generation time.

1. The initial number of cells at the beginning of growth.
2. The number of cells present at the end of a specific time period.
3. The time interval.

Using the above data it is possible to calculate the generation time for any given population with the help of a series of equations. If we consider that the experiment starts with a single cell, the total population (M) of a bacterium at the end of a given period can be expressed as follows :

$$M = 1 \times 2^n$$

In the above 2^n indicates the quantity of the bacterial populationo after *n* number of generations. Generally, however, the experiment not given with a single cell, and it has hundreds of cells. Hence, the expression may be modified as follows.

$$M^n = M^0 \times 2^n$$

Here M^0 = number of organisms initially; M^n = number of organisms after *n* generations and *n* = number of generations. To solve the equation for n, we should calculate as follows :

$$\log M^n = \log M^0 + n \log_2$$

$$n = \frac{\log M^n - \log M^0}{\log_{2},}$$

From the above equation we can calculate the number of generation if we know the initial population and the final population after a specified time *T*. The generation time *G* = T (the time elapsed between initial population Mo and Mn.) divided by the number of generations n as follows :

$$G = \frac{T}{n} \quad \frac{T \log_2}{\log M^n - \text{Log } M^0}$$

An organism in a balanced growth would show a rate of increase which is proportional to the number or mass of microorganisms present at that time. The rate of increase of cells is an index of the growth rate andit is called the exponential growth rate constant (*K*). *K* is defined as number of doublings of population in unit time and it is usually expressed as the number of doubling per hour. Using this, an alternative method of calculation of bacterial growth may be shown as follows :

$$M^n = M^0 \times 2^{KT}$$

M^n = Population at time T

M^0 = Population at time 0

If we take the logarithms $\log M^n = \log M^0 \times KT \log_2$ and, to solve the equation K we have to calculate as follows :

$$K = \frac{\log M^n = \log M^0}{T \log_2}$$

The above equation indicates the exponentail growth rate constant (K) is reciprocal to the generation time, i.e.,

$$G = 1/K$$

For *E.coli*, the generation time is 30 minutes i.e., there are two doubling for every hour (generation time is usually calculated as number of doublings per hour i.e., if 1/2 = 1/K then K = 2 doulbings per hour).

Methods of measuring microbial growth

Measurement of microbial growth and physiology is an important parameter of microbial study. This measurement to a great extent depends on the quantity (number) of microorganisms present in a culture. Therefore, measurement or quantification of microbial population is a must for any meaningful assessment.

Growth of microorganisms can be quantatively measured using a variety of techniques. These techniques fall into the basics categories mentioned below.

1. ***Determination of cell numbers***

    A. Total count (Non viable count)

    i. Bread smear method
    ii. Haemocytometer method
    iii. Proportional count method
    iv. Electronic counter method

    B. Viable count (Indirect count)

    i. Plate count
    ii. Membrane filter count
    iii. Serial dilution count
    iv. Drop technique
    v. MPN count

2. ***Determination of cell mass***

    A. Direct method

    i. Dry weight measurement
    ii. Measurement of cell nitrogen weight

    B. Indirect method

    i. Turbidimetric method

3. ***Determinationo of cell activity***

    A. Measurement of biochemical activity (Indirect method)

TOTAL COUNT (Non viable count)

Total count of microorganisms in any given suspension indicates the total number of cells which includes both living and dead. Hence, it is called a non viable count. The following are various methods of total count.

Bread smear method

In this method, a known volume of (0.01ml) microbial suspension is spread uniformly over a clean, sterile glass slide within a fixed area of a one square cm.

The smear is then fixed to the slide and stained. After placing a cover glass over the smear, it is observed under oil immersion objective and the cells are counted in a microscopic fields. Several microscopic fields are counted and an averages is taken. The total cells per square cm is then calculated by determining the number of microscopic fields per square cm. The total count may then be calculated as follows.

1. Area of microscopic field = πr^2
   r (oil immersion lens) = 0.08mm
   Area of the field under oil immersion lens
   $\pi r^2 = 3.14 \times (0.08\text{mm})^2 = 0.02$ sq cm.
2. Area of smear 1sqcm = 100 sqmm
   Therefore no of microscopic fields 100/0.02 = 5000
3. Average number of bacteria per field

say 50x5000 = 25,0000 cells per square cm (i.e., number of cells per 0.0001 ml of suspension)

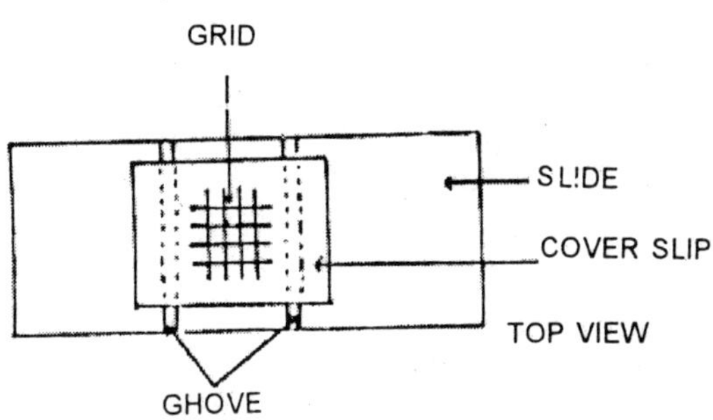

Fig.13.3 Microbial Growth and Reproduction
Diagrammatic representation of Haemocytometer

Haemocytometer method

Also called counting chamber method or the Petroff - Housser chamber method, this consists of the use of a special microscopic slide. This slide is also called haemocytometer as it was originally meant to count the blood cells.

Petroff-Hausser chamber consists of

1. A thick glass slide which has a depression in the centre having a groove on its both sides.
2. A cove slip

The centre of the depression consists of a specific area divided into 9 squares measuring 1 sqmm. The centre square is further divided into 24 smaller squares each with a side of 0.2 mm. Each of these squares is again divided into 16 squares covering an area of 0.0025 sq mm. Counting is done in this small square under the microscope. The procedure for counting is as follows –

1. Clean both the slide and coverslip to make it free of all dust particles.
2. Place a known quantity of dilute bacterial suspension with the help of a pipette syringe.
3. Place the coverslip firmly on the suspension.
4. Allow the suspension to settle. Place the slide on the stage of the microscope.
5. Count the number of cells in the small squares. Take an average of 50 squares.

Calculation of cell number

(a) The area of the square is 0.05mm x 0.05mm = $0.0025mm^2$

(b) The volume of each square is 0.0025 sqmm x 0.1mm = $0.00025mm^2$

(c) Average count per square (say 100)

(d) The sample is diluted 10^4 (10,000 times)

The number of cells in the original square may be calculated as follows

100 x 1/0.00025 x 10,000 cells = 6 x 10^6 cells.

Proportional count method

Standard suspension of materials with a known volume (tiny plastic beads) is taken and mixed with an equal amount of cell suspension. The mixture is spread on the slide and stained. The particles and cells in the microscopic field are counted. A number of fields are counted to take the average. If an average of 5 particles and 20 cells per field is obtained, if the number of particles is one ml of standard suspension is 10,000, then the number of cells per ml of suspension is 20/5 x 10,000 = 40,000 cells/ml.

Electronic counter method

In this method, a special instrument called the coulter counter is used for direct counting of cells from a suspension. The instrument can count thousands of cells in a few seconds. But this system has a disadvantage in that, the counter counts even dust particles. Hence, the suspension must be absolutely free of any foreign particles.

Advantages of total (nonviable) **count**

1. Simple and rapid
2. Morphology of cells also may be studied

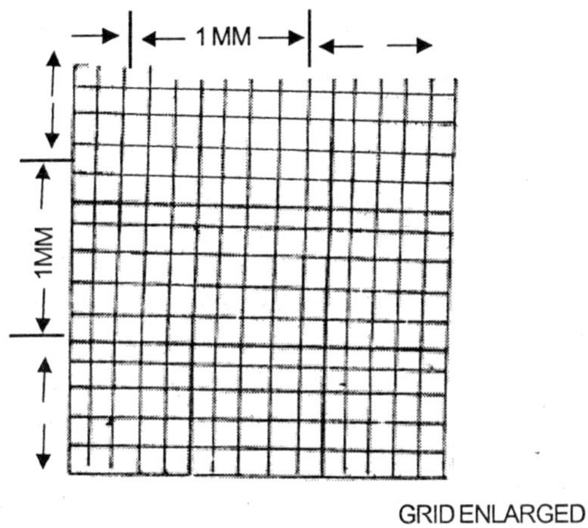

Fig. 13.4 Microbial Growth and Reproduction
A haemocyometer magnified

Disadvantages

1. Includes living and dead cells
2. Accuracy might decline, if the suspension is dense
3. Statistical errors are possible

VIABLE COUNTS

In viable counts, only the living bacterial cells are taken into consideration. The principle behind viable count is that all viable cells or spores, under suitable growth conditions multiply and that (as in an agar medium) each cell or spore forms a colony. The number of colonies therefore is the same as the number of viable cells present in the original sample.

In viable count methods, a knwon quantity of diluted sample is taken in such a way that the number of colonies that develop on an agar plate will be in the range of 30–300.

Viable count techniques are of the following types.

Standard plate count

Also called the spread plate technique, in this method appropriately diluted suspension is inoculated on a series of agar petridishes. The procedure is as follows.

1. A series of ten fold dilution of the bacterial suspension is prepared (10^2 to 10^{-10}).
2. With the help of a sterile Pipette, take 1 ml of the suspension and allow 0.1ml of it to be placed on the melted agar medium. Prepare some replica plates in the same way for this concentration.
3. Repeat the process for dilutions 10^{-4}, 10^{-6}, 10^{-8} and 10^{-10} also, use fresh sterile pipettee for each dilution.
4. Mix the inoculum thoroughly with the melted agar medium by slowly rotating the plates. Allow the medium to solidify.
5. The plates are to be incubated at 30°C for two days in an inverted position.
6. Using a colony counter, count the number of colonies that develop on the plate.
7. If the suspension is of fungus, colonies can be counted visually without a counter.

To facilitate easy counting, the plate may be divided into four sectors by marking on the underside of the plate using a glass marking pencil.

Calculation

For the purpose of calculation, a specific dilution level (say 10^{-4}) may be taken as colony number varies with each dilution. A plate with colonies ranging between 100–300 is deal for the purpose of calculation.

If dilution 10^{-4}, has 200 colonies developed on the plate, the number of cells in the original sample may be calculated as follows.

$$200 \times 10{,}000\ (10^{-4}) = 20{,}00{,}000 \text{ cells } 1/1\text{ml} = 2 \times 10^6 \text{ cells/ml}$$

As each colony develops from a single cell in the original suspension, the number of cells must be 2×10^6.

The plate count has some disadvantages. If the original suspension has a mixture of microbes, not all of them may grow properly on a standard medium. If the suspension contains aggregates of cells, the number of colonies developed may not exactly indicate the number of cells, as many cells may develop into a single colony.

The plate count method is used for the estimation of microbial population in milk, water, soil etc. Plate count method is, however, eminently suitable to count the living cells.

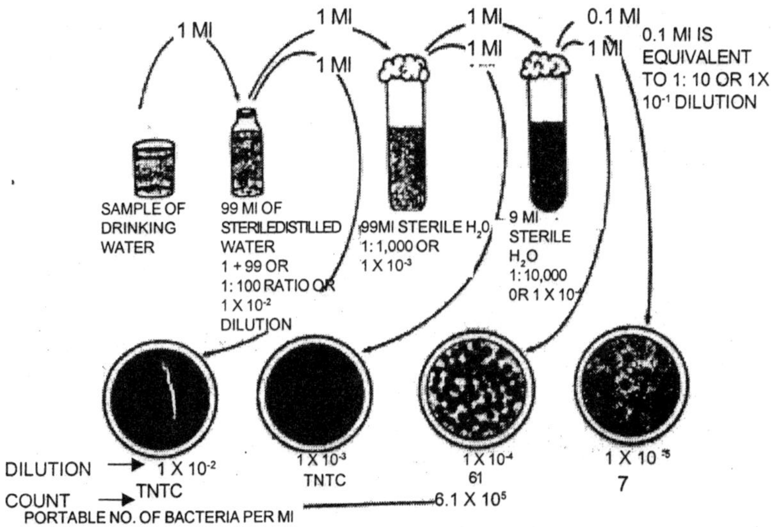

Fig.13.5 Microbial Growth and Reproduction
The viable count technique (standard plate count)

Membrane filter count

This method follows the same principle as in a standard plate count. A diluted suspension of microorganisms is filtered through a millipore filter. The microbes are retained on the filter disc. The filter disc is placed on a culture medium in a petridish. The plates are incubated and the colonies are counted on the filter disc. Following the same method of calculation given for standard plate count, the number of clells may be determined.

Membrane filter method has many advantages over the plate count. A large volume of sample may be analysed when organisms are few in number. This method is better when a variety of microbes are to be analysed. Filter discs with microbes may be kept in different nutrient media suitable for different microbes.

Serial dilution method

In this method, a suspension of microbes is diluted sequentially and plated on agar medium. (For details see Chapter 1. Culture of microbes).

Drop technique

This is an ideal method where a comparison can be made of the number of colonies that develop on a single plate with different dilutions. This is also of help when frequent countings are necessary.

The procedure is as follows.

1. Prepare a series of dilutions of the microbial suspension.
2. Take 2 sterile petridishes (2 pairs), and pour into them melted agar medium. Allow it to dry.
3. Divide each plate into many sectors by marking on the lid with a glass marker. Divide as many sectors as you have dilution levels (10^{-1} to 10^{-6} etc).
4. Take a Pasteur Pipette and find out the volume of drop of calibration. The drop volume is calculated by the formula.

$$V = \frac{2rj}{dg\,\theta}$$

Where V = volume of the drop
r = radius of the pipette
j = surface tension of the liquid
g = acceleration of gravity
θ = correction factor of the neck of the drop

In the above, r and g are constant, and v and θ are almost constant. The value is affected by the volume of j. To keep the value of j constant, the inner side of the pipette must be kept absolutely clean. In order to calibrate, such a quantity of liquid, deliver the solution drop by drop into a measuring cylinder. Count the number of drops required to make the volume to one ml. Suppose, 100 drops make one ml then, each drop has 0.01ml

5. Arrange the plates, take a particular dilution, the lowest and put a drop of the suspension on the sector marked 10^{-1} (by lifting the lid) dilution. Thoroughly clean the pipette (or use a separate pipette) and repeat the process for other dilutions. (10^{-2}, 10^{-4}, 10^{-5} and 10^{-6})
6. Allow the drops to settle and dry. Invert the plates and incubate them at 30°C for two days.
7. Count the number of colonies, as follows.

The volume of each drop is 0.01ml., if the sector at dilution 10^{-4} has 50 colonies, the original sample should have ($50 \times 1/0.01 \times 10^{-4}$) 5×10^{-7} cells/ml.

Most probable number (MPN) count

Also known as the multiple tube method, it is used for estimating the number of viable bacterial cells in a liquid medium. In this method, a set of culture tubes are inoculated with a fixed volume of the suspension (say 5ml for first tube, 4ml for second, 3ml for third tube etc), with the series of tubes getting progressively smaller volume of suspension. Tubes are incubated and observed for bacterial growth. It is assumed that growth will take place in any tube which receives at least one living cell. It should be understood that if the original sample has very few bacteria, then very few series of tubes show positive growth as the series with greater dilution will dilute them out.

For instance, three sets of culture tubes may be taken as follows :

First set - 1ml inoculum
Second set - 1ml inoculum
Third set - 0.1 ml inoculum

The above is said to be triple dilution series. The quantity of the culture medium, however, is same in all the tubes. After incubation, if

1. All the three tubes of first set show growth
2. Only two tubes of second set show growth
3. Only one tube of first set shows growth

The series may be expressed as 3–2–1. This can be checked for MPN from a standard table of MPN. The number may also be calculated as per the Thomas formula.

$$\text{MPN/100ml} = \frac{\text{Number of positive tubes}}{\sqrt{(\text{ml of sample in negative tubes}) \times (\text{ml sample in all tubes})}}$$

DETERMINATION OF CELL MASS

This is a method in which, the weight or mass of the cells is estimated as an indicator of increased growth. There are two methods of measuring the mass direct method in which the cell mass is actually weighed, and an indirect method in which cell mass is determined indirectly as a function of optical density (turbidimetric method).

Direct method

1. Measurement of dry weight

This is a simple and direct method of measuring the cell mass. The culture suspension is centrifuged and the pellet is repeatedly washed to remove all foreign particles. After drying, the pellet which consists of cells is weighed. This method is useful in large scale operations such as industries, where a bulk amount of microbial growth is involved.

2. Measurement of cell nitrogen

A major chemical constituent of cell is protein, of which nitrogen is an important ingredient. A microbial growth can be assessed in terms of the quantity of nitrogen content. The cells are to be obtained by centrifugation as mentioned above, and the cell nitrogen is estimated by chemical analysis. This method is useful with dense cell suspensions where the amount of growth is large.

Indirect method

Turbidimetric method

This is an indirect method of measuring cell mass by measuring the varying intensity of light as it passes through a cell suspension. The apparatus used for this is a photocolorimeter or spectrophotometer. A colonimeter functions on the principle, that subject to certain limitations, the light absorbed or reflected by a bacterial suspension is directly proportional to the concentration of cells in culture. Thus, by applying nepholometry (measurement of reflected rays), or turbidimetry (measurement of percentage light absorption). (For more details see Volume I College Microbiology by the same author).

For turbidimetric estimations, a photocolorimeter is used. This instrument has a source of monochromatic light (light of a single wavelength), which is facilitated by a filter that allows only the desired wavelength of light to be transmitted. The light ray passes through the culture and the amount of light transmitted is measured by means of a photoelectric cell connected to a glavanometer. The percentage of light transmitted/absorbed is usually expressed in terms of optical density (OD), which is directly proportional to the cell concentratin. OD is a function of the negative log of the percentage transmission and is expressed as 2-logG. Therefore OD = log 100 - log of galvanometer reading.

The value of OD reading of a bacterial suspension is only a relative value in that it does not directly indicate the number of cells/mass of bacteria. In order to interpret the OD value in terms of cell mass/number, it has to be compared with a standard curve which indicates the direct relationship of OD and mass.

A standard curve can be prepared by obtaining a series of dilutions of a bacterial suspension. A suitable aliquot from each suspension is then taken and the cell mass (dry weight) or number is measured by plate count or any suitable method. The same dilution level is now used to measure the OD. By measuring the OD and corresponding cell mass/number for series of dilutions, a direct relationship between the two sets of data may be established. This is done by plotting the data on a graph (mass/number on X axis and OD on Y axis), wherein a linear standard curve is obtained.

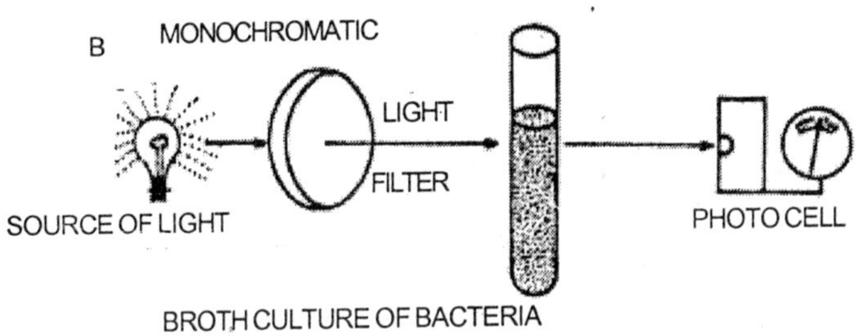

Fig.13.6 Microbial Growth and Reproduction
Method of turbidimetric estimation of bacterial population

Cell mass/ number of an unknown sample can be determined by taking the OD and compmaring it with the corresponding value on the standard curve. Turbimetric measurements, however, are very accurate only with moderate density. If a suspension is very dense it has to be diluted and the dilution factor must be taken into consideration while calculating cell mass.

Disadvantages

1. If the suspension contains any colouring matter, it effects light transmission and the absorbancy of light is not just due to cell concentration.
2. Turbidity measurements are not useful for viable counts as they measure both dead and living cells.

DETERMINATION OF CELL ACTIVITY

Metabolic products produced from a bacterial cell mass can be quantified and expressed in terms of cell mass quantity. The logic behind this is, more number of cells the more the quantity of metabolic product in question. This is an indirect method of assessment of cell mass. For instance, the amount of acid produced in a medium of cell suspension is a measure of the concentration of cells.

Sometimes concentration of specific enzymes also may be assayed to find out the cell mass.

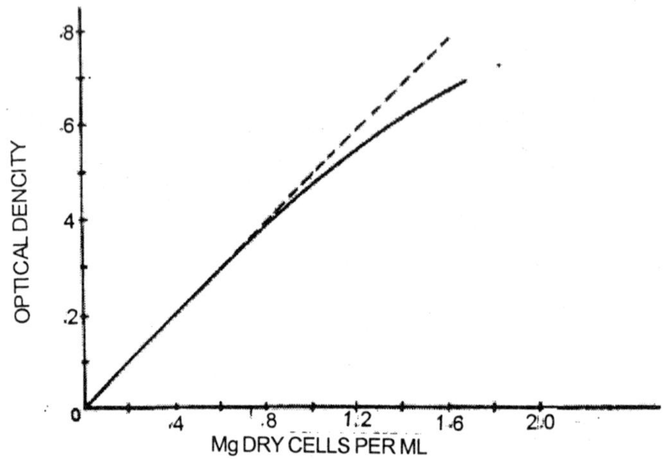

Fig. 13.7 Microbial Growth and Reproduction
Graph showing standard curve and dry weight curve using photocolonimetric estimation (broken line represents standard durve)

REPRODUCTION IN BACTERIA

Bacteria reproduce asexually as well as sexually. Asexuall reproduction takes place by the following methods. 1. Fission (cell division) 2. Budding 3. Endospore formation.

Cell division Generally bacterial cells do not divide mitotically as is true for all prokaryotic cells. There are three aspects of cell division called binary fission. These are 1. DNA duplication 2. DNA partitioning 3. Cross wall formation.

The cell division is completed by the doubling of all the cell components and their precise distribution and partitioning between the two daughter cells. Unlike in eukaryotic cells, there is no spindle formation and no resolution of chromatin into chromosomes.

Electron microscopic studies have revealed the following details in bacterial cell division.

DNA duplication. The two strands separate and each strand then replicates a new strand.

DNA partitioning. This involves the distribution of DNA between the two daughter cells. As the DNA is free and not condensed into chromosomes, it is a complicated task to assure equal distribution of genetic material between the two daughter cells. It is believed that (Kleppe et al, 1979) the DNA molecule attaches itself at some point to the plasma membrane and after duplication, one strand swings towwards one half of the cell keeping itself in the same attached position. Both the havles of the cell receive one double stranded DNA each. The attachment of DNA to the plasma membrane is to assure proper distribution without entanglement.

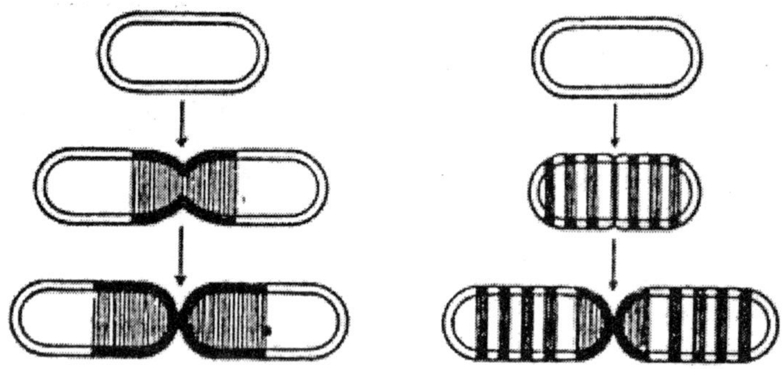

Fig. 13.8 Microbial Growth and Reproduction
Diagrammatic representation of wall formation in gram positive (left) and gram negative (right) bacterium

Some recent studies have indicated that mesosome present in the central portion of the cell plays a significant role in the distribution of DNA to daughter cells. Electron micrographs have shown the attachment of DNA to the mesosome. Along with DNA, the mesosome also bifurcates and each bit of mesosome carries along with it a double strand DNA molecule that ultimately goes to a daughter cell.

Cross wall formation

The cytoplasmic membrane of the bacterial cell grows inward in the middle part of the cell. Mesosome (Chondriod), an intracellular membranous structure continuous with the cytoplasmic membrane plays an importnat role in the formation of septum. The mesosmes of gram positive bacteria are more distinct and may appear as parallel or concentric lamellae, or as vesicles or tubules which get interconnected. Studies have indicated to the contribution of mesosomes towards the synthesis of new membrane. This is followed by the growth of the cell wall and finally the parent cell gets divided into two daughter cells.

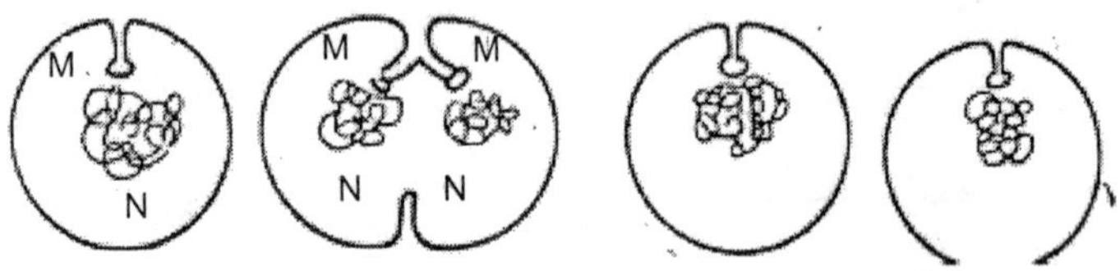

Fig.13.9 Microbial Growth and Reproduction
The role of mesosome in cell division of bacteria

The formation of the septum starts below the equatorial ridge in the cell wall. The newly synthesized cell wall is plastic enough to undergo changes in shape to achieve the final configuration. Hydrolase enzyme is known to bring about the separation of two daughter cells in *Bacillus subtilis* and *Staphylococcus facalis* (Peberdy, 1980).

Mesosomes of gram negative bacteria are smaller and less intricate. They are often seen as small invaginations of the cytoplasmic membrane. Cross wall formation has been studied in detail in *Escherichia coli,* a typical gram positive bacterium. Here, the cell consists of a middle peptidoglycan layer between outer membrane and cytoplasmic membrane. At the site of formation of septum a bud or fold originates very close to the mesosome. This is followed by the invagination of peptidoglycan and cytoplasmic membrane. Continuation of the invagination process ultimately leads to the formation of the cross septum.

Budding

In the process, a protrusion first develops, into which cytoplasm migrates. The nucleus splits into two and one bit enters the protrusion. This is later cut off from the cell by a cross wall. This is the bud. At maturity, this separates from the parental cell and develops into a new individual.

Endospore

Many of the bacterial like *Bacillus, Clostridium* etc., form endospores at certain stages in their life cycle. Strictly, endospore formation cannot be called reproduction because only one spore is formed per cell. Endospores are highly resistant to environmental conditions and may be regarded as an attempt by bacteria to withstand the adverse environmental conditions. Endospores are so highly resistant to desiccation, chemicals, temperature and radiation that they are vialable even for centuries. Some ensospore even withstand boiling temperature also for a long time.

As the spores are formed inside the parental bacterial cell, they (spores) are called endospores. On germination, each endospore develops into a single bacteria cell. The spore wall assures the endospore high resistance due to the presence of compounds like diplolinic acid, calcium and a very low water content.

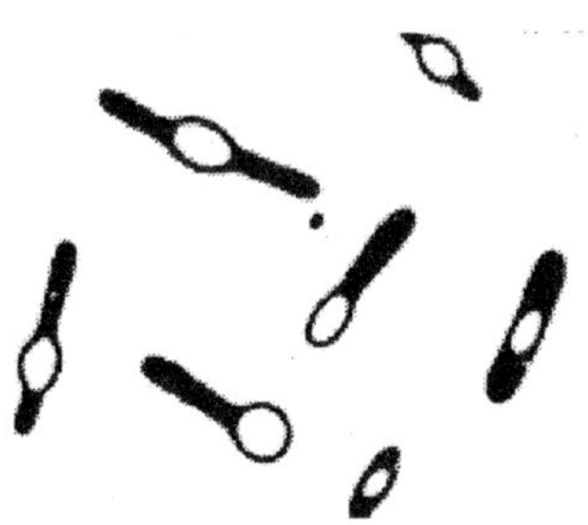

Fig.13.10 Microbial Growth and Reproduction
Various positions of endosporesthe within bacterial cells

Shape and position of endospore

The spores are of various shapes and occupy various positions in the cell. They may be spherical or oval in shape, and in position may be terminal, sub-terminal or central. They may be bulging giving a swollen appearance to the cell, or non-bulging conforming to the outer parameters of the cell.

Ultra structure of endospore

The outermost layer of the endospore is the exosporium which is thick and wrinkled. Chemically, it is made up of lipids and proteins with a low content of methionine and cysteine. Internal to the exosporium are two spore coats–the outer and the inner. The spore coats may be smooth or show various types of ornamentations like ridges, grooves etc. Chemically, the spore coats are made up of lipids, proteins and glycopeptides.

The protein of inner spore coat has a high quantity of sulphur containing amino acids. The spore coats surround a space called cortex which can be divided into outer cortex and inner cortex. Chemically, cortex consists of dipicolinic acid, peptidoglycan and calcium ions. The cortex is internally lined by the spore wall internal to which lies the cell membrane. Chemically, the membrane consists of peptidoglycans but no teichoic acid. The membrane surrounds the central core of the endospore. The core consists of nuclear material and cytoplasm. This region has DNA, RNA and proteins.
Sexual reproduction as shown by genetic recombination is explained in detail in Chapter 6.

REPRODUCTION IN CYANOBACTERIA

While generally one cay say that there is no sexual reproduction and no gametes are formed, there are, however, specific instances of genetic recombination indicating a type of sexual reproduction much the same way as in other bacteria. Genetic recombination has been demonstrated in *Anacystis nidulans* (Kumar. H.D. 1962), and *Cylindrospermum majus,* and other filamentous forms (R.N. Singl & R.N. Sinha 1965).

The most prevalent methods of reproduction are vegetative propagation and formation of non motile spores.

Vegetative propogation

1. Cell division

The cell division starts with the formation of a ring like outgrowth which grows inwards dividing the cell into two. Simultaneously, the chromatin material also divides and is distributed to the two daughter cells. There is no formation of chromosomes or spindles.

2. Hormogones

This is formed in the filamentous species. Hormogones are usually bits of trichomes consisting of few cells. Hormogones are separated from the trichome by the formation of biconcave separation discs. These are very common in Oscillatoriaceae. Each hormogone develops into a new filament.

3. Hormospores or Hormocysts

At the time of formation of Hormocys't a large amount of food is accumulated in the harmogones and they become thick walled. These are usually formed at the tips of the trichomes. They germinate to form a new filament.

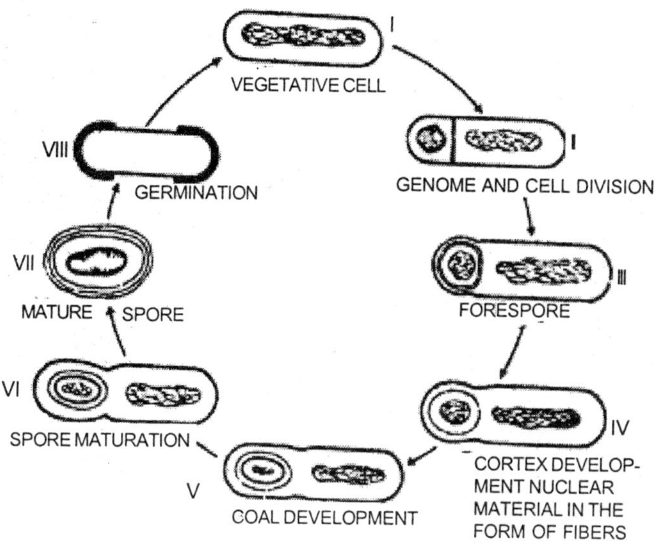

Fig.13.11 Microbial Growth and Reproduction

The sporulation cycle in an endospore forming bacterium

Spore formation

1. Akinetes, Resting spores or Arthrospores

These are non motile spores generally formed in nostocaceae and Rivulariaceae to tide over the adverse environmental conditions. At the time of formation of an akinete' the cell accumulates large amount of food and the cell wall becomes very thick. The wall may be distinguished into outer exospore and inner endospore. This type of spore in which the original cell wall forms the outer most layer is known as "Akinete". Akinetes may be isolated in a trichome or may be formed in a series. These are extremely resistant to environmental hazards and remain viable for long periods. On germination, an akinete develops into a new individual.

2. Endospores

These are also called gonidia and are produced in all the genera of chamaesiphonales and in few members of other orders. These are usually spherical but may be angular also, they become free when the cell wall decays. Unlike akinetes, endospores germinate immediately.

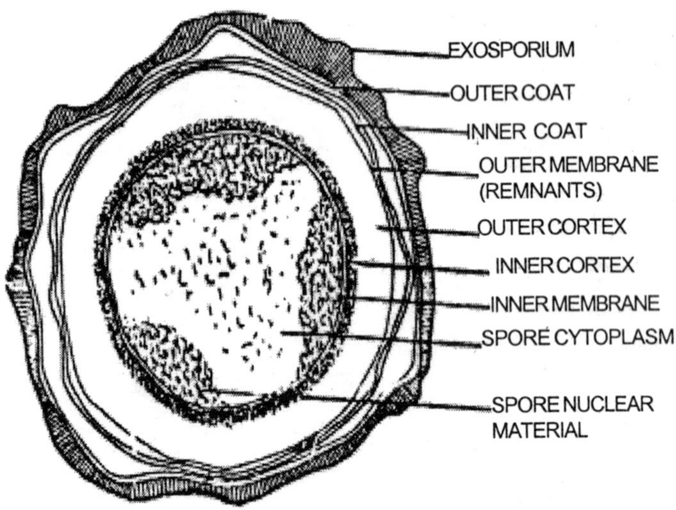

Fig. 13.12 Microbial Growth and Reproduction
Ultrastructure of endospore in *Bacillus cereus*

3. Exospores

In the epiphytic member *Chameosiphon,* exospores are generally formed. After a ceertain period of growth the protoplast abstricts externally cutting off a spore much the same way as a conidium is formed in fungi.

4. Nannospores

In certain forms like *Gloeocapsa,* where the cell division is very rapid the daughter cells are small and are called Nanocysts or Nannospores. They are some what similar to endospore. Eventually, Nanospores develop into new individuals.

5. Heterocysts

All filamentous Cyanobacteria except Oscillatoriaceae have special thick walled cells called Heterocysts. They are formed from the vegetative cells. They may occur singly or in chains. In some general like *Rivularia,* they are always basal while in others like *Nostoc,* they are inercalary.

Each heterocyst consists of a thick two layered wall. At the poles, the wall swells to form a nodule known as the polar nodule. Depending on the position, there may be one or two nodules; one in basal and two in intercalary. As the heterocyst matures, the protoplast inside becomes more and more transparent due to its transformation into a homogenous viscous substance.
Various views have been put forward to account for the nature of heterocysts. One view is that the heterocyst is spore like. Brand (1933), Geitler (1911), Canabouse (1929), and Desikachary (1946), have reported exceptional cases where heterocysts have germinated to give rise to new individuals (Nostoc, Anabena etc). Spratt (1911), has shown heterocysts forming a number of endospores as in *Anabena cycadicae.* These show that heterocysts are archaic reproductive structures which are largely functionless now.

Although functionless, heterocysts have taken a number of secondary functions. They are

1. Helping in the development of Akinetes (Fristch, 1951).
2. Help in the breaking up of a trichome (Kohl, 1903).
3. In genera with false or true branching a relation exists between the origin of the branches and the position of the heterocyst.
4. Help in N_2 fixation.

Genetic recombination

It is generally believed that Sexual reproduction is absent in cyanobacteria, but genetic recombination has been reported in several members. According to

Kumar (1962), biochemical recombination seen in the cyanobacteria is a kind of parasexual phenomenon and it differs from true sexuality in that "it is not attended by syngamy or meiosis", but the true function of sexuality via., gene recombination is achieved. Gene recombination seen in cyanobacteria is similar to what is seen in other bacteria.

Genetic recombination was first demonstrated in *Anacystis nidulans,* (Kumar 1962) *Cylindrospermum majus* (Singh and Sinha, 1965), *Anabena dolislum* etc. Genetic recombination has been indicated with reference to antibiotic resistance and such other markers. Kumar (1962), cultured strains resistant to streptomycin and penicillin on separate media, and obtained a resistant variety as shown below :

Strain 1 - S+P-
Strain 2 - S -P+

Strain 1 is sensitive to penicillin, but resistant to streptomycin. Strain 2 is sensitive to streptomycin but resistant to penicillin. When these strains were kept on a medium containing both streptomycin and penicillin, surpirsingly some cells survived and they possessed resistance to both the antibiotics (S+P+). This is possible only by the recombination of strain 1 and strain 2. Pikalek (1962), raised an objection to the experiment mentioned above by staining that penicillin used by Kumar (1962), was not a stable one.

Brazin (1968), was able to demonstrate unequivocally gene recombination in *Anacystis, nidulans* using single resistant strains with reference to polymixin B and streptomycin. He could isolate double resistant strains to both polymixin B and streptomycin.

Nothing definite is known about the mechanism of genetic system involved in gene recombination. According to Kumar (1982), it is likely that recombination is brought about by conjugation between donor and recipient cells much the same way as in other bacteria.

Lazaroff and other Vishniac (1962), have claimed that under certain conditions, filaments of *Nostoc muscorum* actually fuse, much in the same way as heterokaryotics do in fungi.

14

ENZYMES

A plant cell is virtually a microchemical factory in which hundreds of chemical reactions are taking place at amazingly high speeds and that at normal temperature ranges. These biochemical reactions provide sustenance for a plant to absorb, assimilate, synthesize, degrade, grow and reproduce which are essential for survival. Many of these reactions are simple, but most of them are complex, and involve several sequential steps in order to produce a particular product. Conversion of one compound to another has to be precise and exact, there being no excuse for mistakes. For a long time biologists and biochemists wondered, as to how biological reactions occur *in vivo* in such precise manner and at such rapid rates without involving undue increase of temperature. Very soon they got an answer, when Buchner (1987), accidentally discovered that a juice extracted from a yeast cell can perform fermentation as the living yeast cells can do. This means then that the juice contained some substance which can bring about fermentation. The chemical substance bringing about fermentation was soon isolated and given the name zymase. This started the glorious story of enzymes. Enzymes can be described as *Chemical middlemen* mediating and accelerating a variety of biological reactions. It may not be an exaggeration to state, virtually no metabolic activity in living organisms is participating without enzymes. Each reaction requires the participation one two or sometimes many specific enzymes. What are these enzymes? What is their role in metabolism? How do they accelerate chemical reactions? We shall try to find answers to these in this chapter.

The word enzyme is derived from Greek, meaning – *En* = in; zyme = leaven or living. The term was first coined by Kuhne (1978), while working on fermentation, even though it was Buchner who first made an extract of enzyme from yeast cells. Robert Sumer (1926), was the first scientist to purify and crystallize an enzyme revealing the proteinaceous nature. Even though the scientific discovery of enzymes may be comparatively recent, their use for practical purposes was known to early Greeks and Indians. The *Arthashastra* of Kautilya (4th century BC), mentions the use of extracts of barks in certain trees helpful in the process of fermentation.

Definition of enzymes

(a) Biocatalyst;

(b) Organic catalysts;

(c) Specific proteins, simple or compound in structure acting as specific catalyst;

(d) Biological middlemen.

Structure and chemical composition

All the known enzymes are proteinaceous. These proteins called functional proteins, have a very complex structure and have molecular weights sometimes ranging up to several millions. Even some of the simple enzymes like *pipsin* (35,000), *urease* (4,83,000) etc., also have high molecular weights. While all enzymes are made up of proteins, they also have a non-protein part. The bulk of the enzyme, however, is proteinaceous.

The total eyzyme is given the name *holoenzyme* (Protein + non protein), while the protein part is called the *apoenzyme* and the non-protein part is given the name *prosthetic* group. When the prosthetic group is not tightly bound to the apoenzyme, it is called a *coenzyme* or *cofactor*. The presence of a non-protein part in the enzyme is a must as otherwise the enzyme will be inactive. The following are some of the structural attributes of the enzymes.

(i) Apoenzyme

The protein part constitute the major part of the enzyme molecule. The complex proteins that make up the apoenzyme have specific sequences of aminoacids contributing to the specific reactivity of the enzymes. As the proteins are colloids, they offer a large area in relation to their volume. Very few enzymes are wholly made up of proteins. E.g. *proteolytic enzymes*.

Active site

All enzymes possess an area in their molecular organization where substrate materials bind themselves in order to undergo chemical change. This binding site is called the active site of an enzyme. An enzyme may have one or more active sites. In addition to the active sites an enzyme may also have *regulatory sites*, to which regulatory substances bind and regulate the activity of the enzyme.

Prosthetic group

Most of the enzymes possess a non-protein part tightly bound to the protein part known as the prosthetic group. Many metallic ions like Cu, Zn, Mo etc., constitute the prosthetic groups. Organic compounds such as cytochromes, flavoproteins, pyridoxal phosphate, biotin etc., are also known to be prosthetic groups of certain enzymes.

Coenzymes and Cofactors

Strictly speaking, there is no distinction between a prosthetic group and a coenzyme except that the latter is loosely bound to the enzyme molecule. The term cofactor is employed for inorganic ions such as Cl^- Mg^{++} etc., while organic molecules are called conezymes.

Distribution of enzymes in plant cells

Enzymes are found distributed all over the plant cell. But there is a qualitative distribution i.e., not all enzymes are found in all the areas. Enzymes have a localized distribution with reference to the functions they have to perform. The distribution of some important enzymes are listed below.

1. **Nucleus** : *Polymerases, Transferases* etc.
2. **Chloroplast** : *Carboxydismutases, Phosphatases, Kinases, Dehydrogenases* etc.
3. **Mitochondris** : Enzymes of respiratory pathway - *Dehydrogenases, Cytochrome oxidases etc.*
4. **Cytoplasm** : *Aldolsases, Isomerases, ATP ases, Phosphorylases, Transphosphorylases* etc.

(The list is not comprehensive)

Nomenclature and Classification

The name of the enzyme is usually descriptive and consists of two parts – the first part denotes the substrate on which the enzyme acts, while the second part indicates the type of action of enzyme. For example, *Isocitric dehydrogenase* indicates the enzyme acts on Isocitric acid and removes hydrogen from it. In some of the older nomenclature, this is not followed and the enzymes are named arbitarily. For Eg., *Pepsin, Trypsin, Chymotrypsin* etc.

According to the commission on Enzymology of the International Union of Biochemistry, all enzyme names should end with ase.

Enzymes are classified into six major divisions. These are –

1. Oxidoreductases
2. Transferases

3. Lyases
4. Isomerases
5. Ligases (= synthetases) and
6. Hydrolases.

1. Oxidoreductases

Enzymes which bring about oxidation reduction reactions are included under this group. This group is further subdivided into the following.

(i) *Oxidases* : Reactions involving oxidation of the substrate with molecular oxygen serving as the electron acceptor belng to this category.

(ii) *Peroxidases* : These catalyze the removal of hydrogen from the substrate which combines with hydrogen peroxide (H_2O_2)

$$SH_2 + H_2O \rightarrow \quad 2H_2O + S$$

(iii) *Dehydrogenases*: Enzymes which bring about oxidation of the substrate by removing hydrogen from it belong to this category. The hydrogen acceptor, however, is not molecular oxygen.

$$SH_2R \rightarrow \quad S + RH_2$$

(iv) *Reductases* : These catalyze the incorporation of both the atoms of oxygen into the substrate.

(v) *Oxygenases* : These catalyze the incorporation of both the atoms of oxygen into the substrate.

(vi) *Hydrolyases* : These enzymes catalyze the reaction in which only one atom of oxygen is added to the substrate.

(vii) *Catalases* : These liberate molecular oxygen from hydrogen peroxide.

2. Transferases

These enzymes transfer a particular group from one substrate to another. These are further divided as follows : based on the type of group transfer which they catalyze.

(i) *Transketolases* : Transfer a ketone group
(ii) *Transaminase* : Transfer an amino group
(iii) *Transaldolases* : Transfer an aldose group
(iv) *Transphosphorylases* : Transfer an every rich phosphate group
(v) *Transcarboxylases* : Transfer a carboxyl group
(vi) *Hexokinases* : Transfer an ordinary phosphate group
(vii) *Gransglycorsylases* : Transfer of a glycosyl group

3. Hydrolases

These enzymes catalyze the hydrolysis of complex substrates into simpler ones (starch into glucose). These include *Carbohydrases* (hydrolysis of carbohydrates), *Lipases* (hydrolysis of fats), *Phosphatases* (removal of phosphate groups) etc.

4. Lyases

These catalyze the removal of groups from the substrate without the addition of water (non hydrolytic).

$$\text{Fructose, 1,6 diphosphate} \rightarrow \text{Dihydroxy acetone phosphate} + \text{phosophoglyceraldehyde}$$

Other examples of lyases are - *Aconitase, Isocitrase, Fumarase* etc.

5. Isomerases

They bring about isomerization of one substrate to another (intramolecular shifting of atoms)

$$\text{Glucose 6 phosphate} \xrightarrow{\text{Isomerase}} \text{Fructose 6 phosphate}$$

6. Ligases

Also called synthetses, they help in the synthesis of a new compound generally with the cleavage of ATP or other nucleoside phosphates.

Isoenzymes

Enzymes which perform similar functions but having different molecular structures are called *Isoenzymes* or *Isozymes*. In other words, *different molecular species of an enzyme can be called Isozymes.* Isozymes can be identified based on the appearance of different bands on a polyacrylamid 3 gel. For example, the enzyme *lactic dehydrogenase* has five isozymase.

Zymogens

These are enzyme precursors or inactivated forms of enzymes secreted by cells. These can be activated to the normal enzyme state by chemical modification.

Constitutive and adaptive enzymes

Not all enzymes are present in the site of action always. Some are synthesized at the time of reaction and soon get degraded after the reaction. Such enzymes are known as *adaptive enzymes*. The other category – *constitutive enzymes* are present always and participate in reaction whenever required.

Properties of enzymes (Physical and Chemical)

Enzymes are characterized by the following properties.

1. Catalytic properties

All enzymes are catalysts, and as such they accelerate the pace of reaction. Being true catalysts, they do not undergo any chemical change and do not figure in the end products of a reaction. They also do not disturb the equilibrium of a reaction. Being catalysts, they are required in very small quantities. The number of molecules of substrate converted into product per minute is called, the *turnover number* of the enzyme indicating its efficiency. Turnover number of different enzymes varies from 100-3,00,000.

The velocity of enzyme action like that of a true catalyst is proportional to the concentration of the substrate/concentration of the enzyme.

2. Reversibility of action

Like all catalysts, enzymes can accelerate the pace of a reaction in both directions. The direction depends on the factors present at that particular time. It should be specified here that an enzyme by itself does not initiate a reaction but merely speeds up the rate. Thus, an enzyme *starch phosphorylase* catalyzes the hydrolysis of starch into glucose during day, but brings about the synthesis of starch during darkness. So here, it is not the enzyme that decides the direction of a reaction, but the presence or absence of light, pH etc., which do so. The enzyme merely speeds up the rate of reaction in whichever direction it is proceeding.

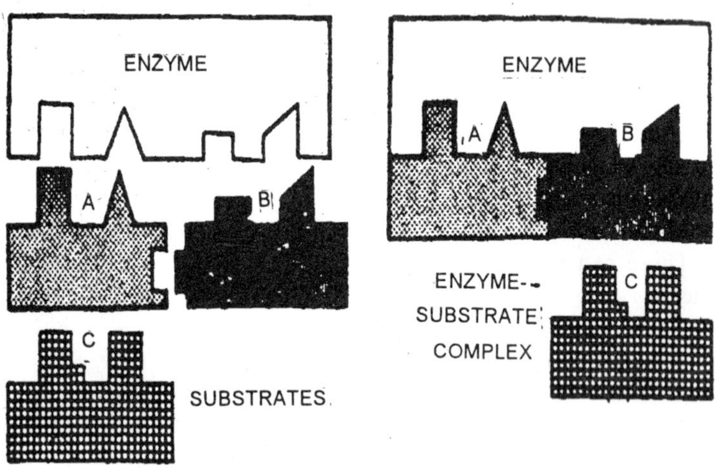

Fig. 14.1 Enzymes
Diagram to illustrate enzyme substrate specificity

3. Specificity

Enzymes are highly specific in their action. A particular enzyme catalyzes only a particular kind of a reaction. Sometimes, if a reaction involves several steps, an enzyme will be specific for each step. Enzyme specificity is of the following types.

(i) *Absolute specific*	:	Acts on only one kind of a substrate. For example, *malic dehydrogenase* acts only on malic acid.
(ii) *Absolute group specific*	:	Acts on a given organic group. For e.g., *alcoholic dehydrogenase* acts only an alcoholic group.
(iii) *Relative group specific*	:	Enzyme catalyses more than one organic group. E.g., *Trypsin* acting on ester bonds involving Arginine or lysine residues.
(iv) *Optical specificity*	:	Certain enzymes can distinguish between optical isomers in the substrate and catalyze any one only. For example *-amino acid oxidase* will not act on D - amino acids and *vice versa.*

4. Heat sensitivity

Enzymes remain active under normal temperature and are denatured under high temperatures (60 - 70°C). Under dehydrated conditions however, as in seeds, they can withstand temperature up to 120°C. Hence enzymes may be termed *thermolabile.*

5. pH sensitivity

Most of the enzymes are active only under a particular range of pH. Any alteration in pH would render the enzymes inactive. Enzymes are denatured by strong acids or alkali. Optimum pH for each enzyme varies depending on the nature of the buffer system (if present), absorbed ions, presence of activators/ inhibitors etc.

6. Inhibition of enzyme action

Most of the enzymes have active sites to which the substrate binds to undergo conformational changes. In some instances, it has been noticed that certain other substances which are partially indentical to the substrates bind themselves to the enzyme and inactivate it. These substances are called inhibitors and their action is called inhibition. There are three types of inhibition *viz.* **competetive inhibition, non competetive inhibition** and **uncompetetive inhibition.**

(i) Competitive inhibition

In this, the inhibitor molecule is structurally analogous with the substrate molecule. It competes with the substrate molecule to occupy the active site of the enzyme. Due to the occupation of the active site by the inhibitor molecule, the enzyme gets inactivated i.e., it cannot accept a substrate molecule (Fig). For example, *succinic dehydrogenase* is inhibited by molecules of malonic acid.

Competitive inhibition can be overcome by the addition of more molecules of substrate to the system which can now favourably compete with the inhibitor molecules for the active site of the enzyme.

(ii) Non competitive inhibition

The inhibitor molecules are not structural analogos of the substrate, hence, they do not bind to the active site of the enzyme, instead they bind themselves to the enzyme in some other area, all the same, the enzyme gets inactivated due to conformational changes. Heavy metallic ions such as Mg ++, Ag++, Pb++ etc. combine with the sulphydryl groups of the enzyme rendering it inactive.

Non competitive inhibition is irreversible. Addition of extra molecules of substrate is of no consequence since the active sites are anyway available to the substrate.

(iii) Uncompetetive inhibition

In competetive inhibition, the inhibitor is supposed to combine with forms of the enzyme but does not combine with the substrate enzyme complex. The form of the enzyme to which uncompetetive inhibitor attaches will always remain substrate non combining, in the sense it can never get back to the substrate combining form of the enzyme. As a result, enzymatic catalyses will stop. This inhibition also cannot be remedied by adding more amount of substrate. Uncompetetive inhibition is very common in multisubstrate reactions, but very rarely in single substrate reactions.

6. Colloidal condition

The enzymes are large molecules of proteins and exhibit all the colloidal properties. They cannot pass through a membrane of *collodion.* The enzymes have the capacity of absorption.

7. Redox potential value

Certain enzymatic reaction which are reversible, depend on the redox potential value for the direction of the reaction

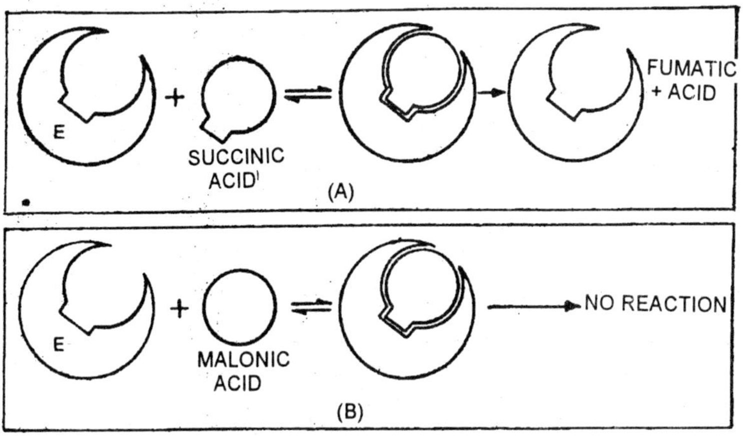

Fig. 14 . 2 Enzymes
Competetive inhibition of enzyme action

MECHANISM OF ENZYME ACTION

How does an enzyme catalyze, or speed up a reaction? The following explanation may be offered by considering the conditions of a biochemical reaction.

Normally, chemical reaction take place when molecules of any substance possessing on an average a higher kinetic energy change into a substance whose molecules have a lower average of kinetic energy. Although on an average all molecules of reactions (substrate) possess equal amount of kinetic energy, a few of them possess higher kinetic energy, and a few of them possess lower than the average due to collisions among molecules. Thus among the molecules of a substrate, there are *energy rich* and *energy poor* molecules. If we, for instance take a reaction where molecules of A get converted into B, at any given time only few molecules of A can get converted into B because of their higher kinetic energy. This extra energy required for the participation in the reaction is called

activation energy. The activation energy is always above the average kinetic energy of the molecules. In a non enzymatic reaction, the average kinetic energy of molecules may be raised to activation energy by increasing the temperature of the medium. This speeds up the rate of reaction. Theoretically, continuous increase of temperature should indefinitely increase the rate of reaction. But this is not possible actually because, a higher temperature beyond a certain point would kill the whole system. The enzymes in living cells do the same things in *vivo,* what the temperature does in *vitro* (they speed up the rate). But the enzymes do not increase the kinetic energy of the molecules. Instead, they combine with the molecules and lower the activation energy necessary for reaction. The combined product of enzyme and substrate known as *enzyme substrate complex,* crosses the activation energy barrier much easily than what the plain substrate molecules can do. All enzymatic reactions are believed to take place as follows.

$$\text{(i)}\quad E + S \rightarrow ES$$
$$\text{(ii)}\quad ES \rightarrow A + B + E$$

(E = enzyme; S = Substrate; ES = enzyme substrate complex; A and B = products)

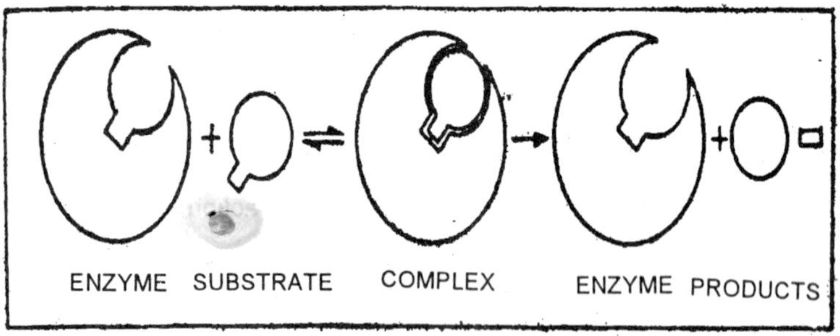

Fig. 14.3 Enzymes
Graph to demonstrate lowering of activation of energy due to enzyme catalysis

Measurement of activation energy has been done for several reactions with or without enzymes. With the participation of enzymes, it has been noticed that more number of substrate particles can participate in a reaction than what is normally possible. For example, hydrolysis of sucrose with only hydrogen ions as catalyst has an activation energy of 25,000 cal/mol., but with the addition of yeast *invertase,* it comes down to 10,000 cal/mol. Similarly, in the decomposition of H_2O_2 (enzyme *catalase*), activation energy comes down from 18,000 cal/mol to 2000 cal/mol; for caesin the figures are 20,600 cal/mol and 12,000 cal/mol without, and with enzyme (*trypsin*).

Enzyme kinetics (Michaelis - Menton equation) : Michaelis and Menton (1913), proposed a theory for enzyme action, in which they believed that the substrate and enzyme form a loose complex, and at the end of reaction the complex breaks releasing the enzyme and products as explained above. In order to explain the rate of substrate conversion, they applied the laws of kinetics which may be represented as follows.

$$E + S \xrightarrow{K_1} ES$$ (E = enzyme; S = substrate; ES = enzyme substrate complex; K_1 velocity constant of the formation of ES and K_2 velocity constant of dissociation of ES)

OR

$$E + S \xrightarrow{K_1} \xrightarrow{K_2} ES \rightarrow P + E$$ (K_2 = velocity constant for decomposition of ES.P = end product)

The equation can be further simplified as $\dfrac{K_1 + K_2}{K_1} = K_M$

K_M is known as the velocity constant, or Michaelis and Menton's constant. Its value is characteristic for each enzyme and substrate and by comparing KM values of an enzyme with different substrates, the affinity of different substrates to an enzyme can be calculated. When the velocity of the reaction is half of the maximum velocity, KM is equal to the concentration of the substrate. However, Michael is Menton equation is applicable only to single substrate reactions.

MODELS FOR ENZYME ACTION

Two models have been proposed to account for the specificity and mode of enzyme action. These are (i) the lock and key theory, and (ii) the induced fit theory.

Lock and key model

This theory originally proposed by Fischer (1898), believes that the enzyme specificity can be accounted for by assuming that the enzyme and the substrate fit into each other structurally like the two pieces of a jigsaw puzzle or the lock and key. Just as only a particular key can fit into a particular lock because of the alignment of levers, enzymes have specific active sites, to the contours of which, the substrate molecules can fit in.

No other molecule can fit in to this slot. When a reaction mixture consists of several substrates, like A,B,C,D etc., only A and B can fit into the active sites and undergo conformational change resulting in the product P (This is an example molecule, into two products, only one active site will be present on the enzyme and only one substrate A or B can fit into the active site.)

The lock and key mechanism assumes a rigid structural configuration for the enzyme molecule and does not allow even minor adjustments between the enzyme and the substrate.

(ii) Induced fit model

Many experimental findings questioned the rigid unchangeable configuration for the enzyme molecule as proposed by the lock and key model. It has been noticed that the enzyme molecule does allow for minor adjustments in its geometry to permit the substrate to fit into the active site. Secondly, it has also been observed that the substrates induce proper orientation of the catalytic groups in the enzyme in order to activate enzyme action.

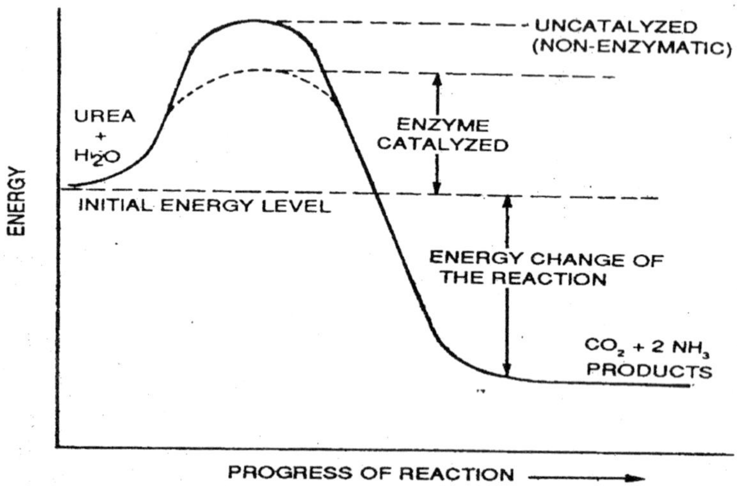

Fig.14 .4 Enzymes
Lock and Key mechanism of enzyme action

In order to make provisions for the above findings, Koshland (1959), proposed a theory called the induced fit theory. According to the induced fit model a substrate after linking itself with the enzyme alters the geometry of the enzyme protein in such a way as to achieve perfect alignment.

Allosteric regulation

In reactions which involve several sequential steps many enzymes participate, each one catalyzing a particular step. Such enzymes constitute the multienzyme complexes. In multienzyme complexes, the product of one step is a substrate for the next step. The intermediate products are transferred from one enzyme to the other. In these enzyme complexes, there is end product inhibition of all the enzymes involved. In otherwords, the end product of the pathway blocks the functioning of the first enzyme (of the complex), by binding itself to a site called the *allosteric site,* which is different from the active site. The end product which binds to the allosteric site is called the *modulator* and the enzyme itself is known as the allosteric enzyme (allosterism means another space). The name allosteric enzyme was first proposed by J. Monod, J.P. Changeux and F. Jacob.

Allosteric regulation is different from non competetive inhibition in that, it is the *end product* of the pathway which blocks the enzyme action of the first in the series of the enzyme complex. Allosteric regulation need not always be inhibition of enzyme action *(negative modulation).* In some cases, the end product may even activate the first enzyme of the series *(positive modulation).*

Mond, Wyman and Changeux (1965), proposed a theoritical model to explain the allosteric regulation. According to them, the allosteric enzymes are oligomers made up of identical units called monomers. The monomers are linked together in such a way that all of them occupy equivalent positions. Each monomer possesses one site for the binding of each molecule of either the substrate or the allosteric effector which are able to bind with the enzyme. The three dimensional configuration of each monomer is dictated by its association with other monomers. The oligomer molecule can exist in atleast two states which differ in the distribution and energy of interpromoter bond. When a transition from one state to another state occurs, there is a change in the affinity of the binding sites towards the molecule.

Double displacement (ping pong) reactions : In enzyme reactions involving a single substrate, it is well known that the substrate concentration influences the rate of the reactions. In many biological reactions however, a single reaction with the enzyme involves two or more substrates participating simultaneously. For example in the following reaction involving *hexokinase,* ATP and glucose are substrates with ADP and glucose - 6 - phosphate being the products.

$$\text{Glucose} + \text{ATP} \xrightarrow{\text{hexokinase}} \text{Glucose 6 phosphate} + \text{ADP}$$

In a situation as mentioned above, what will be the influence of the substrate concentration on the enzyme activity? In several bisubstrate reactions, it has been observed that the enzyme has a separate velocity constant (KM) for each of the substrates involved.

Bisubstrate reactions usually involve the transfer of a functional atom or group from one substrate to the other. This may proceed in one of the two following ways – (a) *Single displacement method* and (b) *Double displacement method.* In the first type, both the substrates bind to the enzyme simultaneously to form the enzyme substrate complex.

$$A + B + E \qquad EAB$$

(A - substrate EAB - Enzyme substrate complex B- substrate E - enzyme)

EAB then breaks up to form the products C and D

$$EAB \longrightarrow C + D + E$$

In the secondary category of bisubstrate reaction called the *double displacement type,* only one substrate can combine with the active site of the enzyme at any given time. After the substrate transfers its functional group to the enzyme molecule, it (substrate) leaves the enzyme as the product. The second substrate now binds to the enzyme (which has the functional group of first substrate), accepts the functional group donated by the first substrate and leaves the enzyme as product.

E + A   EA
EA   E' + C
E + B   E'B
E'B   E + D

(A and B = substrates, C and D = products, E = enzyme EA = enzyme substrate complex, E'=enzyme with the functional group of substrate A, E'B = enzyme substrate complex; enzyme has the functional group of A).
The enzyme which has the functional group of one substrate after it (substrate) has been released as the product, may be called a *partial enzyme substrate complex.*

This type of a reaction in which the substrates alternately bind and get released to the enzyme is called **ping pong reaction** or **double displacement reaction.**

Transamination reactions involving the enzyme *transaminase* provide a typical example of ping pong reactions. In a reaction involving glutamic acid and oxaloacetic acid, the enzyme glutamic transaminase has pyriodoxal phosphate as the prosthetic group. In the first of the reaction, glutamic acid binds to the prosthetic group of the enzyme (pyridoxal phosphate), and donates the amino group and is released as a α keto glutaric acid (product). In the second step, oxaloacetic acid binds to the pyridoxal phosphate (which has the amino group bound to it), accepts the amino group and is released as Aspartic acid (product).

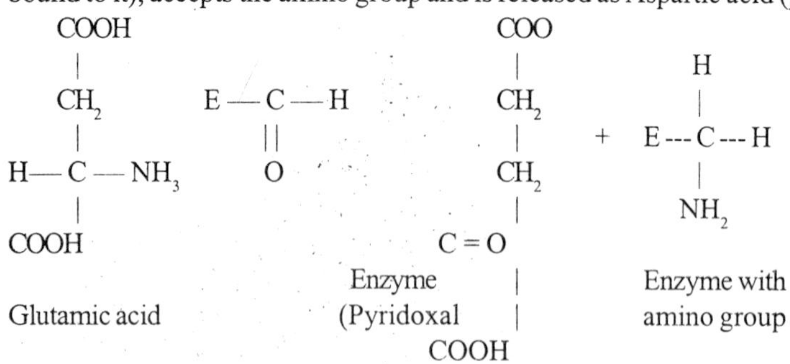

Glutamic acid   Enzyme (Pyridoxal   Enzyme with amino group

(Pyridoxal phosphate)   α ketoglutaric acid

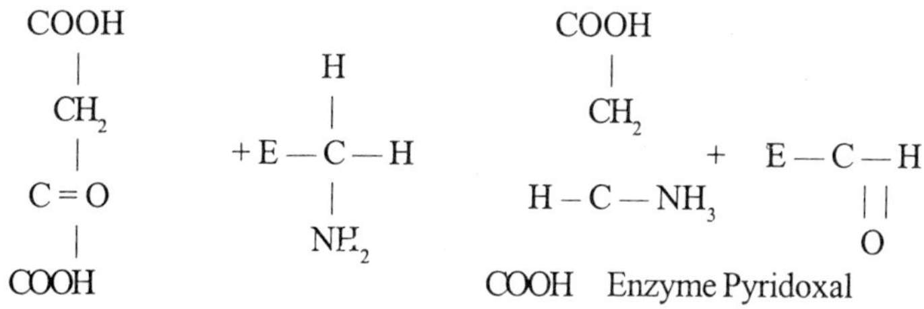

Oxalo acetic acid   Asparticacid   Enzyme Pyridoxal Phosphate)

FACTORS AFFECTING ENZYME ACTION

The following factors influence the enzyme action.

1. Enzyme concentration

At specific concentrations, and in the presence of large number of substrate molecules, the reaction will proceed at a particular speed. Any addition of substrate at this stage is not of any consequence as all the enzyme active sites are already occupied. In such a situation, increasing the concentration of the enzyme will increase the rate of the reaction. This increase stops at a point when substrate molecules become a limiting factor.

Substrate concentration

As the enzyme molecules are very large in size when compared with the substrate molecules, relatively large amount of substrate is required to engage all the enzyme molecules. At a particular concentration of the enzyme, if the substrate molecules are at a low concentration, the rate of the reaction will not proceed at the required pace since many active sites of the enzyme are unoccupied. At this stage, addition of more substrate to the system would accelerate the pace of the reaction as many enzyme molecules can be engaged, indefinite increase of substrate concentration, however would increase the pace of reaction indefinitely, as enzyme concentration would become a limiting factor then.

Temperature

Like every other bilogical reaction, enzyme catalysis is also influenced by temperature. Since enzymes are proteinaceous, they are very sensitive to temperature changes. They can function at a very narrow range of temperature since enzymes are functional even at as low a temperature as 0°C, but the rate of the reaction will be its minimum. A rise in temperature at this stage would bring in a geometric progress in the rate of the reaction. Very soon, this reaches a peak and this temperature level is known as the optimum. Further, increase beyond the optimum temperature would actually retard the rate of reaction. A higher temperature of 50°C and above would incapacitate the enzyme.

The Q^{10} for all enzymatic reactions is about 2.0. The higher value of Q^{10} clearly shows that temperature has a marked influence on enzymatic reactions.

Hydrogen ion concentration (pH)
Most of the enzymes are catalytically active only at a particular pH. Sometimes, they may have a range of pH at which their activity will be sustained. Any increase or decrease from this pH would render the enzymatic activity impossible. The different enzymes, however, would vary in their pH requirement. While some enzymes are active at a low pH, others prefer an alkaline pH; still others require a neutral pH for their optimum activity.

Inhibitors
The presence of certain inhibitor molecules in the system would render the enzyme inactive. Various substances are known to inhibit enzyme action. These are of 3 kinds.

1. Competitive inhibitors
2. Non - competitive inhibitors
3. End product accumulation

In many cases, if the end products of a reaction are not removed from the system they accumulate and bring about inhibition of enzyme action known as feed back inhibition. The feed back inhibition many a time may be due to allosteric effect as can be seen in reactions involving multienzyme complexes. The end product inhibition of enzyme action may also be due to a difference in the pH value of the substrate (at which enzyme is active), and of the end product.

15

BIOENERGETICS

All living cells are virtual biochemical factories where hundreds of reactions take place every second involving conversion, degradation and synthesis of chemical compounds that are vital for the maintenance of life. These reactions require energy or produce energy whic his either obtained from the breakdown of chemical compounds synthesis, or in the bonds of chemical compounds. In other words, these reactions are accompanied by energy changes. Thus, *bioenergetics may be defined as the study of energy changes that accompany biochemical reactions in living systems.* Bioenergetics may also be called biochemical thermodynamics, as the energy changes follow the laws of thermodynamics precisely.

In any biological system, the rectants have a higher energy level and they move **"energy down hill"** after the reaction. The energy thus liberated is wasted as heat either partially, or wholly. The energy that is liberated as heat cannot be put into any biological function and thus is a waste. The living systems, however, have developed means and mechanisms to trap the energy and store them in the form of energy rich chemical bonds. These 'energy currencies' readily donate the energy as and when required to do so, and help perform all the activities of living beings. One such universal energy currency is a molecule of ATP (Adenosine triphosphate). Whenever energy is required for a process, ATP is hydrolysed to ATP and the energy so released is utilized for the necessary activity. Similarly, there are a number of methods in which ATP molecules are synthesized (as during respiration).

The basic rules of bioenergetics are no different from those applicable to physico - chemical energy transformations. The only difference, perhaps, is that biomolecules involved in bioenergetics are more complex and function in a highly specializéd environment. In order to understand the principles involved in bioenergetics it will be necessary first to understand the physiochemical principles involved in energy transformation reactions, and later these can be applied to biological situations.

Energy and matter

The universe is composed of only two components – *energy* and *matter* which were thought to be non interconvertible and fixed in their content. Einstein ,however, has shown that matter can be coverted into energy. What is energy? Energy may be defined as the ability or capacity of a system to carry out some work. This capacity is there in all the molecules of matter to a lesser or greater extent.

Types of energy

Basically, there are two main types of energy. These are – *potential energy* and *kinetic energy*. Potential energy is bound to the molecules. It may also be called as energy at rest i.e., not working. It can be measured in terms of how much work it is capable of doing when it is released. Potential energy is of various types – chemical, light (photic), atomic and even positional (such as that of a boulder on hill top). In biological world, potential energy source is mainly the solar radiation which is trapped, transformed and stored in the form of chemical energy during photosynthesis in greeen plants. All living beings obtain their energy requirements from the green plants directly or indirectly.

Kinetic energy is the other form of energy. It can be called as *energy in motion,* or energy released. Molecules or bodies have this energy as a result of motion. The amount of kinetic energy depends upon the mass of the body and also its velocity of motion. The following equation summaries this relationship.

$$KE = \frac{mv^2}{2} \quad \text{or} \quad KE = 1/2\ mv^2\backslash v^2\backslash v^2$$

In the above equation KE = Kinetic energy, m = mass of the body/molecule, and v= velocity of the body/molecule.

Kinetic energy of molecules also depends on one other parameter viz., temperature. The amount of molecular kinetic energy in a system is proportional to the absolute temperature of that system and at absolute zero (-273.18^0 C), kinetic energy disappears, i.e., it is zero. For every rise of 10^{0}C in temperature, kinetic energy increases by the ratio of 10 to the absolute temperature.

Kinetic energy of molecules may be easily demonstrated through the Browinian movement of the particles. Brownian movement, is the to and fro movement of medium. This can be clearly seen when a beam of light is made to pass through a colloidal medium. The rate of movement of particles increases with the increase in temperature. The mechanism of Brownian movement is attributed to the random motion of molecules colliding with the visible particles, and imparting to them required energy for movement.

In addition to kinetic and potential, energy may also be of two other types. Free energy and Internal energy. Free energy may be defined as the energy that is liberated or consumed in a reaction which can be used or supplied to another system. Free energy is the most important type of energy in the biological system (see later in the chapter for a detailed discussion on free energy).

Internal energy is the energy that is associated with a system based on the constitution and movement of the molecules. Internal energy is again of two types – *Internal potential energy* and *Internal kinetic energy.* The contribution of energy due to its molecular configuration is Internal potential energy, and the energy derived from the motion of its molecules is Internal Kinetic energy. The total internal energy of a system is the sum total of both the above types of energy.

In any given substance, the internal energy is fixed under a given set of conditions. The internal energy of different substances, however, varies. Internal energy is symbolically represented by E or U.

Matter

This may be defined as something that physically occupies space and of which the entire physical universe is composed of. Originally, it was thought that matter can neither be created nor destroyed, but since the epoch making work of Albert Einstein, it has been found out that two atomic reactions *viz; nuclear fusion* and *nuclear fission* can convert matter into energy. But nuclear fusion or fission requires a very high temperature range (Nuclear fission occurs in sun and nuclear fusion is seen in atomic bombs), and they do not occur in ordinary chemical reactions.

The study of matter and its energy, conversions follow the principles of thermodynamics. In order to study the principles of thermodynamics as it relates to matter, an arbitrary position of matter, a system and its surrounding has to be imagined/considered. A system may absorb energy from the environment of liberate energy into the environment. In biological situations, *a system* may be a cell and its surroundings, or an organism and its environment or it may be an ecosystem. But the principles of thermodynamics remain the same irrespective of the size and structure of matter.

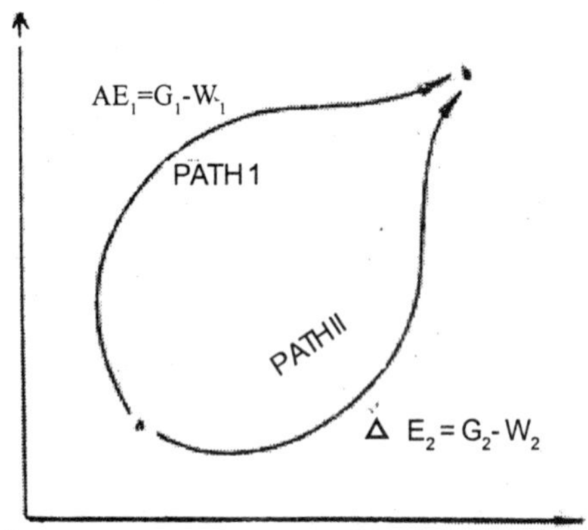

Fig.15.1 Bioenergetics
Illustration of pathways for energy conversion of *a* into *b*

Laws of Thermodynamics

It has already been pointed out that the bioenergy conversions follow the principles of thermodynamics. Hence, an understanding of the laws of thermodynamics becomes necessary before attempting to understand the role of energy in biological reactions.

It has been mentioned above that irrespective of the size and structure of matter, the laws of thermodynamics are the same. Classic thermodynamics in reality deals with reactions in the equilibrium state, and may be regarded as *equilibrium thermodynamics*. Equilibrium thermodynamics is not concerned with the pace of the reaction, instead, it is concerned with the change in the form of energy from the initial stage (before the reaction has begun), to the final stage (when the reaction has been completed). Some of the other salient features of classic thermodynamics are –

(i) The number of steps between the initial and the final are irrelevant as long as the final state is same in different systems.
(ii) Direction of the reaction can be indicated in reversible reactions.

There are two laws of thermodynamics which are universally applicable. These are –

The first law of thermodynamics

This is known as the *Law of conservation of energy*. It states that the total content of energy is fixed and it can neither be created nor destroyed. But during reactions, energy can change form. In other words, the total energy in an isolated system will neither decrease nor increase. For instance when some amount of chemical, light or electric energy disappears (apparently), it would be transformed into equal amount of heat.

Previously, many attempts were made to disprove the principle of conservation of energy by constructing a 'perpetual motion machine', which is supposed to operate without any input of energy i.e., creating energy out of nothing. But these attempts failed.

The first law of thermodynamics may be enunciated in a detailed manner by the two following statements. These are –

(i) The disappearance of energy in one form will result in its reappearance in another form in exactly the same quality.
(ii) Energy of an isolated system will remain constant although it may be transformed from one form to another.

Elaborating the first statement, it may be said that from an isolated system when heat is given out (as in exothermic reactions), the internal energy of the reacting system decreases while that of the surrounding medium increases correspondingly. Hence, the total energy of the system (energy of the system + surrounding) remains unchanged. Alternatively, when there is energy input into the system (absorption of heat), the energy of the reacting system increases while that of the surrounding decreases. In this case also, the total energy remains unchanged.

As an illustration to elaborate the second statement, the burning of coal in a steam engine may be cited. Coal has a lot of internal energy. When coal is burnt, a lot of energy is released as heat, which is converted into mechanical energy of the engine i.e., it is allowed to boil water and release steam to rotate the wheels. In this, the chemical energy of coal gets transformed into mechanical energy of the engine.

Energy exchanges in chemical systems are studied with the help of a 'bomb calorimeter'. In the working principle, a calorimeter is nothing but a slightly modified thermos or Dewar flask. Basically, it has an insulation mechanism to isolate a system with internal energy as in a thermos flask.

A bomb calorimeter consists of an outer jacket (made up of insulating material not to allow heat to escape). This jacket surrounds a chamber which is filled with water. The centre of the calorimeter has a central chamber (combustion chamber) containing a sample which releases heat. The heat released by the sample increases the temperature of the water and its temperature is noted with the help of a thermometer. Since a calorimeter may not have a 100% insulation, corrections must be made for possible thermal leaks. With the help of this, it will be possible to show that the rise in temperature of water corresponds to the quantity of heat released by the sample.

Calculations of total energy (as per the first law)
Calculations of total energy in a reacting system can be made by finding out the quantities of the following :

(i) The work done
(ii) The heat that is exchanged, and
(iii) The energy stored in the system

Mathematically the above may be expressed as follows.

$$\blacktriangle E = q\text{-}w$$

Where $\blacktriangle E$ is change in the internal energy of the system, q is the heat absorbed by the system and w is the work done.
The values of q and w may be positive or negative, depending on the nature of reaction. If heat is absorbed, q has a positive value. If heat is given out, q has negative value. In the same way if the system performs work, w will be positive and if the medium performs work on the system, w will be negative.

The change in total energy does not depend whether q and w are negative or positive. For instance if molecules of *b, the change in energy is* $\blacktriangle E_l$, and the other case $\blacktriangle E_2$ then $\blacktriangle E_1 = \blacktriangle E_2$ The energy change her may also be represented as –

$$\blacktriangle E = Ea - Eb$$

Where Ea and Eb are energy levels of molecules of *a* and *b* respectively. The first law of thermodynamics needed some modification when Albert Einstein discovered that matter is convertible into energy. Accordingly, it is now agreed that matter (mass) can be converted into energy according to the following equation.

$$E = mc^2$$

Where E = energy; m = mass and c is velocity of light in cm/sec.

The reverse of this system i.e., conversion of energy into mass is also possible. Based on these observations, the first law of thermodynamics may be restated as *matter and energy can neither be created nor destroyed but can be inter converted.*

The interconversion of matter and energy however, does not occur in biological systems.

Second law of Thermodynamics

As per the first law of thermodynamics, the total energy in a reaction remains unchanged, but the transformation of energy from one to another proceeds in such a fashion that the capacity of the total energy to do work decreases. This is because all natural processes have a tendency to move towards equilibrium. The second law is actually related to the equilibrium of the process. Stated in very simple terms, it means during every energy change the capacity of total energy to do work decreases.

The second law of thermodynamics states that spontaneous reactions have a direction for the flow of energy. Once a reaction has reached an equilibrium, the reactants cannot return spontaneously to the initial stage.

Potential energy of molecules is often referred to as energy with a high degree of orderliness, whereas the kinetic energy is said to be random or disorderly. When once potential energy changes into kinetic energy, its randomness or disorder increases. This increase in randomness is called entropy. The stage of entropy is energy in that state where energy is incapable of doing any work because of its disorderliness. In otherwords when energy is transformed from one state to another state, the entropy goes on increasing untill it reaches an equilibrium. At equilibrium, entropy is maximum.

In nature, randomness or disorder is more a rule than exception. This is the reason why orderly arrangement which is special, tends to always move towards randomness. Entropy can be zero if a system is fully orderly. For instance, at absolute zero, the entropy of a crystalline substance is zero because there is no randomness.

The second law also stipulates that in any transformation of one type of potential energy into another type, the efficiency of the process is not perfect because some quantity of potential energy gets transformed into kinetic energy. Although theoritically, total kinetic energy can be transformed into potential energy, practically it is not possible because of inefficiency of potential energy conversions. Some transformation does occur as in the case of steam turbine or internal combustion engine, but in these instances only a small fraction of the kinetic energy becomes potential mechanical energy.

Ultimately, all potential energy of the system (whether it be a molecule, body, or even the entire universe) gets converted into kinetic energy or heat and in course of time (in finite), the entire potential energy may get exhausted i.e. Total entropy.

The fundamental principles of the second law of thermodynamics also operates in a food chain, in an ecosystem where starting from producers with a high degree of order (more number of individuals), succeedingly there will be less and less consumersin number of the primary, secondary and tertiary variety respectively. This is due to the fact that the whole of the potential energy of a producer cannot be transformed into the potential energy of a consumer. This is because of the intervening kinetic energy which introduces some amount of randomness (entropy).

Operation of the principles of second law of thermodynamics can be seen in our day to day life. For example water runs downhill and not uphill and when it has flown down its capacity to work gets decreased.

Green plants in this respect can be regarded as the most efficient mechanism to store energy for their own use and and for the use of other life forms. They trap the radiant energy and store an appreciable amount of it in the form of chemical energy. It is this potential energy which forms the fulcrum on which revolve all forms of life.

The concept of free energy

One of, the most important thermodynamic concepts that is of great relevance to biological systems is that of **free energy.** It is the energy that is liberated or consumed in a reaction and can be used or supplied to a system. It is denoted by the letter *G*. It may be regarded as a measure of the potential energy of a substance but it cannot be measured directly. However, change in free energy content that accompanies a reaction can be measured. This change is denoted by ▲ G, may be defined as that quantity of total energy change in a system that is available to do the work. Change in free energy (▲ G) of a reaction is calculated by considering the difference between the sum of free energies of the reactants and products. This may be represented as follows.

A ——— *B*

In the above, *A* is reactant and *B* is in the product. Here, free energy is the maximum energy made available, as *A* is getting converted to *B*. If the free energy content of the product B *(Gb)* is less than the free energy content of the reactant A *(Ga)*, then ▲ G will be negative (Here, reactant has more energy than the product). This may be shown as follows :

▲ G = *Gb-Ga*

Here, *Ga* is greater than *Gb,* hence ▲ G is negative. There is decrease in energy level as A gets converted to *B*. Similarly, when the reaction is reversible, *B* gets converted to *A*. In this reaction the level of free energy is increased (as B has less free energy).

In this case ▲ G will be positive. Based on the fact whether ▲ G is positive or negative (whether it gets reduced or increased), reactions are categorized into two types **endergonic** and **exergonic.**

Exergonic reactions

These reactions do not require the input of external energy. Normally, reactants have more energy than products. Products are *downhill* in energy level while reactants are *uphill* in energy level. Hence, no energy input from the medium is required to carry out the reaction. The reaction is driven by the inherent potential energy of the reactant molecules. Hence, the reaction can take place without any external energy supply. During exergonic reactions energy is released.

Endergonic reactions

Here, the reactant molecules have less free energy than the products. That is, product is *uphill,* and reactant *downhill.* Naturally, the direction of the reaction cannot be from down hill to uphill. The reactant molecules cannot negotiate on their own the uphill migration. Hence, energy from outside has to be supplied to the reactants to make them participate in the process. Here ▲ G is positive *(GB–GA* and *GB* is greater than *GA,* where *GB* is free energy of reactant, and GA is free energy of the product).

The difference between endergonic and exergonic (energy requiring and energy releasing) processes can be explained with an example. Suppose a huge boulder is on top of hill, no extra effort is required to push it down hill; just a shove would do it. It has greater potential energy because of its position on hill top. When it reaches foot hill, its free energy is less because of its different position (location). This can be called exergonic process. Suppose the same boulder has to be now pushed back (from foot hill) uphill it requires an enormous effort (input of energy), which should come from outside and comparable to input of energy into a system. The boulder that is pushed uphill will again have more amount of free energy than the boulder at foot hill. This process is called endergonic, as there is energy requirement.

Photosynthesis and respiration serve as the best examples of endergonic and exergonic processes. For instant the overall formula of the processes.

1. $C_6H_{12}O_6 + 6O_2 + 6H_2O \longrightarrow 6CO_2 + 12H_2O +$ energy (respiration)
2. $6CO_2 + 12h_2O \xrightarrow{\text{light energy}} C_612_{12}O_6 + 6O_2\ 6H_2O$ (photosynthesis)

In the above equation for respiration, reactant (glucose) has more free energy than products. Hence, along with products energy also will be released. Reaction takes place without any energy input into the system as the reactants are *energy uphill,* and products *energy downhill.* The reaction is said to exergonic.

Equation for photosynthesis on the other hand shows that product viz., glucose has more free energy than the reactants viz CO_2 and H_2O. Hence, to drive the reaction *uphill,* there is requirement of energy input into the system which is provided by light energy. The reaction here is said to be endergonic.

Calculation of standard free energy change

In an exergonic reaction where ▲ G is negative, the rate of reaction however, is determined by the surrounding conditions like temperature (heat) etc. For instance suppose 10 molecules of reactants have to become products, not all the 10 molecules may participate in the process as they may have lost some energy because of collision among molecules. Hence, only some molecules which have the required high energy participate in a reaction. This is known as *activation energy.* Activation energy for a process is the higher potential energy that is required to start a reaction. Acvitation energy may sometimes be increased depending on the environment, like a medium having higher temperature. This means then the reactant molecules have varying speeds of the reaction even in exergonic reactions depending on the medium. For example, glucose can be oxidized to CO_2 and H_2O in different surroundings. In a bomb calorimeter, oxidation takes place in a few seconds; in living organisms it takes minutes to several hours. Similarly, even in the presence of O_2 (oxidizing agent), glucose can be kept in a bottle for years without large scale oxidation. In a reaction where A and B are reactant and product respectively.

$$\blacktriangle G = \blacktriangle G^0 + RT\,B/A \blacktriangle$$

Here $\blacktriangle G^0$ is standard free energy. R is the gas constant and T is absolute temperature in Kelvins. This equation shows that when conditions are ideal $\blacktriangle G$ and $\blacktriangle G^0$ are equal ($\blacktriangle G$ = free energy and $\blacktriangle G^0$ is standard free energy). *Standard free energy change may be defined as the free energy change of chemical reaction when reactants and products are in their standard state.* In otherwords, standard free energy change of a substance is that value which it can ideally release when all conditions are optimum. Standard free energy change of a process can be used to find out whether a reaction is proceeding along the ideal path of not. If there is any difference between the standard free energy change and the actual free energy change that is going on in a reaction, it indicates that the conditions are not ideal necessiating corrective measures.

To summarize, the standard free energy change can be calculated as follows :

$$A + B \_\_\_\_\_\_ C{+}D \quad (1)$$

Where A, B, C and D are molecules of substances participating in a reaction. The free energy change $\blacktriangle G$, at constant temperature and pressure is given by

$$\blacktriangle G = \blacktriangle G^0 + RT \text{ in } \frac{C{+}D}{A + B} \quad (2)$$

In the above, $\blacktriangle G$ = is free energy change; $\blacktriangle G^0$ = standard free energy change; R=gas constant and T= absolute temperature.

In the equation (2), above $\blacktriangle G^0$ has a definite (fixed) value characteristic for any given chemical reaction. Thus, we can calculate $\blacktriangle G$ at any given temperature if $\blacktriangle G^0$ is known in addition to concentration of products and reactants.

If the reaction at equation (1) is allowed to proceed towards equilibrium, as the free energy decreases, the reaction will be able to carry out work at constant temperature and pressure. When equilibrium is reached, $\blacktriangle G$ will be at its minimum or zero, and the system halts as it cannot do any more work. The equation at equilibrium is

$$O = \Delta G^0 + RT \ln \frac{C + D}{A = B} \quad (3)$$

If we rearrange the above equation

$$\Delta G^0 = - RT \ln \frac{C + D}{A + B} \quad (4)$$

Since the equilibrium constant K for the reaction would be

$$K = \frac{C + D}{A + B} \quad (5)$$

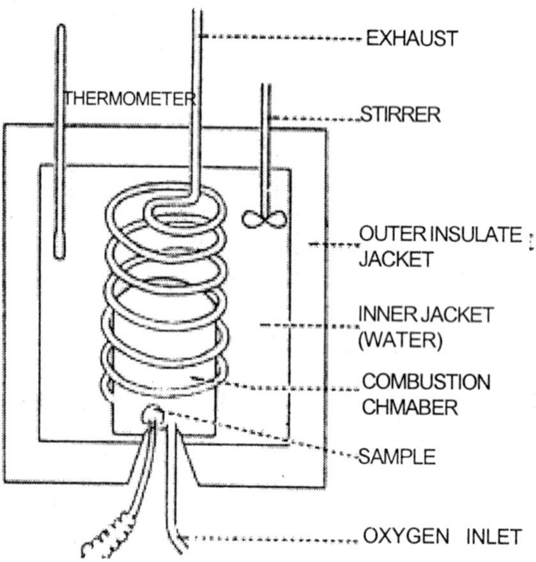

Fig. 15.2 Bioenergetics
A bomb calorimeter for determining heat of combustion of compounds

K can be substitued in the equation (4) to state as follows

G^0 = - RT in K or G^0 = 2.303 RT log K

The above equation shows that if we can calculate the equilibrium constant for any reaction, the standard free energy change can also be calculated. In the same way, if ▲ G^0 is known, ▲ G is calculated either in calories/mole or joules/mole depending on whether gas constant is given in calories (R=1.98 mol^{-1} K-1) or joules (R=8.31 J mol^{-1} k-1)

As a conclusion to the discussion, the following summarized statements may be made about standard free energy change

(a) "The standard free energy change of a given chemical reaction is the difference between the sum of the free energies of the products and the sum of the free energies of the reactants, each reactant and product being present in its standard state" (Lehninger, 1970). The standard condition on the various components in a reaction system are –
   (i) Cencentration of 1.0m
   (ii) Temperature 25°C or 298°k
   (iii) Pressure 1.0 atm

Each compound has a particular architectural pattern that decides its inherent free energy. Hence, the standard free energy ▲ G^0 of a chemical reaction may be expressed as follows

▲G^0 = EG^0 (product) - EG^0 (reactant)

(b) The standard free energy change may also be expressed in terms of the equation A+B______C+D. Here, ▲ G^0 is that quantum of free energy absorbed or lost per mole when molecules of A and B get converted into molecules of C and D, when the conditions of all the components are standard. The following table gives the free energy changes of some chemical reactions.

Standard free energy changes at pH7 and at 25°C of some chemical reactions
(Adopted from Arthur Giese, 1976)

Oxidation	**▲F**
Glucose + 60_2 ----- $6CO_2$ + $6H_2O$	-6,86,000
Lactic acid + 30_2 ----- $3CO_2$ + $3H_2O$	-3,26,000
Palmitic acid + $23O_2$ ----- $16CO_2$ + $16H_2O$	-2,338,000
Hydrolysis	
Sucrose + H_2O ----- glucose + fructose	-5,500
Glucose 6 phosphate + H_2O ----- glucose + H_3PO_4	3,300
Glycyglocine + H_2O ----- 2 glycine	-4,600

Rearrangement

Glucose 1, Phosphate ----- glucose 6. Phosphate	-1,745
Fructose 6, Phosphate ----- glucose 6. Phosphate	-400

Ionization

$CH_3COOH + H_2O$ ----- $H_3O_3+ + CH_3COO$ -	+6,310

Elimination

Malate ----- fumarate + H_2O	+750

A sample calculation of ▲ G^0

In order to elucidate the calculation of ▲ G^0 from the data obtained from equilibrium constant, the reaction in which the enzyme phosphoglucomutase can be considered. This enzyme catalyses the reversible reaction of glucose 1. Phosphate to glucose 6. Phosphate. We can take 0.020m glucose 1. Phosphate add the enzyme allowing forward direction change or take 0.020m glucose 6. Phosphate and allow the reaction to go in reverse direction. In either case, at equilibrium the medium contains 0.001m glucose. 1. Phosphate and 0.019m glucose 6. Phosphate at 25°C and at photosynthesis 7.0. The equilibrium constant can be calculated as follows

$$k = \frac{\text{glucose 6. Phosphate}}{\text{glucose 1. Phosphate}} = \frac{0.019}{0.001} = 19$$

From the value of k ▲ G^0 can be calculated as follows

▲ G^0 = RT in K

= - 1.987 x 298 In 19

= - 1.987 x 298 x 2.303 log 19

= - 1.745 cal mol^{-1} or - 7301 KJ mol^{-1}

Energy rich compounds

Living systems must get free energy from their environment in order to conduct life processes. Autotropic organisms obtain energy from sun light and their metabolism is linked to the exergonic process (respiration) to obtain energy. On the other hand, heterotrophic organisms couple their metabolism to the breakdown of complex to say that an endergonic process is always linked to an exergonic process.

In the living systems, a number of organic compounds are present which are energy rich (high energy compounds) because they undergo a large decrease in free energy change during their breakdown. These are also called energy currencies because they readily give out energy by undergoing hydrolysis. Some of these compounds are –

1. Pyrophosphate compounds
2. Acyl phosphates
3. Enolic phosphates
4. Thiol esters
5. Guanidine phosphates

Pyrophosphate compounds

Among the pyrophosphate compounds, the most important ones are ATP (Adenosine triphosphate) and ADP (Adenosine diphosphate), from the point of view of energy changes.

ATP

Historically, ATP was first discovered by C.F. Fiske and Y. Subbarow, in USA, and by K. Lohmann in Germany in 1929. After its discovery, it was thought that ATP was primarily involved in only muscle contraction reactions. Subsequent findings of Otto Warburg and Otto Meyerhof, indicated that ATP is generated from ADP during the anaerobic breakdown of glucose to lactic acid. Later, Kalckar of Denmark, and Belister of Russia, showed the generation of ATP (from ADP) during aerobic respiration also. Hydrolysis of ATP to ADP was demonstrated by Engelhardt and Lyubjmova during energy requiring reactions.

Finally, it was Fritz Lipman in 1941, who organized all the known facts of ATP into a general hypothesis concerning energy transformation reaction in living systems. Lipman opined that – ATP functions in a cyclic manner as a carrier ATP is used for various cellular activities. When fuel molecules undergo degradation, ATP is generated by the coupled phosphorylation of ADP. The ATP so generated can donate its third phosphate group for energy requiring processes such as biosynthesis of macromolecules, transport of substances across the membrane against the concentration gradient etc. As the energy from ATP is released, it undergoes cleavage to form ADP. ADP will be rephosphorylated to ADP through the energy obtained by degradation of fuel molecules thus completing the energy cycle. The terminal (third) phosphate group is the key substance that undergoes continuous cleaving and regeneration.

Fig. 15.3 Bioenergetics
Space filling model of ATP

Structure of ATP, ADP and AMP

The three compounds are basically built up on the structure of ribose of deoxy ribose nucleoside. Coupled with a Phosphate group, these occur in three states – 51 monophosphate, 51 diphosphate and 51 triphosphate. Here, 51 indicates that the phosphate group(s) is attached to the 5th carbon position.

For Adenosine nucleoside also, there are three types. These are –

(a) Adenosine monophosphate (AMP)
(b) Adenosine diphosphate (ADP)
(c) Adenosine triphosphate (ATP)

The phosphate groups of these are designated as α, β, γ. AMP has only phosphate group, ADP has α, and β, while ATP has α, β, and γ phosphate groups. The γ phosphate group is energy rich and undergoes cleaving and regeneration.

The structure of ATP was first deduced by Lohmann (1930), and subsequently confirmed through total chemical synthesis by Alexander Todd *et al,* (1948). The three compounds (ATP, ADP and AMP), are of universal occurrence in all living beings.
The sun total of these three is constant (between 2 to 10mm) in a given species. In actively dividing cells, the amount of ATP exceeds those of ADP and AMP.

At pH 7.0 in aqueous solution both, ATP and ADP are highly charged ions. ATP has four anions while ADP has three protons. Three of the four protons of ATP get fully ionized at pH 7.0 while the fourth one will be dissociated about 75% as it has a pH of 6.95. The high concentration of closely placed negative charge around the triphosphate group in ATP has a significant bearing on its ability to carry out energy transformations.

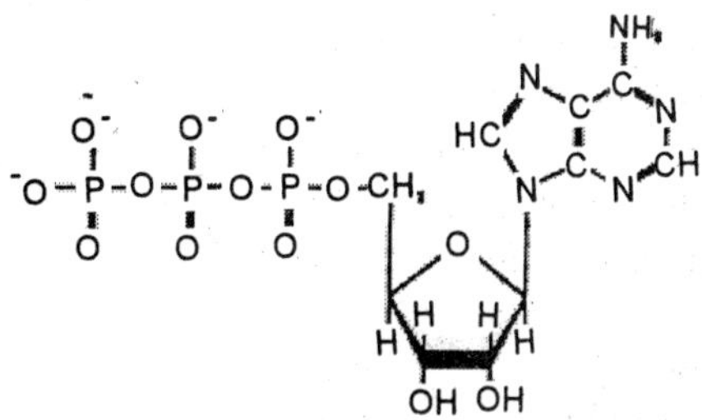

Fig. 15.4 Bioenergetics
Molecular structure of ATP

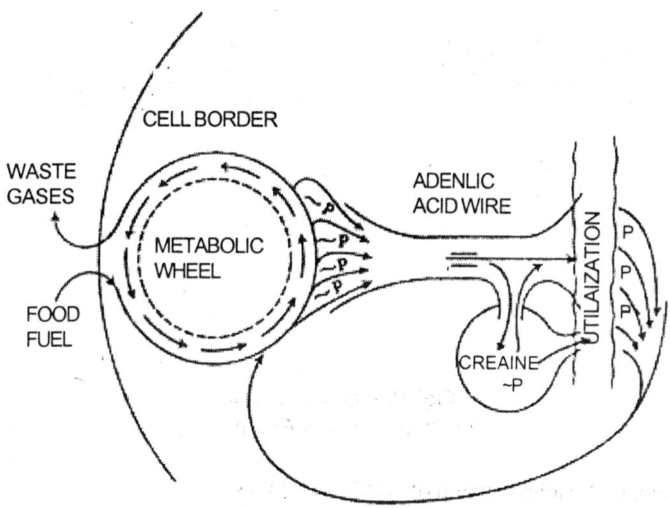

Fig. 15.5 Bioenergetics
The ATP cycle as depicted by Lipmann (1941)

Hydrolysis of ATP and Standard free energy

Theoretically, ▲ G^0 for ATP can be calculated by determining the equilibrium constant at pH 7.0. However there are several practical difficulties in this method as it is not easy to find out when exactly equilibrium has reached.

One of the easier methods is to find out of the additive values ▲ G^0 when ATP participants are in consecutive reactions. For instance, reaction of ATP with glucose and its subsequent breakdown into glucose + phosphate will help us to calculate the total standard free energy of hydrolysis of ATP.

1. ATP + glucose ———— flucose 6. Phosphate + ADP
   Here k = 661 therefore ▲ G^0 = 4.0 cal mol·
2. Here k = 171 ▲ G^0 = 3.3 k cal/mol

Adding 1 and 2 of the above (-4+3.3) we get - 7.3 k cal/mol as the total standard free energy available during the hydrolysisn of ATP.

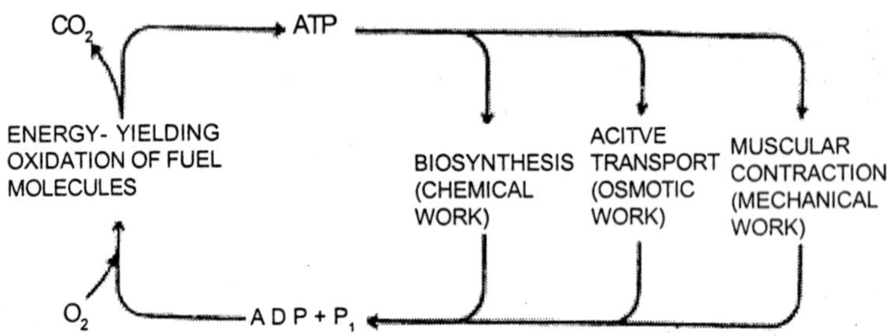

Fig. 15.6 Bioenergetics
The major uses of ATP energy

The second phosphate group of ADP also has a similar value (-7.3 k cal mol.), while the single phosphate group of AMP has G^0 value of - 3.40 k cal/ mol. This is due to the fact that the first phosphate group is linked to the nucleoside with an ester linkage, while the other two are anhydride linkages. It is well known that anhydride linkages have a great G^0 than ester linkages.

In the above discussion on energy rich pyrophosphates, attention is given to only ATP and ADP, as these participate in almost all energy transformation reactions, hence called universal energy currency. However, there are other energy rich pyrophosphate compounds such as GTP, GDP, CTP, CDP, UTP, UDP, dATP, dGTP,dTTP and dCTP (These correspond to the nucleotides of RNA and DNA). These, however, participate only in specific reactions. Thus, UTP is used in the biosynthesis of carbohydrates - polysaccharides); GTP is used in protein synthesis and CTP is utilized in lipid metabolism. These three along with ATP are involved in RNA synthesis, while dATP, dGTP, dCTP and dTTP are involved in DNA synthesis.

Another pyrophosphate compound that is energy rich is cyclic AMP (CAMP). Eventhough on hydrolysis, CAMP has a large negative free energy, it is not known to function due its high negative energy and unstable anhydride ring, instead, it acts as an allosteric effector and second messenger. Cyclic AMP is known to stimulate inactive enzymes (phosphorylase) into the active state.

2. Acyl phosphates

1.3 diphosphogylceric acid is an example of acyl phosphate. Its standard free energy on hydrolysis is – 11.8 k cal/mol. The strain of the C = 0 bond is responsible for a very high ▲ G^0 in this kind of compounds. High energy is required (during synthesis) in these bonds to overcome the repulsive changes of carbon and phosphorous atoms and consequently on hydrolysis a large amount of energy is released.

3. Enolic phosphate

An example of enolic phosphate is phosphoenol pyruvic and PEP. This is formed during the glycolysis when glucose gets converted intp pyruvate. On hydrolysis, it has a standard free energy of - 14800 cal/mole at pH 7.0

The large negative ▲ G^0 observed on hydrolysis of this compound, is due to stabilization of enolic form of pyruvic acid in PEP by the phosphate ester group. On hydrolysis, the unstable enol will instantly isomerize into the much more stable keto structure. It has been in ▲ G of about 8000 cal/mol thus bringing the total upto - 14800 cal/mole. This tautomerization of PEP makes it one of the most 'energy rich' compounds in biological systems.

4. Thiol esters

An example of thio ester is Acetyl CoA. This is also an energy rich compound which can be utilized to generate ATP from ADP. The ▲ G^0 of this compound on hydrolysis is approximately - 75v00 cal/mole. The higher negative value for ▲ G^0 is due to the fact that thioesters do not exhibit the resonance forms that is exhibited by ordinary O_2 atom. The sulphur in thio esters do not readily release the electrons for double bond formation and hence, resonance *forms possible* for oxygen ester. Usually, oxygen ester atom is knocked to the carboxyl acid carbon by means of a double bond. As the thio ester linkage is unstable (to S), the ▲ G is higher on hydrolysis.

5. Guanidine compounds

These represent a fourth type of energy rich compounds. A typical example of guanidine compound is guanidine phosphate that plays an important role in energy transfer and storage. Also known as *phosphagous,* these compounds are formed by the phosophorylation of creatine to arginine with ATP in the presence of appropriate enzyme. Hydrolysis of these compounds has a standard free energy of – 1–300 cal/mole i.e., about – 3000 more than that of ATP.

The guanidine phosphates are not inherently less stable because of bond strain as in ATP and ADP. There are no ionization or tautomerization processes which account for greater stability of reactants than their products.

In addition to the above mentioned 5 compounds which are energy rich, there are other compounds'. Some of these are, glucose 6. Phosphate, fructose 6. Phosphate etc. These have a standard free energy less than that of ATP (i.e., less than - 7300 cal/mole). The table below gives the standard free energy (on hydrolysis) of some important compounds.

Standard Free Energy on Hydrolysis of some important metabolities

	▲ G' at pH 7.0 (Cal/mole
Phospheoneolypyruvate	-14,8000
Cyclic - AMP	-12,800
1,3-Disphosphyglycerate	-11,800
Phosphocreatine	-10,300
Acetyl phosphate	-10,100
S - adenosylmethionine	-10,000
Pyrophosphate	-8,000
Acetly - CoA	-7,500
ATP to ADP and Pi	-7,300
ATP to AMP and pyrophosphate	-8,600
ADP	-6,500
UDP - glucose to UDP and glucose	8,000
Glucose - 1 - Phosphate	-5,000
Fructose - 7 - Phosphate	-3,800
Glucose - 6 - Phosphate	-3,300
sn - Glycerol - 3 - Phosphate	-2,200

Coupling reactions

In biological systems, ATP is the source of energy for most of the endergonic reactions. ATP itself is synthesized in exergonic reactions by the process of phosphorylation. The endergonic and exergonic reactions are always coupled so that the energy synthesized from one (exergonic) is used to drive the other (endergonic). The net effect of this coupling is the transfer of kinetic energy from one system to the other. This coupling is mediated by a number of specific enzymes. Coupling reactions or energy coupling for efficient energy transfer is a common phenomenon. Let us consider an example in energy coupling.

The conversion of phosphoenol pyruvate to pyruvate is an exergonic process (energy releasing).

The liberated energy is used for the formation of ATP from ADP. This ATP can be hydrolysed to provide energy for other reactions. For instance, formation of glucose 6. Phosphate from glucose and phosphoric acid requires energy to the tune of – 3k cal/mole. Phosphorylation of ADP to ATP however, has free energy of –4k cal/mole. Participation of ATP in this reaction provides energy for the formation of glucose-phosphate bonds and the extra energy (of about -1k cal/

mole) is liberated into the system (waste). The above reaction may be shown as follows :

1. Glucose + Pi ——————— glucose 6, Phosphate + H_2O
(This is an endergonic reaction requiring energy of about –3k cal/mole)
2. ATP ——————— ADP + Pi

(This is an exergonic reaction releasing energy of about –4 k cal/mole)
Here reaction 1 cannot take place unless there is energy input. In order to do this reaction 1 gets coupled with 2 mediated by the enzyme hexokinase as follows :

Glucose + ATP hexokinase glucose 6. Phosphate + ADP

Energy coupling reactions produce complex compounds apparently against the second law of thermodynamics. But it has to be noted that the algebraic sum of all the components of coupled reactions is negative i.e., the sum of energy liberated is more than the sum of energy consumed (-3 k cal/mole is used while the energy liberated -4k cal/mole). In biological systems, hydrolysis of energy rich compounds is always associated with an endergonic reaction in order not to waste the energy. This requires effective energy management which is provided.

Redox potential

The ability of any particular atom, or molecule, or ion to eject an electron (getting oxidized), is measured by its oxidation – reduction potential (ability to lose an electron - oxidation, and also receive an electron - reduction). This is known as redox potential and may be defined as *"The quantitiative measure of the affinity of a compound to lose or gain electrons"*.

During oxidative phosphorylation in mitochondria ATP is produced when electron are tossed from one acceptor to another. This flow occurs according to the relative affinities of the substrate to gain or lose electrons. This is accompanied by the alternate oxidation - reduction of the system. A compound losing an electron is oxidized, while the one accepting an electron is reduced. For instance, ferric ion is oxidized to ferrous ion by losing an electron, and when ferrous ion accepts an electron, it gets reduced to ferric ion.

$$Fe^{+++} \underset{\text{reduction}}{\overset{\text{oxidation}}{\rightleftharpoons}} Fe^{++} + e$$

A redox system can be compared to a dry electric cell. In the cell, the electrons are then transferred through a wire from one electrode to another, generating electric current. In such a system the capacity to lose or gain electrons is called the electrode potential. This can be measured through a standard hydrogen electrode. The electrode potential of a reducing oxidizing system also can be measured is a similar way. The electrode potential is here called the redox potential.

16

PHOTOSYNTHESIS

Ever since its origin, sometime in the remote past, life has been able to maintain itself in myriads of forms mainly because of its capacity of reproduction. An organism, however, has to exist,maintain itself in the healthy condition and only then can it reproduce. The existence of all forms of life on this planet has been made possible mainly by means of energy conversions. Organisms trap energy from environment which they utilize for their activities, and after a series of interconversions it (energy) is returned to the environment.

The phenomenon of energy input from the abiological environment to the biological system is always through the plants – green plants. Green plants thus, can be regarded as the basic fulcrum on which revolve all forms of life. In fact it is impossible to imagine what course the evolution of life would have taken but for the green plants. All forms of life are dependent on green plants (directly or indirectly) for their sustenance.

This unique phenomenon of energy input into the biological system is known as photosynthesis. The energy input has to be mediated only through the green plants, for they alone are capable of absorbing the energy from solar radiation.

Source of energy

There are only two sources of energy on this planet. One is, atomic energy and the other is solar energy. Of these two–atomic energy is not easily negotiable and available. Solar energy, however, is easily available in large quantities and living organisms have developed in the course of evolution mechanisms to trap and utilize this unique source of energy.

All forms of life, however, cannot directly harness solar energy. The capacity to trap solar energy is unique only for green plants and some bacteria. Green plants possess the green pigment viz., chlorophyll which can capture, transform, translocate and store energy which will be readily available for all activities of life.

Photosynthesis may thus be explained as the process in which the light energy will be converted into chemical energy. Except for green plants and some bacteria, all other organisms cannot directly utilize the solar energy; hence they are dependent on green plants for their nutritional requirements. Green plants and some bacteria are called autotrophic (they can prepare their own food), while all other organisms are called heterotrophic (they cannot prepare their own food). There is however one group of exceptional organisms which are autotrophic in spite of being non-green. These are the Chemosynthetic bacteria, which obtain energy by the oxidation of inorganic substrates such as ferrous ions and sulphur dissolved from the earth's crust of H_2S released from volcanic action. But considering the overall energy budget of lifeforms, the energy produced by chemosynthesis is very meagre and is of least significance quantitatively.

Definition

Photosynthesis (Photo = Light), may be defined as the process in which CO_2 and H_2O are put together (by green plants) into an organic molecule (sugar) utilizing the solar energy.

$$CO_2 + H_2O \xrightarrow[\textit{Green Plants}]{\textit{Light}} \text{Sugar (Carbohydrates)}$$

Photosynthesis may also be defined as the reduction of CO_2 to glucose ($C_2H_{12}O_6$) with the **H** necessary for reduction coming from water.

Photosynthesis involves both oxidation as well as reduction ultimately in synthesize energy rich compounds. Water is oxidized (electrons are removed), and the **H** released from water is used to reduce CO_2 to carbohydrates. Oxidation of water and the units of light energy (absorbed) produce energy intermediary compounds called ATP (Adenosine Triphosphate), and reduced coenzymes called NADPH + $H_{+|}$ (Nicotinamide adenosine diphosphate). These two compounds help in the fixation (reduction) of carbon dioxide into carbohydrates. These carbohydrates in turn constitute the starting point for the synthesis of various other metabolities necessary for the survival of the organism.

The definition and brief explanation of photosynthesis given above sums up the significance of the process. A few other definitions (different ony in the usage of the words) of photosynthesis are given below.

1. "Biosynthesis of simple carbohydrates from CO_2 and H_2O in the presence of sunight inside chlorophyll containing cells".
2. "Production of carbon containing compounds from CO_2 and hydrogen down by illuminated green cells".
3. "Photosynthesis is a redox process"
4. "Photosynthesis is a sensitized photochemical oxidation and reduction mechanism between water (H_2 donor) and CO_2 taking place at biological temperature".
5. "Photosynthesis is the formation of carbon containing compounds from CO_2 and water by illuminated green cells, water and oxygen being the by products".
6. Photosynthesis deals with the capturing and transforming of light energy into chemical energy".

The overall reaction of photosynthesis may be represented as follows

$$6CO_2 + 6H_2O \xrightarrow[\textit{Chlorophyll}]{\textit{Light}} C_6H_{12}O_6 + 6O_2$$

Oxygen which is released as a byproduct originates from water (Ruben and Kamen, 1941). In order to account for this, the equation should be changed as follows –

$$6CO_2 + 12H_2O \xrightarrow[\textit{Chlorophyll}]{\textit{Light}} C_6H_{12}O_6 + 6O_2 + 6H_2O$$

MAGNITUDE OF PHOTOSYNTHESIS

As has already been pointed out, photosynthesis is the most important biological process that supports all forms of life. Either directly or indirectly, all forms consume photosynthetic energy. From the point of view of human welfare, the contribution of photosynthesis is immense. Coal, gas oil etc., represent the photosynthetic products of the plants belonging to early geological periods.

The significanse of photosynthesis as a vital process contributing to the survival of life can be understood when we take into consideration the overall; rate of photosynthesis.

According to an estimate, the overall area under green vegetation on this earth is approximately 510 million sq.km. Of this, oceans alone account fo r361 million sq.km.

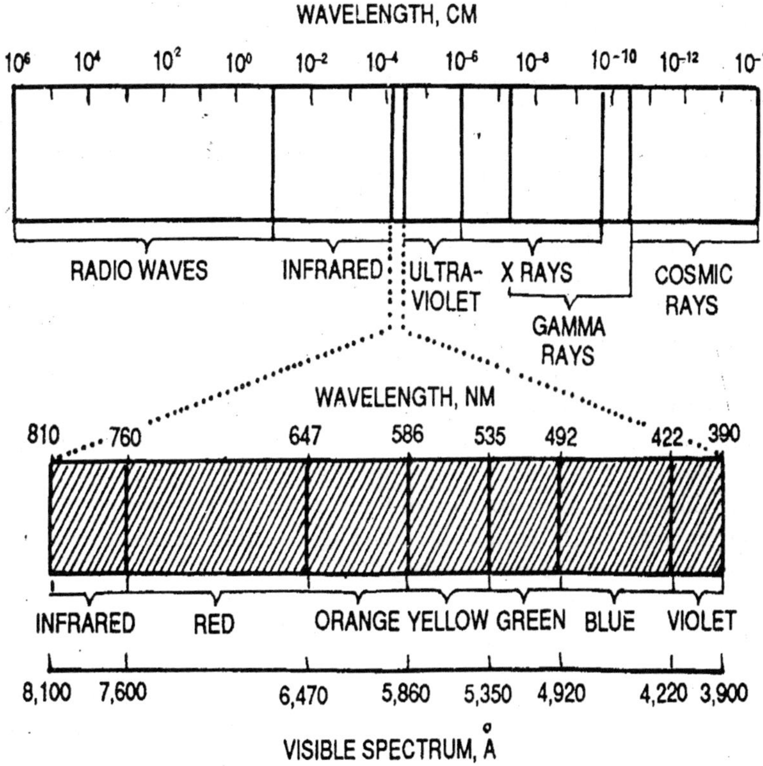

Fig. 16.1 Photosynthesis
Electromagnetic Spectrum

Regarding the availability of light, solar radiation which is emitted from the sun in the form of electromagnetic radiation reaches the outer atmosphere. In terms of energy, the outer atmosphere receives about 1300 x 10^{21} cal/year. Much of this is shielded by the atmosphere and the surface of the earth receives light equivalent to 650 x 10^{21} cal/year. Considering the distribution of water and land masses, marine plants receive about 90 x 10^{21} cal/year and terrestrial green vegetation receives about 25 x 10^{21} cal/year. If we assume photosynthetic efficiency is 2% of this energy, it amounts to 50 x 10^{9} tons of organic carbon for terrestrial green plants and 180 x 10^{9} tons for marine plants.

Rabinowitch (1951), calculates that photosynthesis produces approximately 155 x 10^{9} tons of organic carbon of which 90% is contributed by the oceans. However, some of the recent estimates (Rhyer and Wood Well, 1970) suggest a ratio of 1:3 fixation of carbon between terrestrial and marine plants. Of the total amount of light received on earth's surface (650 x 10^{21} cal/year), about 60% is in the form of very short and very long wavelength (far and infra red) radiation and does not participate in photosynthesis. This leaves a total of 180 x 10^{21}cal/year for oceans and 80 x 10^{21} cal/year for terrestrial area. Added to this, is the fact that more than half of the land area is devoid of vegetation. As a result (efficiency of photosynthesis being 2%), this represents only 0.5 x 10^{21} cal/years. Sheoder estimated the production of organic carbon by land plants to be about 16 x 10^9 tons, while Leibeg and Ebermayer estimate the carbon production by land plants at 30 x 10^9 tons and 24 x 10^9 tons respectively. In oceans, where planktons and other plants absorb about 50% of the incident light, the estimates come to 1.8 x 10^{21} cal/year or 180 x 10^9 tons of organic carbon.

HISTORICAL ACCOUNT

Eventhough scientific study of photosynthesis is very recent, from time immemorial man has looked upon the Sun as the source of ale light and sustenance. Early writings of Indians, Romans and Greeks bring out the idea very clearly. The Rig veda says that the 'Sun is the soul of the entire world'. *Bhagavadgita* also mentions that all forms of life are dependent on the Sun. Apart from these early writings, the following may be mentioned as significant names in the development of photosynthetic knowledge.

Aristotle (257 - 80 B.C.), **Theoprastus** (370 - 285 B.C.), and others up to the 17th century believed that plants obtained their nutrition from the soil, and the Sun had nothing to do with plant nutrition.

Van Helmont (1648)

In this experiments with willow seedlings, he observed that water, and not the soil was the principle source of nutrition.

Wood Ward (1699)

Opined that the plant body is made up not of water, but a peculiar soil substance which is absorbed along with water.

Stephen Hales (1927)

Who is regardes as the **Father of Plant Physiology** proposed that green plants obtained their nutrition partly from the air through the leaves, and sunlight may have a role in it. He, however, was not aware of the part played by carbon dioxide.

Joseph Priestley (1972)

Who conducted several experiments on mint plants stated that plants can purify air. He also observed that impure air (CO_2) would get purified (vital air - O_2) if kept in contact with the green mint plants. Priestley also knew nothing about the role of either CO_2 or light in photosynthesis.

Jan Ingen Housz (1771)
A Dutch physician, demonstrated that plants purified air only in the presence of light. He also stated that plants too contributed to 'bad air' during darkness.

Jean Zenebier (1782)
A Swiss minister agreed with the findings of Ingen Housz, and commented that plants absorb CO_2 from the atmosphere during daytime and exhale it (CO_2) during darkness. The role of two gases (O_2 and CO_2) had been implicated in plant nutrition by 1782. Works of Lavoisier and others identified that these gases were O_2 and CO_2.

N.T. de Saussure (1804)
Identified that water is an active participant in photosynthesis together with CO_2. He also made the first quantitative measurements of photosynthesis and opined that approximately equal volumes of CO_2 and O_2 are exchanged during photosynthesis.

Mayer (1842)
Who formulated the law of conservation of energy, observed that during photosynthesis solar energy is stored in the form of chemical energy.

In 1871, two French Chemists, Joseph Bienaime Caventou, and Piere Pelletier isolated the chlorophyll pigment from green plants. Richard Willstatler obtained chlorophyll in the pure form. He also identified two forms – **Chlorophyll a** and **Chlorophyll b** which differed in the manner of their light absorption.

Leibeg (1843)
Stated that the source of all carbon in plants is the CO_2 of the atmosphere.

Julius Sachs (1846)
A German plant physiologist, observed the growth of starch grains in choloroplasts, only in the area of the leaf exposed to light. Thus, he was the first one to note the increase in organic content due to photosynthesis.

Boussingault (1846)
Calculated the ratio of CO_2 intake and O_2 release, and came to the conclusion that it is 1:1.

Baeyer (1870)
Proposed the formaldehyde hypothesis. He argued that CO_2 gets reduced and

forms CH_2O which on polymerization forms the glucose molecule. The formaldehyde hypothesis, however, has been abandoned not only because it has not been detected in plants, but it is also poisonous (to the plants), if it is present.

Blackmann (1905)
Identified that photosynthesis involves two distinct phases if it is present.

Blackmann (1905)
Identified that photosynthesis involves two distinct phases *viz.,* light phase and dark phase.

Van Niel (1931)
Showed that photosynthesis is a biological oxidation reduction reaction.

Robin Hill (1937)
Showed that photosynthesis is localized in the chloroplasts. In the presence of certain electron acceptors (oxidizing - reducing agents), isolated chloroplasts accept light energy and release O_2. This phenomenon of illuminated (isolated) chloroplasts splitting water and releasing O_2, has come to be known as Hill's reaction.

Samuel Ruben, Randall Kamen and Hyde (1941)
Showed (using radioisotopes of oxygen) that the oxygen released during photosynthesis comes only from water and not from CO_2.

Consden, Gordon and Martin (1944)
Developed paper chromatography technique for the separation of amino acids. This technique later was employed to detect the various intermediate compounds during the dark fixation of CO_2.

Arnon (1951 - 60)
Developed the concept of photophosphorylation and showed that coenzyme II picks up the H released by splitting up of water.

Benson, Bassham, Melvin Calvin and others have worked out in detail the carbon reduction cycle with the help of techniques like chromatography and radioautography.

RAW MATERIALS FOR PHOTOSYNTHESIS

There are some basic requirements (raw materials) for the process of photosynthesis. These are CO_2, a hydrogen donor (water), and light energy, besides of course the structural framework of the green plant in the form of chloroplasts.

Carbondioxide

The only source of carbon for plants for the purpose of assimilation is the atmospheric carbon dioxide. Approximately about 20 x 20^{11} tons of carbon dioxide is used for the production of 50 x 10^9 tons of organic carbon. Besides atmosphere, other sources of carbon are biological respiration decaying organic matter etc. Carbon dioxide is in the gaseous form in the atmosphere, while it is in dissolved condition in water.

Carbon dioxide is present in trace quantities in the atmosphere (o.0-35%). There has been a gradual increase in the concentration of the atmospheric carbon dioxide since the middle of 19th century perhaps due to increased pollution (Wood Well, 1970).

Hydrogen Donor

In almost all photosynthetic organisms, water is the hydrogen donor. Water splits into H and OH ions in the presence of light. H is picked up by the enzyme nadph for reduction of CO_2, while OH recombines to form water and oxygen.

Certain chemosynthetic bacteria, however, use compounds of Iron and Sulphur as hydrogen donors.

Light

Solar radiation is the only source of light for all forms of life. The radiation from Sun is emitted in the form of electromagnetic radiation with radiowaves at one end and cosmic rays at the other. Much of this radiation light shielded by the outer reaches of the atmosphere and only visible Light may be defined as the visible (to human eye) part of the electromagnetic spectrum. The visible spectrum is confined between wavelengths of 390mμ and 760 mμ. Spectral analysis of the visible light shows the following rays with their respective wavelengths given in parenthesis.

Violet	(390 - 430 mμ)
Blue	(430 - 470 mμ)
Bluegreen	(470 - 500 mμ)
Green	(500 - 560 mμ)
Yellow	(560 - 600 mμ)
Orange	(600 - 650 mμ)
Red	(650 - 760 mμ)

Rays of the radiation with wavelengths shorter than violet are called U.V. (Ultra violet - 100 A^0 - 390 A^0), X rays (0.1 A^0 - 100 A^0), Gamma rays (0.001 A^0 - 0.1 A^0) and cosmic rays (less than 0.001 A^0). Radiation with wave lengths longer than red are called infra red (7600 A^0 - 1000,000 A^0). Beyond this, are electric and radio waves, which are measured (in terms of wavelength) in kilometers.

Radiation travels in the form of discrete corpuscles or particles called **Quanta** or **Photons.** Each quantum represents a unit of energy. Energy quantity of each ray depends on its wavelength; shorter the wavelength, higher is the energy. The wavelengths of radiation are so small that they are measured not in terms of millimeters or centimeters, but in terms of microns (μ), millimicrons (mμ) or nanometers (nm) or Angstrom (A^0) units.

$1\ A^0 = 10^{-10}$ or 10^{-8} cms.

Each photon has an energy constant $E = hc/\lambda$, where h is planck's constant (=6.625 x 10^{-27} erg/sec) c is velocity of light (3 x 10^{10} cms/sec), and λ is the wavelength as given by Einstein.

The energy value of different light rays is given in the following table.

Energy Value of Different Wavelengths of Light

Sl.No.	Wavelength in mm	Approximately colour	Energy in cal
1.	300	Ultra Violet	95,200
2.	420	Violet	68,000
3.	470	Blue	60,760
4.	530	Green	53,890
5	620	Orange	46,060
6.	700	Red	40,800
7.	683	Red	41,810
8.	672	Red	42,490
9.	650	Red	43,940
10.	850	Infra Red	33,590
11.	1000	Infra Red	28,560

THE PHOTOSYNTHETIC APPARATUS

Within the cell of the green plants are present certain unique organelles called plastids. These perform many vital functions in cell physiology. Among the plastids, the most important from the point of view of photosynthesis are chloroplasts. All the physiological reactions, starting from the absorption of light energy to the fixation of CO_2 take place in the chloroplasts. No wonder then, chloroplast is aptly called the photosynthetic apparatus.

CHLOROPLAST

These are by far, the commonest and the most important among plastids. As the primary sites for trapping and converting solar energy, they are very vital for the existence of not only the green plants, but for the whole living world.

Shape

Chloroplasts have varied shapes. The algal chloroplasts exhibit an array of shapes. They may be cup shaped (*Chlamydomonas*), girdle shaped

Size

There is a great variety in size. Normally, they are about 1 μm thick and 4–6μm in length. Chloroplasts of polyploid cells are generally larger than in the diploid cells. Cells in the shaded area have longer chloroplasts but with less intensity in colouration.

Distribution

They are uniformly, distributed all over cytoplasm, but in some instances they cluster towards the nucleus. The concentration of chloroplast will also depend on light intensity.

Number

The number of chloroplast per cell varies. In some algae, like *Chlamydomonas,* there is only one chloroplast per cell, or two as in

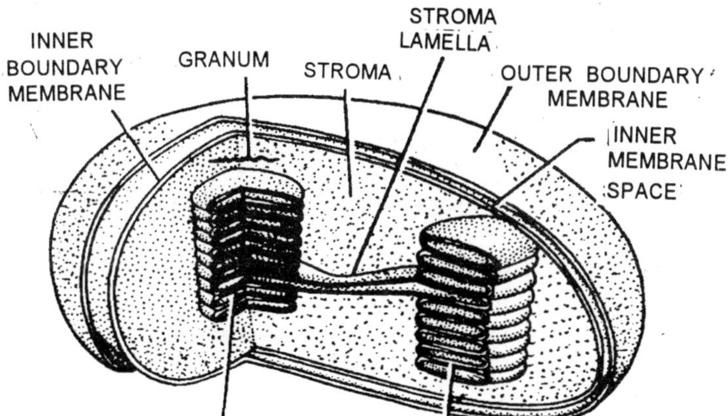

Fig. 16.2 Photosynthesis
Model of Ultrastructure of chloroplasts

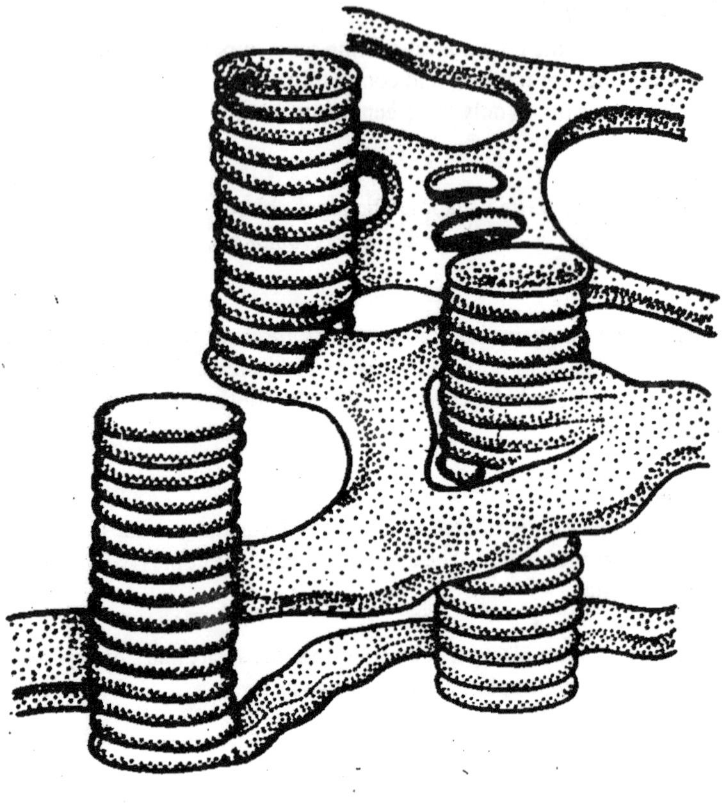

Fig.16.3 Photosynthesis

Three dimensional view of Grana (enlarged view)

Zygnema. In higher plants, cells have a chloroplast. The leaf of *Ricinus communis* has nearly four lakh chloroplasts per square millimetre area.

Chemical composition

The chloroplasts contain carbohydrates, lipids, proteins, chlorophyll, Carotenoid and xanthophylls. The protein lipid ratio is 40:30. Besides the above, nuclei acids have also been demonstrated in the chloroplasts.

Ultra structre of the chloroplasts

The chloroplast has a covering of two membranes with an inner membrane space. These membranes are smooth and there are no perforations or particles. The membranes are differentially permeable (Mudrak 1969).

A section of the chloroplast reveals an intricate system of membranes enclosed in a granular matrix. These membranes are called *lamellae* and the surrounding matrix – the *stroma.* In a sectional view, the lamellae can be seen packed and these stacks are called *thylakoids.* In the chloroplasts of higher plants, the thylakoids themselves form highly compact bundles called **grans.** Some thylakoids of granum extend into the stroma and maintain contact with other grana. These are called – Stroma thylakoids or stroma lamellae or intergrana.

The chloroplasts of algae lack the granum arrangement and in cyanobacteria (blue green algae), the thylakoids lie naked in cytoplasm without any envelope. In such instances, pigments are uniformly distributed on or in the lamellae.

The stroma has in addition to the lamellae – granules (globuli), lipid droplets, starch grains and vesicles.

Structure of the lamellae

All the photosynthetic pigments are concentrated in the chloroplast, lamellae as well as the outer membranes are composed of lipoprotein, sub units. Each sub unit according to Weir and Benson, has a protein core surrounded by a lipid sheath. The chloroplast membranes have a single layer of sub units, whereas in thylakoids, they are double due to the appressing of two membranes. The appressed areas are called partitions. These are hydrophobic in nature. The space between the two membranes of a thylakoid (fret membrane), enclose a space called fret channel. The space between two thylakoids is called loculus while the end of disc shaped thylakoids are called margins. These areas (loculus, fret channel) are hydrophilic. The chlorophyll molecules are concentrated in the fret cannels i.e., in the space between the two membranes of thylakoids. The chlorophyll molecules may be entirely in space, or the heads may be in the fret channel, while the tails are buried into the sub units. Occassionally, the entire chlorophyll molecule may be found in the fret membrane. Other pigments like cytochromes, carotenoids etc are also found in the fret membranes.

Four sub units of the partition consitute a photosynthetic unit or a *quantosome.*

Pigments in chloroplast

Some of the important pigments present in chloroplasts are chlorophylls, Carotenoids, cytochromes etc. In algal cells, the thylakoids also have phycobilins.

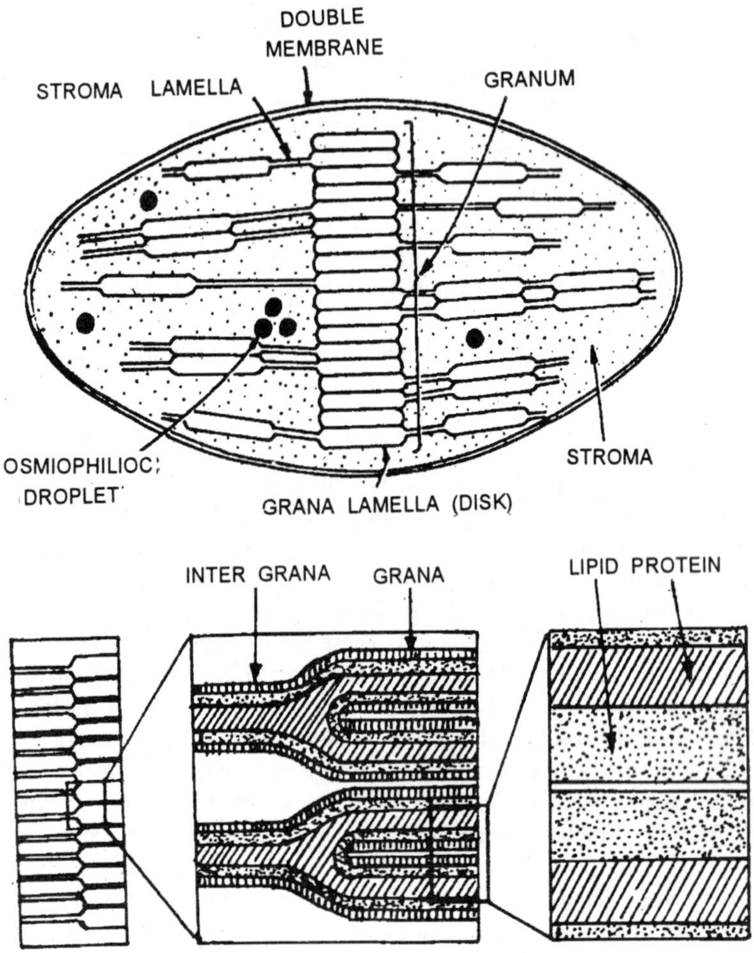

Fig. 16.4 Photosynthesis

Diagrammatic representation of Ultrastructure of chloroplast showing the lamella, **A.** Entire chloroplast, **B-D.** Show successive enlargement of a part of the granum

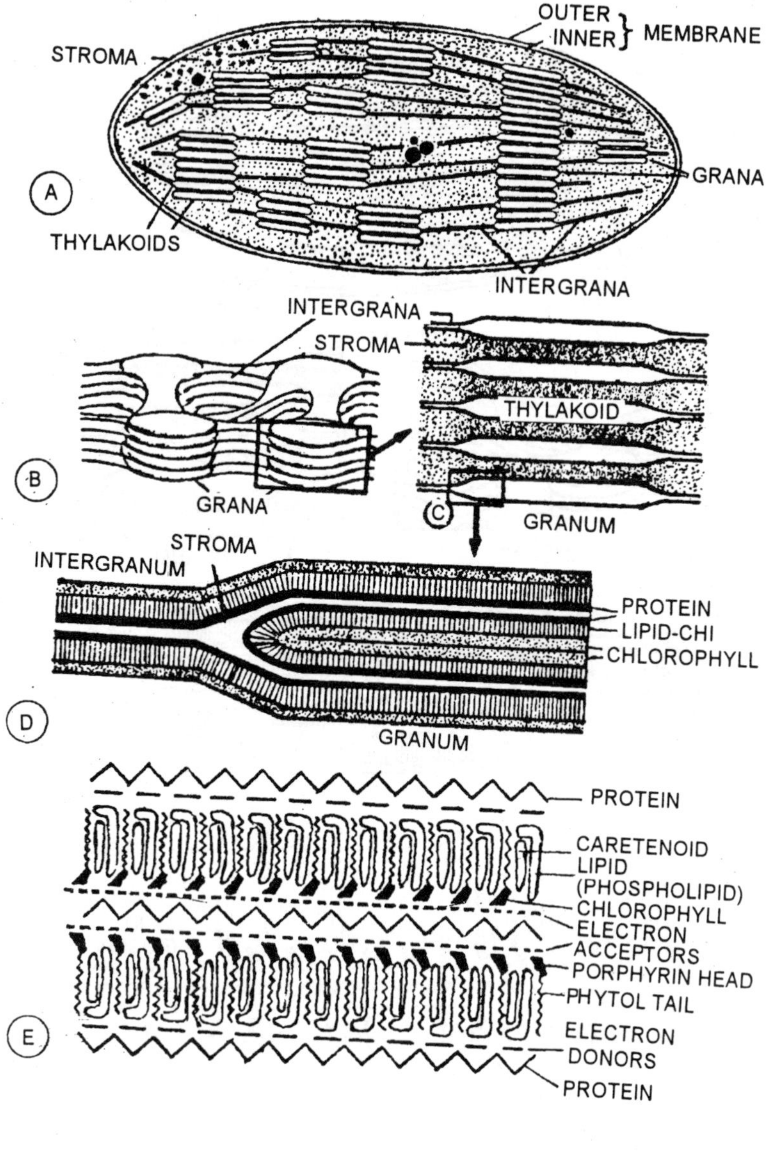

Fig. 16.5 Photosynthesis

Stages in the development of chloroplast**A.** Entire chloroplast, **B.** Grana and Intergrana (diagrammatic), **C.** Section through a granum, **D.** Lamellar structure (in detail), **E.** Model for the arrangement of pigments in a granum

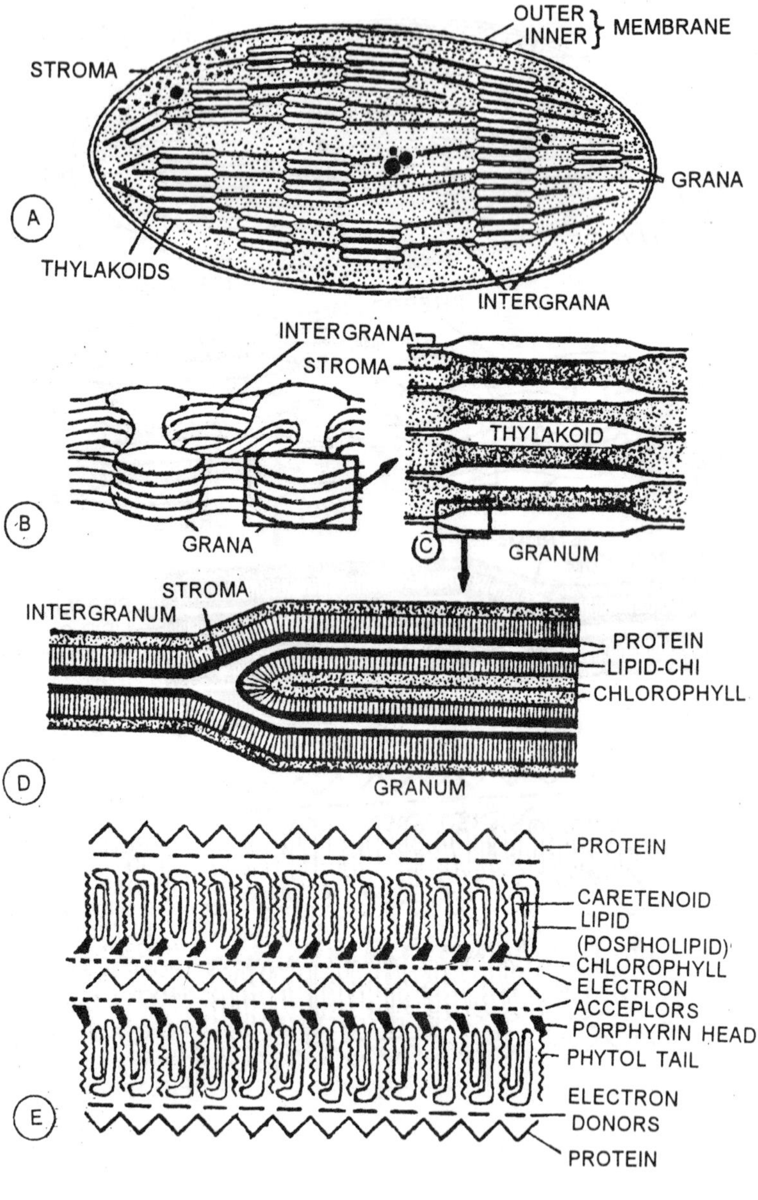

Fig. 16.6 Photosynthesis

A. Entire chloroplast, **B.** Grana and Intergrana (diagrammatic), **C.** Section through a granum, **D.** Lamellar structure (in detail), **E.** Model for the arrangement of pigments in a granum

Chlorophylls

These are the most important pigments involved in the basic photochemical reaction that sustains life. Atleast nine types of chlorophylls have been distinguished so far – chlorophylls *a,b,c,d* and *e.*, bacterio chlorophylls *a* and *b,* and the chlrobium chlorophylls 650 and 660 μm.

Among the chlorophylls, chlorophylls, and b are best known and most widely distributed. These are absent in pigmented bacteria. Chlrophylls *c*, *d* and *e* are found only in algae in combination with chlorophyll a. Bacterio chlorophylls and chlorobium chlorophylls are found only among photosynthetic bacteria.

The chlorophyll molecule is a complex structure basically made up of a 'head' and a 'tail' resembling a tennis racquet. The head is a porphyrin structure made up of four Pyrrole rings attached to each other at the centre by an isocyclic ring containing Mg atom at the centre. Extending from one of the pyrrole rings is the tail–the alcoholic chain (Phytol). The emperical formula for the chlorophyll molecule is $C_{55}H_{72}O_5N_4$ Mg. The phytol chain is esterified on the C atom of one of the pyrrole rings and has only one double bond.

Chlorophyll *a* and *b* differ structurally in having different atomic groupings at the C_3 atom in the pyrrole ring. In chlorophyll *a,* C_3 atom has a methyl group, while chlorophyll b has an alehyde group; besides the two pigments have different absorption spectra. The peaks are as follows.

Chlorophyll *a* 429, 410, and 66nm
Chlorophyll *b* 430, 453 and 442 nm

(The above spectra are of chlorophyll *in vitro*)

Carotenoid pigments

These are lipid compounds ranging in colour from yellow to purple. They are widely distributed in both plants and animals. They are also present in microorganisms including red algae, cyanobacteria, photosynthetic bacteria, fungi etc. Wackenroder (1931), isolated the first Carotenoid – carotene from, the carot tissue.

Carotenoids are derivatives of lycopene, a red pigment found in many plants. They (Carotenoids) are known to have eight isoprent (CH2=C [CH3] - CH = CH2) like residues.

The main Carotenoid found in plants is the orange yellow coloured β carotene with some quantity of α carotene.

The Carotenoids are also located in the chloroplast. According to Goodwin

(1960), Carotenoids and chlorophylls may be combined with the same protein to form a complex known as photosynthein.

Carotenoids protect the chlorophyll from photooxidation and transfer the light energy they absorb to chlorophyll *a*.

Phycobilins

These are found only in algae. There are two known types of phycobilins – the phycoeroythrin (red), and phycocyanin (blue). These pigments are strongly associated with a protein and consequently it is difficult to isolated the pigment in the pure state.

Like Carotenoids, phycobilins are also involved in the transfer of light energy (they absorb) to the chlorophyll.

The absorption spectra of phycobilins presents an interesting study considering the fact that they act as accessory pigments in photosynthesis. R - phycoerythirn has peaks at 495, 540 and 545 nm, while R-phycocyanin has peaks at 550 and 615 nm.

Photosynthetic units

A photosynthetic unit may be defined as "the smallest group of coordinating pigment molecules necessary to effect a photochemical act, i.e. absorption and transportation of a light quantum to a trapping center where it causes excilation and release of an electron".

Emerson and Arnold (1932), who experimented in Chlorella plants, observed that about 2,500 molecules of the chlorophyll pigment are necessary to fix a molecule of CO_2 and they termed this number (2,500), as a photosynthetic unit. They also observed that to release one molecule of O_2, 10 quanta of light are necessary. Some of the recent works, however have shown that on the basis of absorption of 10 quanta, photosynthetic unit should consist of 230 molecules of the chlorophyll pigment. Park and Beggins (1964), working on the chloroplasts opined that photosynthetic units are distinct morphological identities and called them quantasomes. The quantasomes are 180 A^0 broad and 100 A^0 thick (`8 x 16 x 10nm). Each quantasome has 230 molecules of chlorophyll (160 chlorophyll a and 70 chlorophyll b), 48 carotenoids, 46 quinone compounds, 116 phospholipids, 144 digalactosyl diglycerides, 346 monogalactosyldiglycerides, 48 sulfolipids, unidentified lipids and many sterols. Quantasomes have a molecular weight of about two millions and are arranged in a monolayer on the thylakoid bodies. Electron microscopic studies have revealed that the individual quantasome may have four or more sub units. This middle region of the quantasome is called the **reaction centre.**

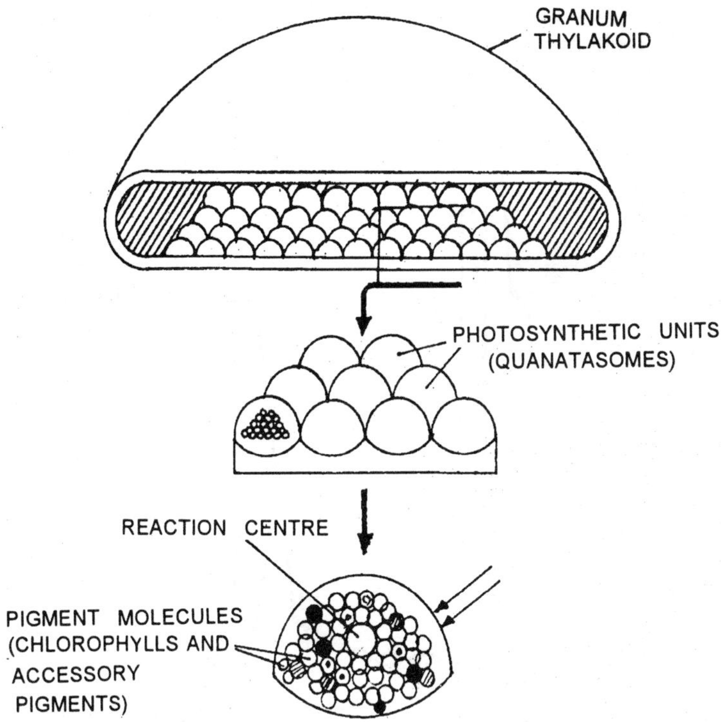

Fig. 16.7 Photosynthesis
Diagrammatic representation of a thylakoid, quantasomes and the reaction centre

Robert Huber, Johann Deisenhofer, and Jartmutt Michel were awarded the 1988 chemistry Nobel Prize for unravelling the secret of the reaction centre. Working at the Max Planck Institute in Martinsried, Germany, these scientists, made their breakthrough by extracting the reaction centre from the membranes of certain photosynthetic bacteria. They then crystallized it and analysed the structure using X - ray diffraction.

Photosynthetic pigments

The most important pigments involved in photosynthesis are chlorophylls and carotenoids. Other pigments present, but not involved in photosynthesis are biliproteins or chromatoproteins.

Chlorophylls

There are five important types of chlorophylls. These are Chlorophyll *a,b,c d* and *e*. In photosynthetic bacteria, two Chlorophyll–like pigments–bacteriochlorophyll and bacterioviridin have been reported.

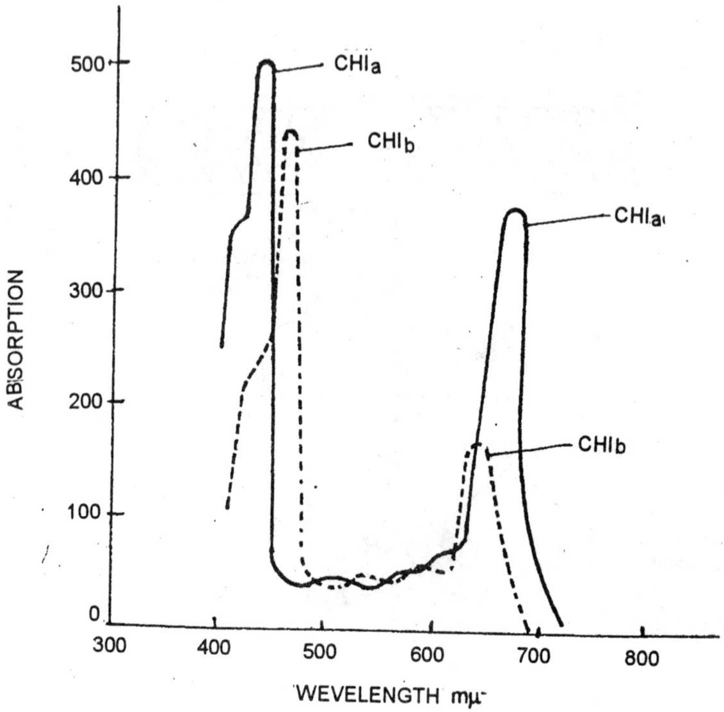

Fig.16.8 Photosynthesis
Absorption spectra of chlorophyll a and b

Chlorophyll *a* is present both in higher plants and algae. The pigment is insoluble in water but readily soluble in either. It appears blue green in transmitted light, but reddish in reflected light. Chlorophyll *a* is the principal pigment involved in the trapping of light energy.

Chlorophyll *b* is also found in green and higher plants. It is yellowish green in transmitted light but reddish in reflected light. It is similar to chlorophyll a in solubility. Chlorophyll *b* is absent in Cyanophyceae, Rhodophyceae Phaeophyceae and diatoms.

Chlorophyll *c*, *d* and *e* are present in brown, red and golden brown (Xanthophyceae) algae respectively.

Carotenes and Xanthophylls. They are insoluble in water but soluble in organic solvents.

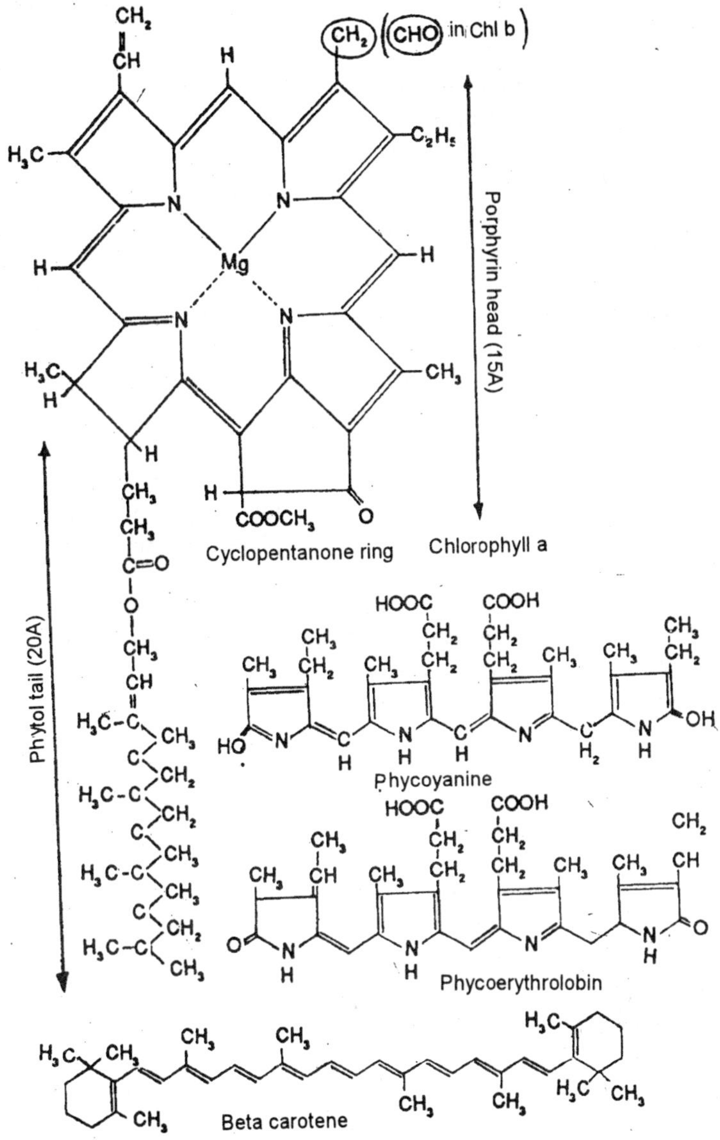

Fig. 16.9 Photosynthesis
Molecular structure of photosynthetic pigments

Photosynthetic apparatus in prokaryotes

In photosynthetic prokaryotes, the pigments are not localized into definite organelles as in eurkaryotes. But they do have some organization in some membrane systems.

Photosynthetic apparatus in bacteria

Phototrophic purple bacteria are rich in intracytoplasmic membrane system. In ultra thin sections of bacterial cells these appear as tubules, vesicles and plates. In *Rhodospirillium ruburm* and *Chromatium* spp, the cell lumen is almost packed with closely placed spherical vesicles. The vesicles seem to be formed by invagination and tubular accretion of the plasma membrane. The tubular structures have constrictions which at intervals disturb the vesicular arrangement, but the tubes do not break completely. On lysis of the cells (during investigations), and differential centrifugation, the tubular structures are liberated as isolated vesicles. These membranous vesicles have a diameter of 500-600 A^0, and can be isolated easily. Schachman, Pradec and Stainer (1952), considered these particles and named them chromatophores. The chromatophores are capable of photophosphorylation and contain chlorophylls, carotenoids, cytochromes, quinones, proteins, lipids and many other compounds that are necessary for carrying on photosynthesis.

Chromatophores appear in various shapes such as disc, shaped, cup shaped, hemispherical or spherical depending upon the organism. Gibson (1965), who has made a study of bacterial chromatophores, reports that they are spherical or nearly so and have a diameter of 540A^0 and contain core pigments surrounded by 25 to 55A^0 thick lipid layer. The chromatophores are not distinguished into grana or stroma and the pigment molecules are not specifically localised as in chloroplasts. Some bacterial species are very rich in carotenoids.

In some marine purple bacteria as *Eclothiorhodospira mobilis,* the organization of photosynthetic pigment is somewhat more advanced in comparison to chromatophores. Here the vesicular bodies are flattened and stacked into regular plates. These are called thylakoid plates as they bear a resemblance to thylakoids of eukaryotes. Here, 10-16 thylakoid plates are stacked into bundles resembling the granum of chloroplasts.

In the cells of green sulphur bacteria. the photosynthetic pigments are bound to two separate cell structures. The antenna pigments are concentrated in the chlorosomes, whereas the pigments of reactions centre are in ther cytoplasmic membrane.

The photosynthetic pigments of phototrophic purple bacteria are bound to intracytoplasmic membranes.

Photosynthetic apparatus in cyanobacteria

In cyanobacteria, the peripheral region of cytoplasm contains the photosynthetic apparatus. In unicellular forms, this commonly consists of a few membrane layers which extend in concentric sheets below the cell membrane and surround the central nucleoplasmic zone. In filamentous members, the cells have more elaborate membrane systems formed due to expansion and invagination of the membranes. The basic structure of the membranes looks like a flattened sac and is designated as the thylakoid. The membranes which are closely thylakoid sac.

The thylakoid lamellae have a lipid bilayer incorporating the lipophilic photosynthetic pigments – chlorophyll a and various carotenoids. In addition, it also incorporates cytochromes, plastocyanin and other components of the electron transport chain. Chlorophyll *a* which is the only chlorophyll present in cyanobacteria exists in three spectral variations (with peaks at 670, 680 and 700nm). Among the carotenoids, which absorb in the range between 480 and 520nm, β carotene appears to be widely distributed in cyanobacteria. Xanothophylls, however, differ in their abundance as well as presence in various species of cyanobacteria. Some of the xanthophylls seen in cyanobacteria are echineone, zeaxanthin, oscillaxanthin, myxoxanthophyll etc.

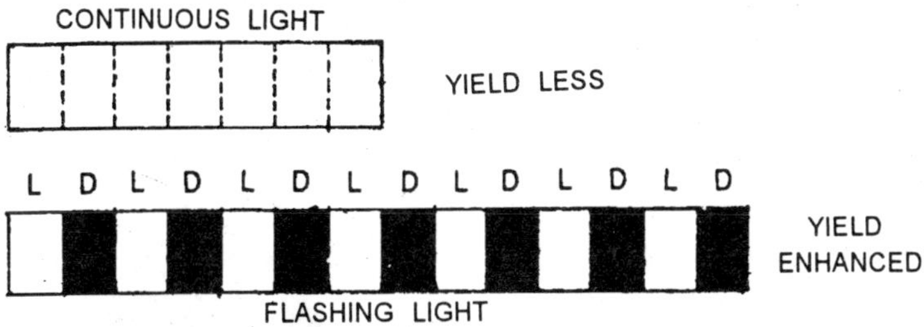

Fig.16 .10 Photosynthesis

Red drop and Emerson effect in Chlorella

A significant part of the photosynthetic pigment complement of cyanobacteria is located in some granular supermolecular complexes such as phycobilisomes. These contain the chromophore bearing water soluble phycobiliproteins namely blue phycocyanin, allophycocyanin and phyceorythrins. The relative quantities of these pigments may vary according to species and the spectral composition of light. It is that which determines the varied colour appearance of the organisms.

The table belwo gives an account of the various classes of pigments present in cyanobacteria.

The major photosynthetic pigments of blue - greens.

Group	Class	Major absorption bands (nm)
Chlorophylls	Chlorophylla	435, (670), 680, (700)
Carotenoids	β -carotene	(431), 450-454, 478-480
	Echinenone	455-459, (475)
	Zeaxanthin	(430), 453, 479
	Canthaxanthin	466
	Myxoxanthophyll	504-508, 474-476, 450-452
Phycobiliproteins	Allophococyanin	650
	Phycocyanin	610-625
	Phycoerythrin	555-565

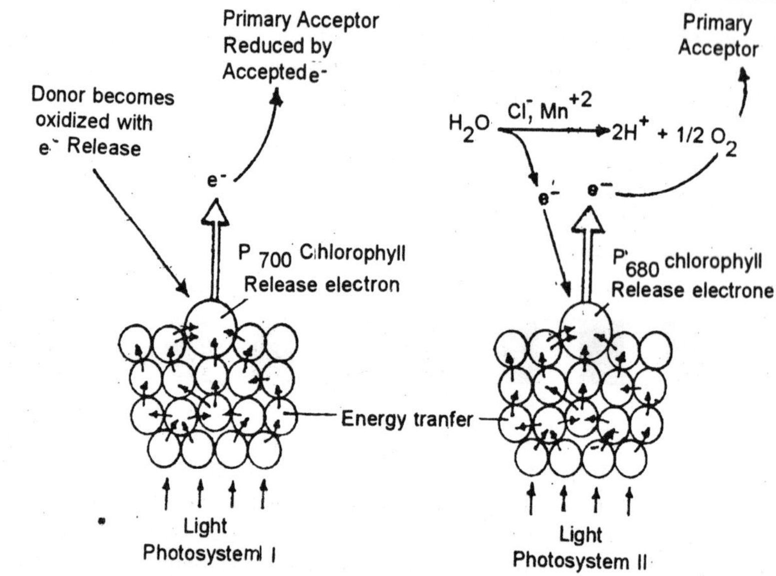

Fig. 16.11 Photosynthesis
Energy transfer in Photosystem I and Photosystem II in green plants

Phycobiliproteins represent, the primary light harvesting or antenna pigments, for photosystem II in cyanobacteria. Light energy trapped by these pigments is transferred to chlorophyll with a very efficiency. Each phycobilisome consists of a triangular core surrounded by six peripheral rods and is made up of disc shaped subunits. The triangular core which is in direct contact with the thylakoid membrane has the pigments allophycocyanin, while the rods contain phycocyanin and phycoerythrin. The phycobilisomes are arranged in parallel rows on the outer surface of the thylakoid membrane. Close proximity between phycobilisomes and thylakoid membranes is necessary for maximum energy transfer in the primary act of photosynthetic.

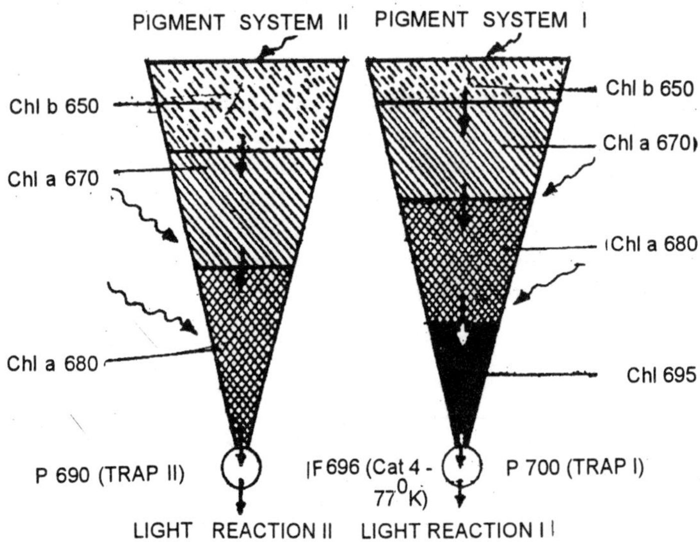

Fig.16.12 Photosynthesis
Distribution of different pigments in PS I and PS II

Complementary chromatic adaption

Environmental conditions such as mineral nutrients, and intensity and quality of light is known to affect the pigment composition in cyanobacteria. It has been suggested that the changes enable the most efficient utilization of available light energy. The synthesis of phycobilin pigments and even the formation of thylakoids is susceptible to environmental influence. Change in pigmentation is largely due to change in phycocyanin phycoerythrin ratio. Certain strains appear distinctly bluish-green when grown under red light because phycocyanin which absorbs minaly red light is present in large quantities. On the other hand, strains grown under green light will appear reddish due to the preponderance of green light absorbing phycoerythrin in light cells. Thus, the organisms produce a pigment which can absorb best the existing light. The shift between red light and green light effects was found to occur at 590nm wavelength. The phenomenon of this shift is called *complementary chromatic adaptation* or *Gauidukov phenomenon* (named after the discoveror). This phenomenon is seen in only those cyanobacteria which synthesize phycoerythrin.

Photosynthetic pigments is bacteria

Suspensions of photosynthetic bacteria appear green, purple, violet, red, brown or salmon coloured to the eye based on the pigment composition. The pigment components can be readily recognised by their absorption spectra of intact cells. Absorption peak in the blue (450nm), and the red as well as infra red (650-1000nm) is due to chlorophylls; absorption in the range of 400–550nm is mainly due to carotenoids.

The various chlorophyll types differ from each other mainly by the presence or absence of a double bond between carbon atoms 3 and 4 and by substitutions at different positions in the poryphyrin skeleton. These changes are reponsible for the differences in absorption spectra (see fig).

The following are the pigments present in bacteria.

1. Bacteriochlorophyll *a* and *b, c d, e* and *g*
2. Chloroboium chlorophyll (bacteriviridin)
3. Carotenoids

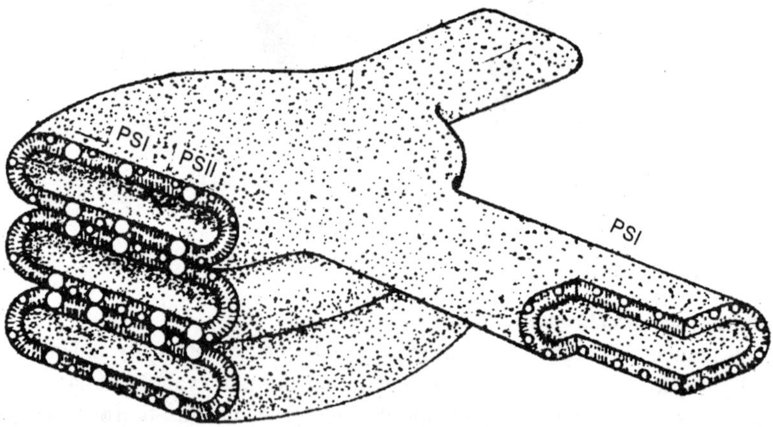

Fig.16.13 Photosynthesis

Model of the structure of Granum thylakoid and a stroma thylakoid showing the distribution of PS I and PS II

Bacteriochlorophulls

These are found only in purple sulphur and non sulphur bacteria. The types of bacteriochlorophylls differ in their absorption spectra which is given in the following table :

Types of Bacteria Chlorophylls

Designation of Jenson et. Albugo, 1964	Former Designation	Characteristic Adsorption Maxima in living cells (nm)
Bhcl a	Bacteriochlorophyll	375, 590, 805, 830-890
Bhcl b	Bacteriochlorophyll *b*	400, 605, 835-950, 1020-1040
Bhcl c	Chlorobium chl 660	Long wavelength absorption max 745-755
Bhcl d	Chlorobium chl 650	– do – 705-740

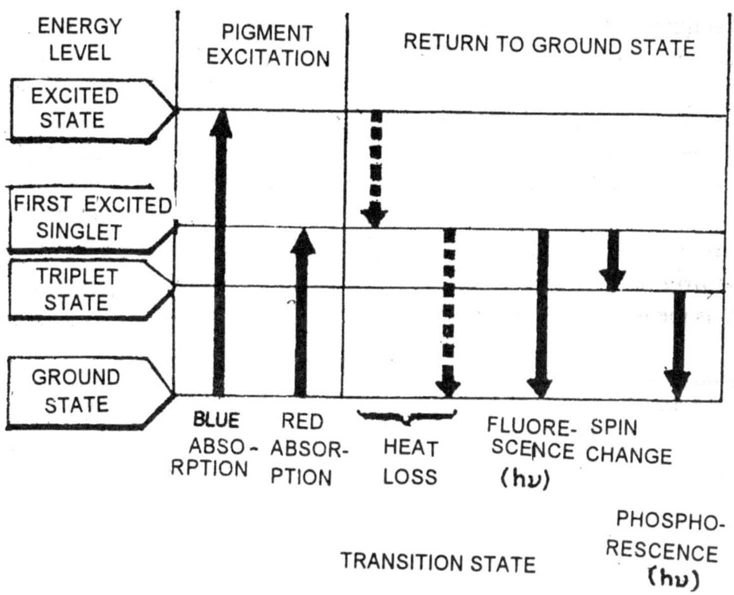

Fig.16.14 Photosynthesis
Diagram illustrating the absorption of light and changes in the electronic configurations in chlorophyll

Bchl (Bacteriochlorophyll) *a* found in purple sulphur bacteria has a peak between 850–890nm. *Bchl b* which is found in *Rhodopseudomonas viridis, Ectothiorhodospira halochloris* and *thiocapsa pfennigii* has an absorption peak at 1020–1040nm.

Bchl a of purple bacteria occurs in four spectral forms – B 800, B 820, B 850 and B 870–890. The differences in absorption spectra are caused by the kind of binding and the position of the *Bchl* molecule in the pigment protein complexes of the photosynthetic apparatus. The number of peaks and the height varies between species and also cultural conditions.

Chlorobium chlorophyll

These are present in green sulphur bacteria of the family chlorobiaceae. Three chlorophylls– chlorobium chlorophylo 660, chlorobium chlorophyll 650 and chlorobium chlorophyll 715 have been repored in green sulphur bacteria such as *Chlorobium, Chloropseudomonas, Chloroflexus* etc.

According to the recent nomenclature, however, chlorobium chlorophylls of three types mentioned above are called *Bchl c, Bchl* and *Bchl e.*

Like higher green plants and algae, in bacteria also, pigment molecules may aggregate to form photosynthetic units. Light may be absorbed by any pigment molecule but it is directed to a special chlorophyll in a reaction centre which serves as energy trap. Many bacteria absorb lights of longer wavelength. Based on the maximumof absorption, this chlorophyll has been called *chl*890 for *Rhodospirillum rubrum* and P 870 for *R.Spheroides.* In green sulphur bacteria *chle* is the reactive chlorophyll.

Carotenoids

These are the so called accessory pigments, and absorb light in the region of 500–550nm. Most of the bacterial carotenoids are open chain aliphatic molecules (unlike the bicycle and β carotens of algae and higher plants). In plants bacteria, they are minaly aliphatic C40 compounds with tertiary hydroxy or methoxy groups (lycopene, rhodopin, spirilloxanthin, spheriodene). Carotenoids with aromatic rings are also found in bacteria. These are restricted to a few families such as chromatiaceae, chlorobiaceae etc.

Green bacteria possess monocyclic γ carotenoids. Lycopene is the most widely distributed carotene in bacteria. Bicycle carotenes are rare in photosynthetic bacteria. The table below gives various groups of carotenoids found in bacteria. Carotenoids fulfill two functions. On the one hand, they are photosynthetic on the other, they protect the chlorophylls from possible damage due to photoxidative action. It has been seen that carotene free mutants of purple bacteria can grow only in dim light and are killed by higher light intensities.

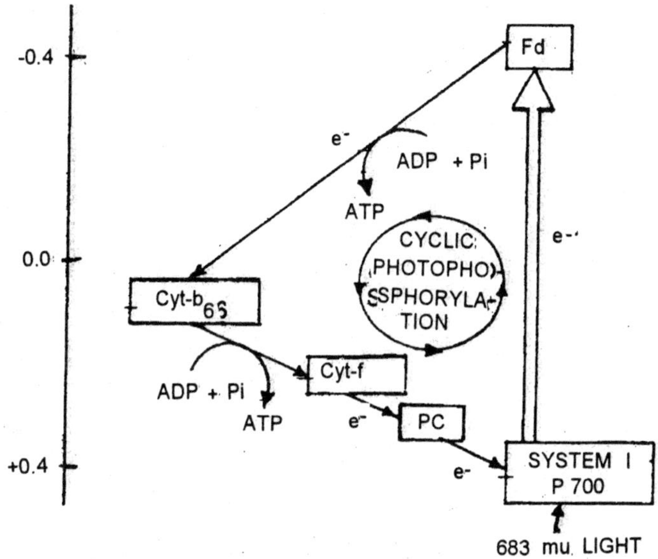

Fig. 16.15 Photosynthesis
Systematic representation of cyclic photophosphorylation

The differences in the absorption spectra of green algae, cyanobacteria, green bacteria and purple bacteria leads to the conclusion that different phototrophic organisms use different spectral rays for their photosynthetic activities. The differences correlate well with the light conditions in their natural habitats. These differences can be used for selective culture of these organisms

Carotenoid groups of phototrophic bacteria

Group	Name	Major components
1.	Normal aspirilloxanthin series Rhodopin Spirilloxanthin	Lycopene
2.	Alternative spirilloxanthin series and ketocarotenoids of spheriodenone type Spheriodenone Hydroxyspheriodenone Spirilloxanthin	Spheroidene Hydroxy-spheriodene

3.	Okenone series	Okenone
4.	Rhodopinal series	Lycopenal Lycopenol Rhodopin Rhodopinal Rhodopinal
5.	Chlorobactene series	Chlorobactene Hydroxy- chlorobactene β-Isoernieratene Isorenieratene

Types of Photosynthetic Bacteria

The prokaryotic phototrophs include a number of Autotrophic pigmented organisms. These are –

1. Purple non sulphur bacteria - Rhodospirillaceae
2. Purple sulphur bacteria - Chromatiaceae
3. Green sulphur bacteria - Chlorobiaceae Chlofoflexaceae
4. Cyanobacteria - formerly blue green algae.

Types of Bacterial Photosynthesis

Phototrophic bacteria normally utilize CO_2 of the environment as a source of cellular carbon. The energy required for the reduction is obtained from inorganic or organic sources. In photolithotrophic organisms, inorganic substances like H_2S or thiosulphate are the source of reducing power. In photoorganotrophic organisms, organic substances like malate or succinate yield reducing power. The general equation for reduction of carbon in photosynthesis is as follows :

$$2H_2A + CO_2 \text{ ——— } C(G_2O) - H_2O + 2A.$$

In algae and other green plants, which grow in the presence of air H_2A is H_2O. In obligatory anaerobic photosynthesizing bacteria. H_2A is H_2S or organic molecules. In chromatiaceae and chlorobiaceae, reductive power comes from H_2S or thiosulphate. In Rhodospirillaceae organic molecules provide reducing power. The following is an outline of different types of reactions involved in bacterial photosynthesis.

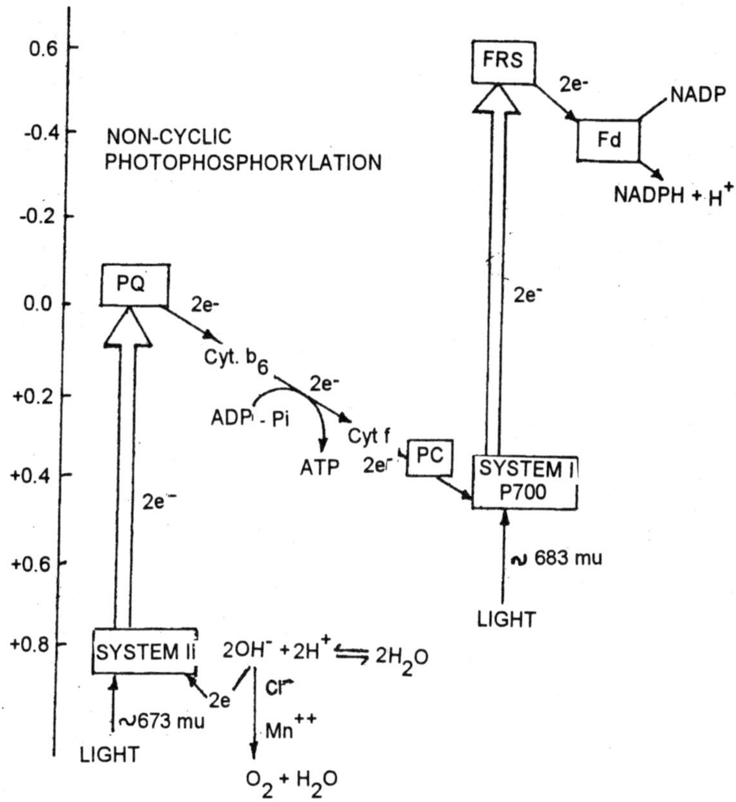

Fig. 16.16 Photosynthesis
Systematic representation of non-cyclic photophosphorylation

1. Purple sulphur bacteria

Here, H_2S is the substrate that provides reducing power. Sulphur is produced instead of oxygen.

$$CO_2 + H_2S \text{ —— } C(H_2O) + 2S + H_2O$$

Purple sulphur bacteria can also utilize thiosulfate as the source of reducing power

$$2CO_2 + Na2S2O_3 + 5H_2O \text{ —— } 2C(H_2O) + 2NaHSO_4$$

2. Non sulphur bacteria

Here, the reducing power is provided by organic compounds like ethanol, succinate etc.

$$CO_2 + 3C_2H_5OH \text{ —— } C(H_2O) + CH_3CHO + H_2O$$

It will be seen that in the above types of reactions there is no evolution of O_2.

In cyanobacteria and other green plants, water is the source of reducing power. In this respect, cyanobacteria differ from green and purple bacteria. As a result, in cyanobacteria oxygen is produced as a byproduct in photosynthesis. In purple and green bacteria however, O_2 is not produced as a byproduct as H_2O is not the source of reducing power. Based on this, photosynthesis is divided into two types – **Oxygenic** and **Anoxygenic.** In the former O2 is evolved during photosynthesis and in the latter O_2 is not evolved. Oxygenic photosynthesis is found in cyanobacteria and higher plants, while anoxygenic photosynthesis is characteristic of purple and green bacteria. In the paragraphs below, oxygenic and anoxygenic photosynthesis is discussed in some detail.

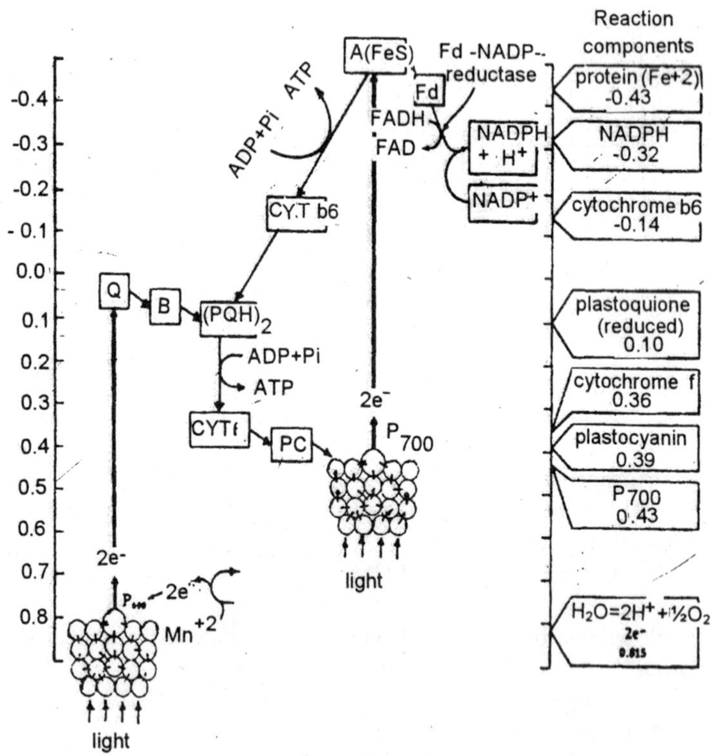

Fig.16.17 Photosynthesis

Systematic representation of overall electron flow in light reaction combining both cyclic and non-cyclic photophosphorylation

MECHANISM OF PHOTOSYNTHESIS

Photosynthesis in both green plants and bacteria can be understood by dividing the process into two–light reaction and dark reaction. In light reaction, molecule(s) of ATP is generated along with reducing power. While ATP is generated in all phototrophic organisms using light energy as the source, the reducing power comes from a variety of sources, inorganic compounds or water. It is based on the source of reducing power (H_2O or otherwise), O_2 released or not. Photosynthesis is divided into two types – anoxygenic and oxygenic.

Light reaction in anoxygenic photosynthesis

Purple sulphur, and green sulphur bacteria conduct anoxygenic photosynthesis. Light reaction here involves the following steps.

1. Source of electron
2. The electron transport system
3. Photophosphorylation

Source of electron

Evolution of oxygen has not been demonstrated in bacteria, and hence these are incapable of using water as a source of reducing power. The source of electron are therefore, hydrogen sulphide and reduced inorganic and organic compounds.

In sulphur bacteria oxidize sulphide to sulphate and the reducing power is utilized to reduce CO_2 as follows :

$$CO_2 + H_2S \text{ ——--- } 3\ CH_2O + 2S$$

$$3CO_2 + 2S + 5H_2O \text{ -—— } 3CH_2O + 2H_2SO_4$$

Chromatium Okenii, a purple bacterium can utilize a variety of sulphur compounds and molecular hydrogen.

Electron transport system

In photosynthetic bacteria. the electron transport system consists of a primary electron acceptor (x), a secondary acceptor (y), and *b* and *c* type cytochromes. Cytochrome *a* is absent as molecular oxygen does not participate in the electron transport. Some scientists believe that there are two separate electron pathways–one for photosynthesis and another for respiration. But as is seen in *R.Rubrum,* only one electron transport mechanism is considered enough for both the processes.

The electron transport carriers are asymmetrically located in the membrane to help maintain the hydrogen ion gradient.

The primary electron acceptor (x) receives the electron from the bacteriochlorophyll P870. Eventhough the exact nature of the primary acceptor is not known, researches have shown that it contains nonhaeme iron and it is quite likely that it is ferredoxin, an iron protein complex.

The events in electron movement can be summarized as follows. The electron driven from bacteriochorophyll is accepted by ferredoxin, which then releases an electron to be accepted by FMN (a flavoprotein). From FMN, the electron travels down and is accepted by NAD which gets reduced providing the necessary reducing power (to be utilized for reduction of CO_2). Ubiquinone may receive an electron either from FMN or FAD, and the electron released from it is accepted by *cyt b* and *cyt c*. Finally, the electron moves back to *Bchl.*

Photophosphorylation

The process of conversion of light energy into chemical energy is called photophosphorylation. Two types of phosphorylation are known to exist in bacteria in the same way as in green plants. These are cyclic phosphorylation and non-cyclic phosphorylation. There is one difference between bacteria and green plants. In the case of green plants, the photosynthetic units have two reaction centres to trap light rays of two different wave lengths. In the case of bacteria, however, there is only one reaction centre and only one PS (photo system) is involved.

It was thought earlier that in bacteria there is only cyclic photophosphorylation. This was based on the assumption that evolution of oxygen is a must in non-cyclic photophosphorylation. But some discoveries have clearly shown that non cyclic photophosphorylation occurs in bacteria and that evolution of oxygen need not necessarily accompany non-cyclic photophosphorylation.

Works of Losada, Whatley, and Arnon (1961), have shown that non-cyclic photophosphorylation reaction does not depend on oxygen evolution. The difference between cyclic and non-cyclic photophosphorylation is based on the dependence of the latter on the presence of an electron donor and acceptor which is necessary, produce the reducing power. Occurrence of non-cyclic photophosphorylation in the chromatophores of *Rhodospirillum rubrum* has been shown by suppressing cyclic photophosphorylation with actinomycin *a* and supplying DPIP ascorbate as an electron system, and NAD as electron acceptor.

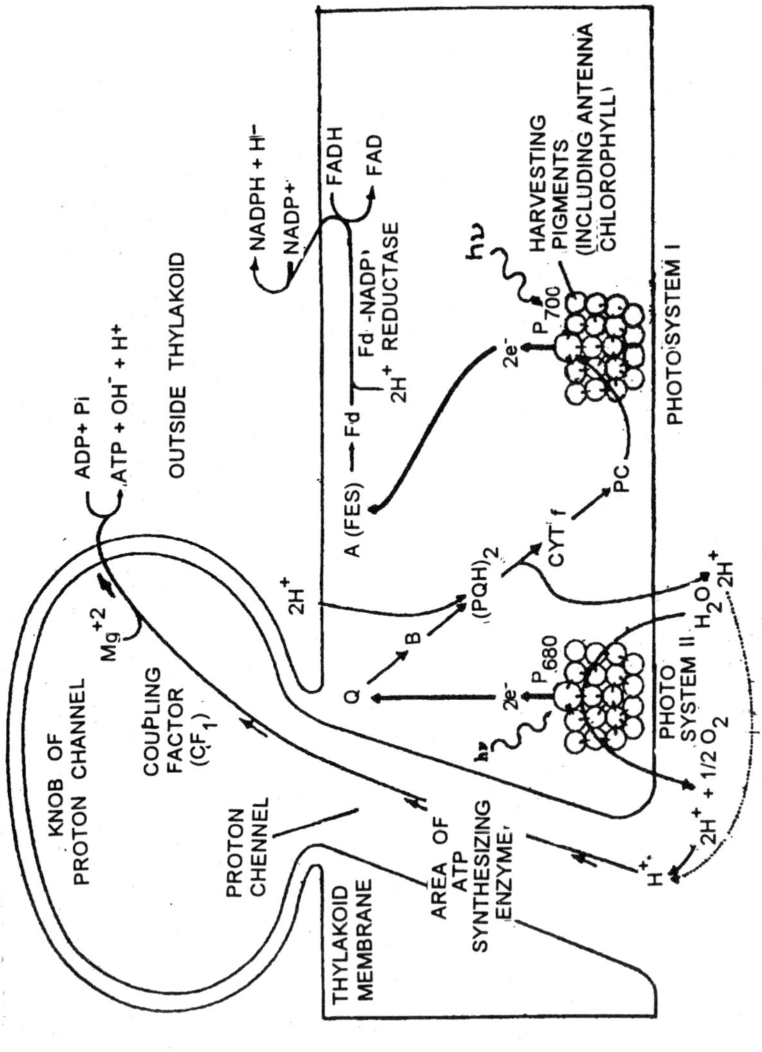

Fig. 16.18 Photosynthesis

Photophosphorylation and ATP synthesis in granum thylakoid according to Chemiosmotic hypothesis

In bacterial pigment system also, the photosynthetic units consists of two parts namely: an *antenna* consisting of light harvesting molecules and the reaction centre where the actual energy conversion takes place. Light energy harvested by the antenna molecules is channelised to the reaction centre. In the green and purple photosynthetic bacteria about 40 or more bacteriochlorophyll molecules make up the photosynthetic unit. The reaction centre complex is commonly refered to as P870 or P890, although the wavelength band of peak absorption might vary in different species.

In composition, the reaction centre generally consists of *Bchl a* or rarely *Bchl b.* In the purple non sulphur bacterium, *Rhodospirillum rubrum,* the principle light harvesting pigment is *Bchl a* and the reaction centre molecule is P890. In *Rhodopseudomonas sphaeriods,* the light harvesting *Bchl* absorbs at a peak of 850nm and reaction centre *Bchl* is P870.

In green sulphur bacteria, the principle light harvesting pigment is *Bchl c* with a peak 650nm or Bchl *d* with a peak of 660nm. Some of the brown species of green sulphur bacteria have *Bchl e* in the harvesting pigment. The reaction centre chlorophyll is *Bchl a* in both the green and brown species.

The following components are found in the reaction centres of *Rhodopseudomonas sphaeriodes, R.Virids* and *Rhodospirillum rubrum.*

1. Four molecules of *Bchl* a
2. Two molecules of Bacteriopheophytin (*Bph*)
3. 1 molecule of carotenoids
4. One or more quinones
5. One atom of non haeme iron

The reaction centres in most of the photosynthetic bacteria contain three chains polypeptides having a molecular rate of 20,000, 24,000 and 30,000. The *Bchl.* Bacteriopheophytin and Darotenoids are located on the first two polypeptides. In the primary photochemical reaction, the light quantum absorbed by the light harvesting pigments are transferred to the reaction centre (P870). In the reaction centre, the absorbed photons rais the chlorophyll to excited state and the electron is expelled from the molecule. The P870 now becomes positively charged. The charged P870 transfers the electrons to an intermediate electron acceptor which is reduced. This reduced acceptor then transfers its electron to another acceptor and ultimately the electron comes back to chlorophyll bringing the P870 to its normal state.

Cyclic photophosphorylation

The process of electron vibration here is the same as what has been explained above. When the electrons are moving from the ferrodoxin to cytochromes, a molecule of ATP is synthesized (ADP + Pi = ATP). This is called cyclic photophosphorylation, because the electron comes back to the chlorophyll in a circular fashion. The net turn of the cycle yields one molecule of ATP.

Non-cyclic photophosphorylation

Here the electron from ferrodoxin instead of moving down towards cytochromes is accepted by NAD, and the reaction is catalysed by ferrodoxin reductase. The reduced NAD now provides the necessary reducing power to fix the carbon dioxide.

The *Bchl* P890 now is in an excited state as it has donated its electron to ferrodoxin. In order to come to the normal state, it must receive an electron from some other source. The electron comes from the oxidation of organic acids, thiosulphates or H2S. These compounds get oxidised as a result of the elimination of an electron. The electron thus released moves via cytochromes back to *Bchl* P890. The chlorophyll molecule now comes back to the normal state and is ready to receive photons from light.

The scheme of photophosphorylation is called non-cyclic because the electron that comes back to *Bchl* is not from its own source but by the reduction of other compounds such as H_2S etc.

Light reaction in oxygenic photosynthesis

The light reaction or photochemical decomposition of water takes place in the grana region of the chloroplast. According to the current understanding, light reaction (Photochemical excitation) takes place in about 10^{-9} seconds. Within this short span of time, there will be the synthesis of molecule(s) of ATP and a molecule of NADPH. These two are utilized in the dark fixation of CO_2. In this section we will study the following aspects in detail with respect to light reaction.

(a) Source of oxygen released during light reaction
(b) Hill's reaction
(c) Arnon's work
(d) Quantum requirement
(e) Emerson effect and two pigment system
(f) Primary photochemical reaction (mechanism of pigment excitation)
(g) Non-cyclic Photophosphorylation.
(h) Cyclic Photophosphorylation
(i) Pseudocyclic photophosphorylation and
(j) Mechanism of ATP formation

SOURCE OF OXYGEN

The overall reaction of photosynthesis upto 1930 was believed to be the exact opposite of respiration and oxygen evolved during photosynthesis came from carbon dioxide.

$6CO_2 + 6H_2O \text{---} C_6H_{12}O_6 + 60_2$

The discovery by Van Neil (1931), that in certain purple photosynthetic bacteria photosynthesis takes place in the presence of CO_2 but without releasing oxygen. Surely, if the source of O_2 were to be CO_2 even in the absence of water O_2 should have been released. These photosynthetic bacteria used H_2S as hydrogen donor instead of water. Accordint to Van Neil (1931), in purple bacteris H_2S, splits releasing H to reduce CO_2 and in the process elemental sulphur is being formed as follows :

$CO_2 + 2H_2S$ —— $(CH_2O) + H_2O + 2S$

Van Neil (1931) further argued that, if elemental sulphur comes from H_2S in purple bacteria, in green plants which use water, O_2 should come from water only and not from CO_2. In order to account for the release of O_2 from water the equation should be modified as follows

$6CO_2 + 12H_2O$ —— $C_6H_{12}O_6 + 6H_2O$

According to Van Neil (1931), water is split up by light (photolysis of water) into H^+ and OH^- ions. The H + ions are used to reduce CO_2, while the OH ions react with one another to form water and oxygen as follows :

$4H_2O$ —— $4H + 4(OH)^-$

$4(OH)^-$ —— $2H_2O + O_2$

$CO_2 + 4H^+$ —— $(CH_2O) + H_2O$

Here (CH_2O) represents not the actual photosynthetic product but the empirical formula of a bigger molecule (glucose).

The work of Ruben, Kamen and Randall (1941), using radio active oxygen (O_{18}) clearly showed that O_2 came only from water and not from CO_2. When the experimental plant was supplied with labelled water (H_2O_{18}), the released oxygen was the O_{18} type, and when the plant was supplied with labelled CO_2 18, the O_2 released was the normal type (O_{16}).

HILL'S REACTION

Hill (1937), and Hill and Scaribrick (1940), showed that the initial reaction of photosynthesis can take place even in isolated chloroplasts in the cell-free environment. When isolated chloroplasts (obtained from ground leaves) are supplied with light, water and a suitable hydrogen acceptor, O_2 will be released and the H_2 acceptor gets reduced. A similar release of O_2 may be observed, when grana from disintegrated chloroplasts are suspended in water with ferric salts and benzoquinones acting as hydrogen acceptors. Besides these, chromates, indophenols etc. also can act as H_2 acceptors. The various H_2 acceptors used by Hill are called Hill oxidants. Hill attributed the release of oxygen to the splitting of water into H and OH by light. Hill's reaction proves the following :

(a) The entire photosynthetic activity including the reduction of CO_2 takes place within the chloroplasts, and

(b) O_{21} released during light reaction comes entirely from water.

ARNON'S WORK (production of assimilatory power)

The products of light reaction provide the necessary inputs for the reduction or assimilation of CO_2. That the assimilatory power (input) is utilized for the dark fixation of CO_2 was demonstrated clearly by the work of Arnon (1951).

According to Arnon (1951), the hydrogen released by water was accepted by coenzyme II (NADP). He found that in isolated chloroplasts, the release of O_2 takes place at a faster rate in the presence of NADP. Arnon also opined that ATP molecules will also be synthesized and he termed this as photophosphorylation.

$$\text{ADP} + \text{Pi} \xrightarrow[\textit{(inorganic phosphate)}]{\textit{energy}} \text{ATP}$$

Arnon also showed that CO_2 fixation takes place in the pigment free portion of the chloroplast (stroma), while the assimilatory power (ATP)/molecules and NADPH), is produced in the grana portion.

QUANTUM REQUIREMENT

The solar radiation that reaches the earth consists of two portions – the visible (400nm – 800nm), and UV, and infrared (less than 400 nm and more than 800 nm). Whether the light consists mainly of waves or particles, as far as photosynthesis is concerned, it is obsorbed in units of energy called photons or quanta.

The quantum yield may be explained on the basis of molecules of CO_2 reduced and amount of O_2 released. Consequently, a minimum quantum is the one required to fix one molecule CO_2 and release one molecule of O_2. According to Warburg and and Negelein (1922), the minimum quantum requirement is four i.e. 4 quanta are required to fix one mol of CO_2 and release one mol of O_2. This (requirement) concurs with the fact that every CO_2 molecule requires $2H_2O$ molecules for reduction i.e., 4 hydrogen atoms. Emerson and his co-workers (1938), however opine that eight light quanta is the minimum requirement. This has been further confirmed by the work of Govindjee (1968), on *Chlorella* plants. Thus two quanta of light are required for the transfer of every atom of hydrogen during photosynthesis.

EMERSON EFFECT AND TWO PIGMENT SYSTEMS

Emerson and Lewis (1943), worked on different wavelengths of light (monochromatic light) as to their quantum yield and came to the conclusion that 8 quanta of light are required for the reduction of one molecule of CO_2 (or to release one molecule of O_2). Accordingly, the quantum yield is 1/8 or 12%. Since 4 electrons are required in the release of one molecule O_2, Emerson opined that 2 light quanta are required to excite one electron. Proceeding further, Emerson *et al* (1943), determined the quantum yield of different beams of monochromatic light and found that the average quantum yield (12%) suddenly dropped near the far red end of the spectrum. This decline in quantum yield is called red drop. Red drop begins at wavelengths greater than 680 nm in green plants and greater than 650nm in red algae. The phenomenon of red drop was also confirmed by the works of Haxo and Blinks (1950), on the red alga *Porphyra neregtis*.

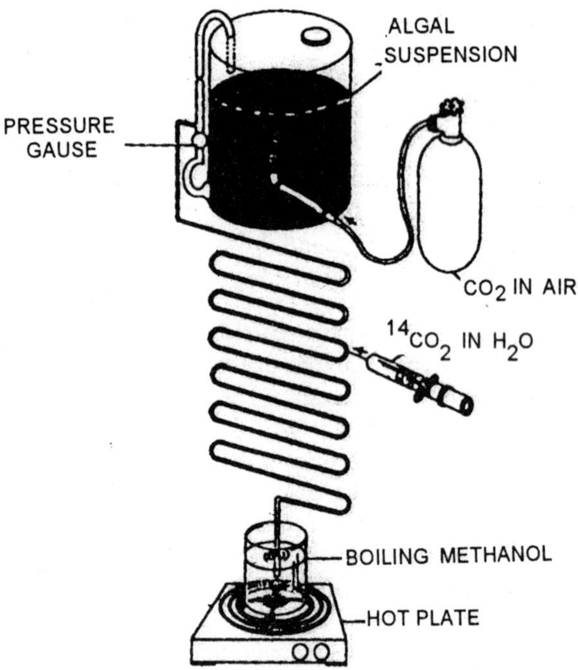

Fig.16.19 Photosynthesis

Apparatus used by Calvin et al for short time exposure to C^{14}

The phenomenon of red drop was studied further by Emerson and Chalmers (1958), who found out that the reduction in the rate of photosynthesis due to red drop can be rectified by providing a beam of monochromatic light of shorter wavelength. In other words, by providing shorter wavelengths of light either simultaneously or quickly alternating with a longer wavelength of light, the rate of photosynthesis can be enhanced (effect of red drop is reversed).

This enhancement of photosynthetic rate under the influence of two wavelengths (long and short) of light is called Emerson effect. This may be represented as follows :

$$E = \frac{\text{Rate of evolution of } O_2 \text{ in double beam - Rate of Evolution } O_2 \text{ in red beam (short)}}{\text{Rate of evolution of } O_2 \text{ in far red beam (long)}}$$

Reason for Emerson effect

The discovery of the Emerson effect led to the idea that there must be two different pigment systems in the plants which differed in their absorption peak (of light). While one pigment system was more efficient in longer wavelength, the other was efficient in shorter wavelength.

Obviously, under a specific wavelength (either short or long), the rate of photosynthesis will not be maximum as one or the other pigment, system will be ineffective. Butler (1966), discovered two forms of chlorophyll viz *chl a* 673 and *chl a* 683 (with absorption peaks at 673nm and 683 nm respectively). Clayton (1966), discovered another form of *chl a* with absorption peak at 700nm which he named P700.

It has now been confirmed by various researches that two pigment system PS I and PS II are involved in photosynthesis, each pigment system having different absorption peaks and consisting of *chl a* and other accessory pigments. The two pigment systems may be removed from the thylakoids, by treating the chloroplast with mild detergents. These can be separated on polyacrylamide gel when they appear as discreet green bands. The details of the two pigments system are as follows :

According to Malkin (1982), Anderson (1982), and Barber (1983), PS I consists of Chl *a* 670, Chl *a* 680 and Chl *a* 710 (P 700), Chl *b* and some β cartoenes. P700 is the reaction centre in PS I into which light energy is transferred by all other pigments. Besides, PS I also consists of iron containing proteins called Fe-S proteins and Cytochromes. PSI is weakly fluorescent.

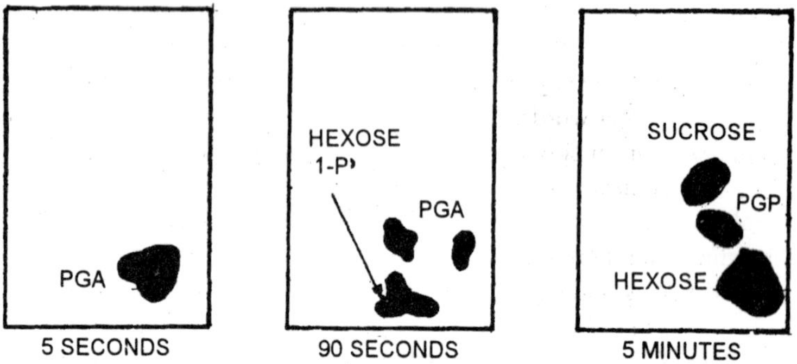

Fig. 16.20 Photosynthesis
Auto radiograph (Diagrammatic representation of chromatograms showing the fixation of C^{14} after different periods of exposure to light

Pigment System II (PSSII) :
It has several forms of Chl *a* with P680 (Chl a 680) being the reaction centre. A little amount of Chl *b* and β carotenes are also present. There is also a colourless Chl *a* called Pheophyrin (Klima and Krasnovskii, 1981), which acts as the primary electron acceptor together with these are present quiones (Q) and manganese proteins.

Light Harvesting complexes
Electrophoretic separation of pigments from Chloroplasts, reveals the presence of two other green bands besides PS I and PS II (Salisbury and Ross, 1986). Each band contains Chl *a* + Chl *b* but very little β carotene. All these pigments are protein bound. According to Salisbury and Rosss (1986), these green bands represent light harvesting complexes of pigments and proteins; one band functions in association with PS I and the other with PS II. The function of the light harvestng complexes is to absorb light energy and to transfer it to the appropriate pigment system.

Co-ordination between PS I and PS II : According to Junge (1982), each granum has about 200 units of PS I and II, while the number is variable in stroma thylakoids. It was originally believed that chloroplasts contain equal quantities of PSI and PS II, but according to Melis and Brown (1980), the ratio of distribution of PS I and PS II varies with the species and other environmental conditions. Anderson and Melis (1983), believe that grana consists mainly PS II, while stroma thylakoids have mainly PS I.

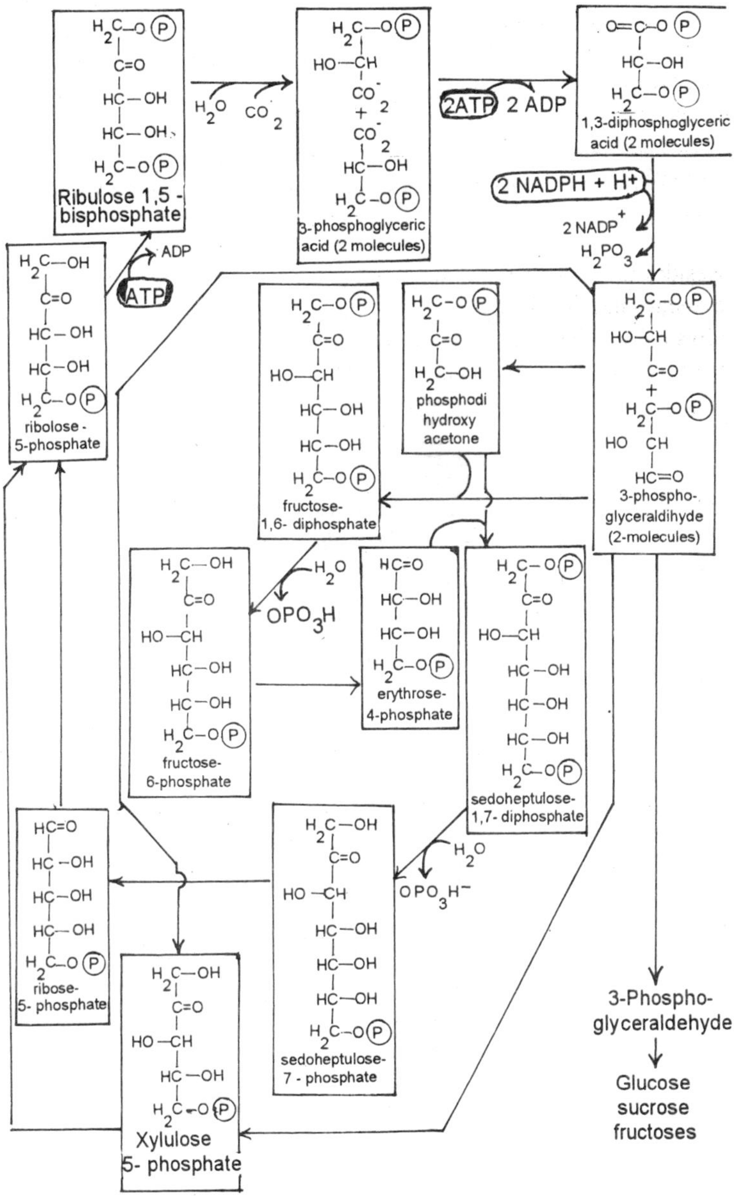

Fig. 16.21 Photosynthesis

Systematic representation of C3 pathway (Calvin Cycle

PS I and PS II should function in close co-ordination because they jointly function to transfer electrons from water to NADP (see later - non - cyclic electron transfer). Since they are located a little apart, certain intermediaries help to carry electrons from PS I to PS II. These intermediaries or carriers are of two types – a) copper containing proteins called *Plastocyanins* bound loosely to the inside of the thylakoid membranes and b) a group of quinones called *Plastoquinones* (PQ). Plastocyanin in its reduced form can carry an electron from PS II to PS I. Plastoquinones carry two electrons and two H+ from PS II to PS I.

Besides the above, additional electron transport components in the form of cytochromes also occur in thylakoids. *Cytochrome* b6 and *Cytochrome f* lie physically in between PS I and PS II and help in electron transport between them. Another cytochrome - cyt b3 and an Fe-S protein (Ferredoxin) also participate in the light mediated electron transport.

Also present in thylakoids and necessary for light mediated ATP synthesis is, a group of complex proteins called *coupling factor* (CF) complexes of ATP ases (see later under mechanism of ATP formation for more details). The CF complex consists of a spherical head attached to the thylakoid membranes towards the stroma and a stalk that extends across the lipid bilayer.

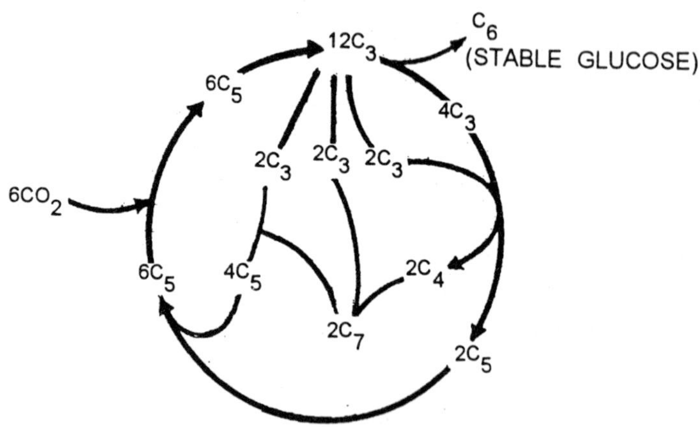

Fig. 5.22 Photosynthesis
Balance Sheet of carbon Calvin's Cycle

MECHANISM OF PIGMENT EXCITATION

Normally, the chlorophyll molecule wil be in its stable state (no excitation), in the sense the two electrons of its atoms will be spinning in opposite directions and the molecule has no magnetic movement.

When the chlorophyll molecule absorbs light energy in the form of photon one of the electrons is raised to a higher orbit (higher energy). However, the spin of the electrons continues to be in opposite directions, and hence to magnetic movement. The unexcited state of the chlorophyll molecule (before the absorption of light) is called the *ground state* or the *single state.* Here, the energy level is zero, hence it is also called the E_0 state. After the excitation, the stage is known as the first excited state or the S_1 state, with the energy being E_1. The life span of the S_1 state is very short and very soon (10^{-9} sec), the excited electron returns to the ground state by losing energy in the form of heat, light, or by the transfer of ėnergy. The chlorophyll molecule reaches the S_1 state by absorbing the light at the red level (660 nm). If the light is absorbed at the blue level (400 nm), the chlorophyll molecule isknown to reach the S_2 level (with energy level being E_2), because the 'blue photon' has more energy. However, even here the magnetic movement is nil as the spin of the electrons is still in the opposite directions. The electron of S_2 state reaches the ground state via the S_1 state by losing energy either as heat or as heat of light (fluorescence or phosphorescence). Since the electron reaches the ground state very soon (10^{-8} or 10^{-9} see), the energy cannot be put into use (it canot be converted into chemical form).

Under certain circumstances, an electron at the S_1 state may lose some energy in reversing the spin and drops into a low orbit. At this state, there will be two electrons in two different orbits, but spinning in the same direction. The molecule is now said to possess magnetic movement with electrons oriented in three different ways relative to the axis. This is called the *triplet state* or the *metastable state* (T_1 state). The conversion of S_1 state electron to T_1 state is called *Inter system crossing.*

The lifespan of T_1 state electron is longer (10^{-2} or 10^{-3} sec). The electrons do not dissipate energy quickly, for they have to reverse their spin before reaching the ground state. Since the electron in the T_1 state remains energised for a long time, the energy can be made use of for chemical reaction (conversion into chemical energy). The electron at the T_1 state is dislodged from the chlorophyll molecule and is shuttled between a series of electron acceptors, the energy is utilized to produce ATP and (reduced) NADPH molecules. The series of electron acceptors constitute the electron transport system (ETS).

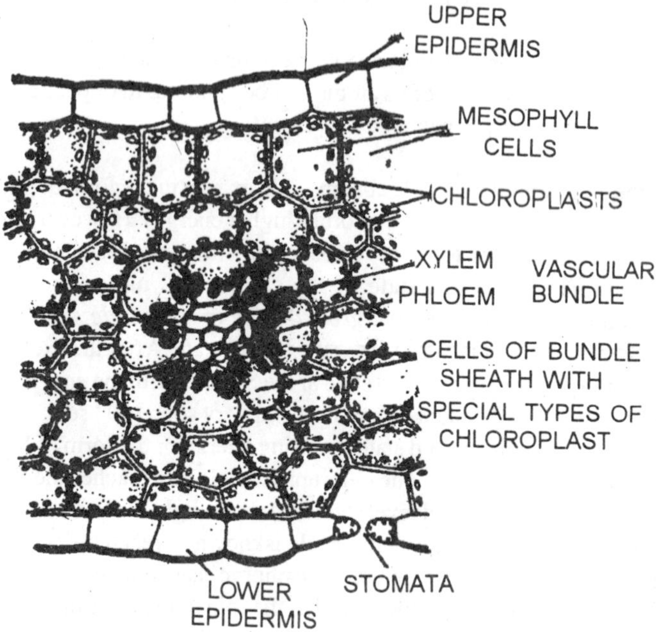

Fig. 16.23 Photosynthesis
TS of a monocat leaf (grass) showing Kranz anatomy

Photophosphorylation

Arnon and Frankel (1954), showed that isolated chloroplasts, under illumination can produce ATP molecules. The reaction is believed to take place as follows.

$$ADP + Pi \xrightarrow[\text{Chlorophyll}]{\text{Light}} ATP$$

(ADP = Adenosine Diphosphate)
ATP = Adenosine Diphosphate)
Pi = Inorganic phosphate

Aron *et al* (1954), named this phenomenon as Photophosphorylation (addition of phosphate group to ADP under the influence of light energy). These ATP molecules are utilized to reduce PGA to PGAL during the fixation of CO_2. They are also required to phosophorylate Ribulose 5 phosphate to Ribulose 1.5 diphosphate (For details, see dark reaction). The following types of photophosphorylation have been discovered in plants.

1. Cyclic Photophosphorylation
2. Non - Cyclic Photophosphorylation, and
3. Pseudocyclic Photophosphorylation

Cyclic Photophosphorylation

Here only PS I participates. The light energy absorbed by the light have harvesting complex which is transfered by inductive resonance to P700. P700 can accept light energy only of a wavelength longer than 680 nm. The excited electron of PQ700 is first accepted by an unknown acceptor (x), then to ferredoxin. This causes ferredoxin to release one of its own electrons (energy rich), which shuttles through the chain consisting of Cyt b6, Cyt f. state and now ready to receive one more unit of radiant energy.

The electron released from the chlorophyll (P700) is returned in a continuous chain i.e., it travels in cyclic fashion, hence the name, cyclic electron transport or cyclic photophosphorylation.

Synthesis of ATP during cyclic photophosphorylation

The electron ejected from P700 has an energy of 0.42 v which gets reduced to 0.417 when it reaches ferredoxin. Here the electrons passes through 3 down hill migration (Cyt b6, Cyt f and Pi) steps before it is returned to P700. The potential gap between Ferrodoxin to P700 (ground state) is about one volt which is sufficient to produce at least two molecules of ATP. ATP molecules are formed at two stages – a) between ferredoxin and Cyt b6 and b) between Cyt b6 and Cyt f

In cyclic photophosphorylation, water does not participate; hence there is neither evolution of O_2 nor is there any formation of reduced NADPH molecule (for there is no hydrogen donor). Further, PS II is not involved in any stage.

According to Park and Sane (1970), there are two types of PS I, PS I*s* and *PS1g*. *PS 1s* is present in the stroma thylakoids and *PS 1g* is present in the granum. *PS 1s* participates in cyclic photophosphorylation, while *PS 1g* participate in non-cyclic photophosphorylation together with PS II.

Ramreiz et al (1968), opine that cyclic photophosphorylation is of limited occurrence in plants, and a high rate of cyclic reaction may even retard the CO_2 fixation as NADPH will not be available. They argue that cyclic reaction is not involved in the main path of carbohydrate synthesis as it can only provide part of ATP requirement for CO_2 fixation.

Non-Cyclic Photophosphorylation

This is the major pathway of light reaction involving both PS I and PS II. Non-cyclic photophosphorylation is also called the 'Z' scheme electron transport because of the zig-zag fashion of electron travel. It takes place as follows :

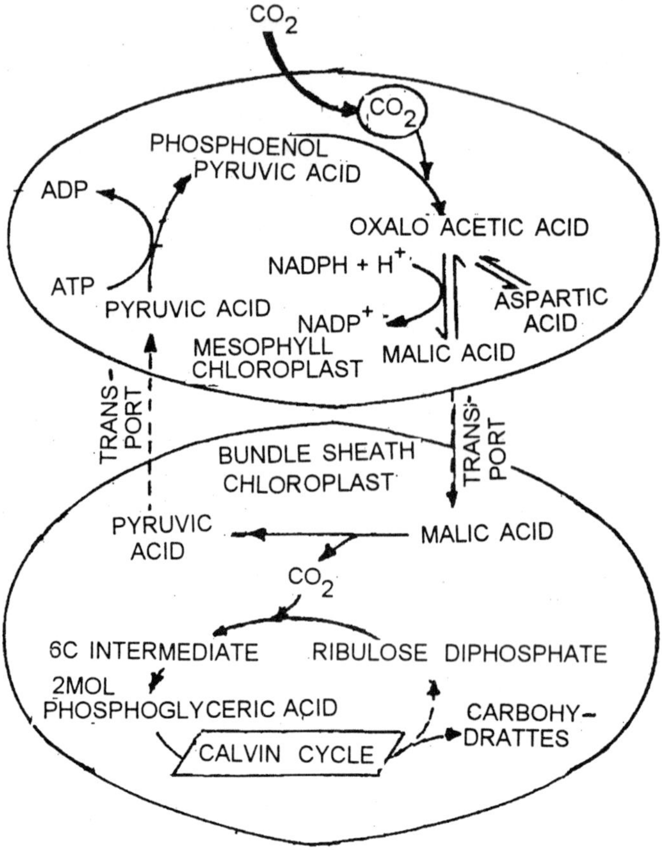

Fig. 16.24 Photosynthesis
Systematic representation of Hatch and Slack pathway

(a) The radiant energy is trapped by PS I (P700), which ejects one if its electrons and is trapped by X, FRS, and finally goes to ferredoxin. The reduced ferredoxin now transfers its electrons to NADPH. NADPH gets reduced. PS I, however remains in the excited state as it has not got back its electron.

(b) Meanwhile, PS II absorbs light energy (680nm) and ejects one of its own electrons which is trapped by an unknown quinone acceptor (Q). From Q–the electron moves downhill to cyt, b6, plastoquinine, cyt f, Plastocyanin (PC), and finally to PS I. PS I which is in an excited state, now returns to the ground state as it has got an electron from PS II.

How does PS II come back to the ground state now that it has sent its electron to fill up the 'hole' in PS I?. Because, unless the PS II returns to the normal state, the system cannot contunue functioning. The answer lies in the dissociation of water molecules. Water splits into H^+ and OH^+ ions. What is the mechanism of this splitting of water? Some physiologists call it photolysis of water, others, the breakdown of water. In any case, the excited PS II returns to the normal state by getting an electron from the OH^- ions of water. According to Salisbury and Ross (1986), the excited PS II returns to the ground state by attracting an electron from an adjacent Mn^- protein, which in turn gets an electron from the OH^- of water. The H^+ ions released by water are accepted by NADPH to become NADPH + H^+. Thus, a reduced NADPH is formed.

The electron transport system or photophosphorylation is called non-cyclic because the excited PS I returns to the normal state by getting an electron from PS II, while the excited PS II returns to the ground state by receiving an electron from water. Hence, the movement of electron is not in a cyclic fashion.

ATP Synthesis during non-cyclic photophosphorylation

The electron (+ 0.8 v) released by PS II is accepted by a quinone from where the downhill migration begins to reach PS I via Cyt b6, Cyt f and PC. One molecule of ATP is synthesized when the electron shuttles between PQ and Cyt f.

Release of oxygen and formation of NADPH + H^+

Four molecules of water split up for every turn of non-cyclic photophosphorylation. This is necessary (see fig) to provide four electrons to the excited PS II to bring it back to the ground state. Similarly, four H + ions are required to reduce 2 molecules of NADPH.

These reactions take place as follows :

$$4H_2O \text{ ——— } 4H^+ + OH|^-$$

$$4OH^- \text{ ——— } 2H_2O + O_2$$

(4 electrons from OH reach the excited PS II)

$$2NADPH + \text{ — } 4H^+ \text{ ——— } 2nadph + 2H^+$$

(four electrons are required to reduce 2 molecules of NADPH and these come) from the excited PS II via ferredoxin)

Thus the end products of non-cyclic photophosphorylation are :

(a) One molecule of ATP

(b) Two molecules of reduced NADPH
(c) 2 molecules of water and
(d) One molecule of O_2.

While a and b are used for the dark fixation of CO_2, c and d are byproducts.

Comparison between cyclic and non-cyclic photophosphorylation

	Cyclic	Non-cyclic
1.	Only PS I participates	PS I and PS IIparticipatre
2.	PS I gets back the electron in a cycle fashion	PS I gets back the electron from PS II from water
3.	Only ATP molecules are formed	Both ATP and NADPH molecules
4.	Water does not participate, hence no evolution of O_2.	O_2 evolved as water participates
5.	Found predominantly in bacteria	Found predominantly in green plants.

Pseudocyclic Photophosphorylation

Arnon and his co-workers demonstrated another kind of Photosphosphorylation, when illuminated chloroplasts produced ATP molecules in the absence of CO_2 and NADP, but in the presence of FMN or Vit K and oxygen. This involves the reduction of FMN by water with the evolution of oxygen. The reduced FMN is reoxidizable by oxygen. Hence, Arnon (1954), called it oxygen dependent FMN catalysed photophosphorylation or pseudocyclic photophosphoyrlation.

MECHANISM OF ATP FORMATION

The source of energy for the synthesis of ATP molecules from ADP comes obviously from the electron transport chain. There are many evidences which point out to this relationship brought about by some common agents. For instance, if some uncoupling agents are introduced into the system, electron transport continues, while the ATP synthesis is inhibited. If the uncoupling agent is removed, ATP synthesis is resumed. Inhibition of electron transport also inhibiting ATP synthesis is another evidence to show that both are coupled.

While there does not seem to be any doubt about the coupling of electron flow and phosphorylation, the mechanism is as yet not fully understood (according to some physiologists). The following are the three hypotheses proposed for the mechanism of ATP synthesis.

(a) Conformational coupling
(b) Chemical coupling
(c) Chemiosomotic coupling

(The mechanism of ATP synthesis is the same for both photophosphorylation during photosynthesis as well as for oxidative phosphorylation during respiration.)

Conformational coupling

During electron transport, both in mitochondria as well as in chloroplasts, some conformational (structural) changes occur which facilitate the release of energy resulting in the production of ATPase mediated ATP synthesis, ATPase is a reversible action enzyme which decomposes ATP into ADP and Pi under less energy conditions, while facilitates synthesis of ATP during high energy conditions. According to Devlin and Witham (1983), electron micrographs of membranes indicate structural variations during peak activity. But there does not seem to be any definite relationship between activated membrane and ATP synthesis.

Chemical coupling

This hypothesis believes in the presence of a coupling factor (most probably a protein), mediating the energy transfer between electron transport and ATP synthesis. Initially, the coupling factor (CF) forms a high energy CF complex with one of the electron carriers. Later, an inorganic phosphate replaces the electron carrier in the CF complex. This results in the formation of high energy rich phosphate to ADP resulting in the formation of ATP.

Chemiosmotic coupling

Among the hypothesis proposed for the mechanism of ATP synthesis, this is by far the most acceptable, Originally proposed by Michell during 1960s for oxidative phosphorylation in mitochondria, it has recently been applied to photophosphorylation in chloroplasts also (Jagendorf, 1975).

As it relates to oxidative phosphorylation, the flow of electrons from NADH to atmospheric oxygen through the respiratory electron transport, influences the extrusion of protons from the mitochondrial matrix. As a result, a proton concentration gradient is produced across the mitochondrial membrane. In effect, the energy of electron flow in the respiratory chain is partially conserved as the *proton concentration gradient* or *proton motive force* across the inner mitochondrial membrane. The protons return to the mitochondrial matrix down the electrochemical gradient through ATPase resulting in the formation of ATP from ADP and Pi.

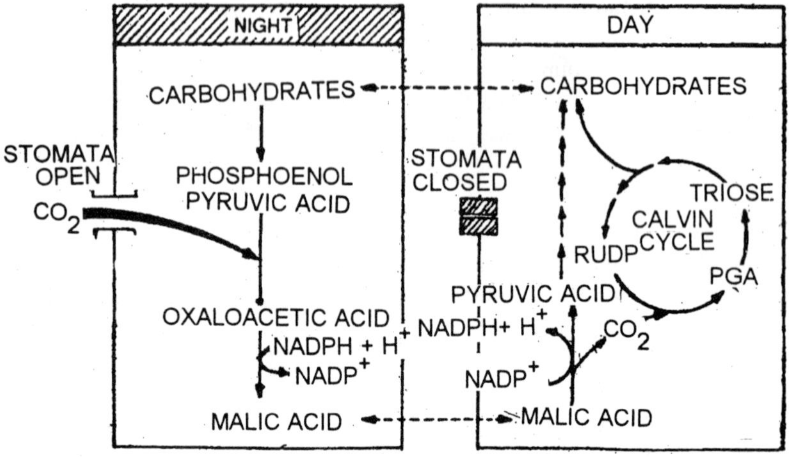

Fig. 16.25 Photosynthesis
Systematic representation of CO_2 fixation in CAM Plants

Ubiquinone, one of the components of the respiratory chain known to transfer protons across the mitochondrial membrane through alternate oxidation and reduction. The flow of protons across the membrane is due to a force called *Proticity* (Noggle and Fritz, 1983), which is comparable to electricity. According to Noggle and Fritz (1983), the word chemiosmotic was proposed to emphasize the fact that chemical energy (oxidation energy) is converted to osmotic energy (difference in the proton concentration gradient between the inner and outer faces of the membrane).

Oxidative phosphorylation may be halted by treating the membrane with uncouplers. These are weak acids which dissolve in the protoplasmic membrane and disturb the proton concentration gradient by allowing the free flow of protons. Thus the proton uncouplers (protonophores) disturb the ATP synthesis.

The Chemiosmotic theory also explains ATP formation in chloroplats. Jagendorf (1975), demonstrated that a pH gradient established across the thylakoid membranes stimulates the ATP synthesis. The electrons are transported across the membrane to the interior through the electron transport chain. Light energy provides the necessary drive the flow of electrons from water to NADP and for transport of protons across the membrane. Plastoquinone, an electron acceptor transfers electrons to Cyt f and picks up H + ions on the outside of the thylakoid membrane and releases protons to the interior of the channel. This results in the building up of a pH gradient (the interior of the thylakoid becomes acidic due to accumulation of protons) across the membrane. The accumulated protons flow from the inside to the stroma side of the membrane through pathways of coupling factor (which include stalks and knobs), which are the sites of photophosphorylation. The downhill migration of protons provides the necessary energy for the addition of Pi to ADP to form ATP, with ATPase acting as a mediator.

DARK REACTION OF PHOTOSYNTHESIS (carbon assimilation)

The source of carbon for reduction is atmospheric CO_2 for green plants. For photosynthetic bacteria, carbon may come from Co_2 or from organic compounds. The process of CO_2 assimilation in photosynthetic bacteria is of the following types :

1. Reductive pentose pathway or Calvin cycle
2. Reductive carbosylic acid cycle or pyruvate synthetase pathway
3. Organic carbon assimialation pathway

CALVIN CYCLE (reductive pentose pathway)

Calvin cycle or the C_3 cycle, wich

is the predominant pathway of CO_2 fixation in higher plants is seen in photosynthetic bacteria also. Researches carried out on *Rhodospirillum ruburm* and *Thiobacillus* spp using radio isotopes of carbon (Fog, 1968), have confirmed this. Sugar phosphates are, however, less as these bacteria are unable to utilize or store them.

The phenomenon of thermochemical reduction of CO_2 was first established by Blackmann in 1905. He opined that light is not necessary for the reduction of CO_2. In dark reaction the products of light reaction (ATP and NADPH + H) are used to reduce CO_2 into glucose. But the precise manner in which CO_2 is absorbed, and reduced resulting in the formation of a molecule of glucose was understood fairly recently, thanks to the availability of radioactive isotopes. Utilizing the isotopes of carbon dioxide, one can trace the different stages of entry of

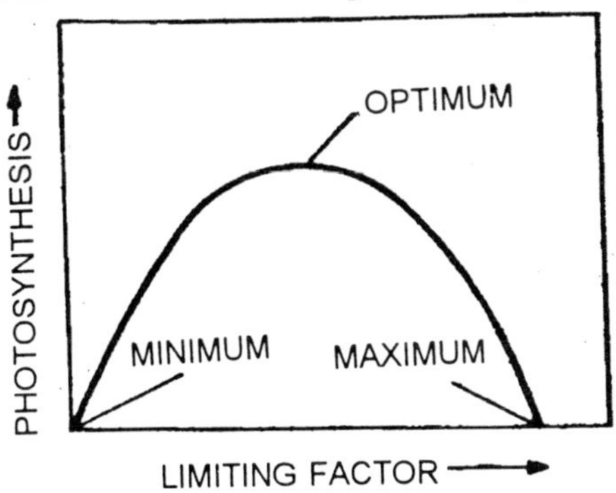

Fig. 16.26 Photosynthesis
Sach's concept of 3 cardinal points

CO_2 into the plant system untill it gets converted into glucose.

Many physiologists, biochemists etc. have contributed a great lot towards the understanding of the path of carbon in Photosynthesis. The names of Bassham (1957 - 1964), Melvin Calvin (1961), Benson (1951) etc are worth mentioning in connection with the untravelling of the secrets of CO_2 fixation. Melvin Calvin was awarded the Nobel Prize in 1961 for tracing the CO_2 reduction cycle, which is also known as the Calvin cycle.
In CO_2 reduction cycle, what we have to precisely understand is how the CO_2 enters the plant system, which is the compound that accepts the CO_2 molecule and how a molecule of glucose is synthesized and how the compound (that originally accepts the CO_2) is regenerated. In addition, the system should explain the acceptance of 6 CO_2 molecules, their reduction and formation of only one molecule of glucose for every turn of the cycle.

CALVIN CYCLE

This is the major pathway for the fixation of CO_2 in green plants and in some bacteria. It is called Calvin Cycle owing to its discovery by Melvin Calvin of the university of California, USA. It is also called C_3 pathway because, the first stable compound formed after the entry of CO_2 into the plant system is a triose sugar (C_3) viz phosphoglyceric acid (PGA).

Some experimental details of Calvin's work

In order to understand the significance of the discovery of CO_2 reduction cycle, we should understand the experimental methods followed by Calvin.

To trace the various steps of conversion of CO_2 after its entry into the plant system until the formation of glucose, Calvin allowed the experimental plants to photosynthesize for various periods of time and analysed the various compounds formed after different periods of photosynthesis.

Clavin chose for his experimentation two green algae – *Chlorella pyrenoidosa* and *Senedesmus olbiqus*. The plants have the same pigmentation as higher green plants and can be easily cultured and harvested. To make sure that the compounds are already present in the plant, the experimental plants were starved of CO_2, and later supplied with a radioactive isotope of CO_2 (C_{14}). Only those compounds in which C_{14} was incorporated were taken into consideration because they were produced after the experiment began. By varying the time of photosynthesis (1/16 sec to 5 sec), and then killing the plant to stop photosynthesis and analysing the intermediate compounds formed, Calvin and his co-workers were able to map the whole of the CO_2 reduction cycle.
Two, very simple but efficient tools is namely *Chromatography* and *Radioauthgraphy*, to trace the path of carbon in photosynthesis.

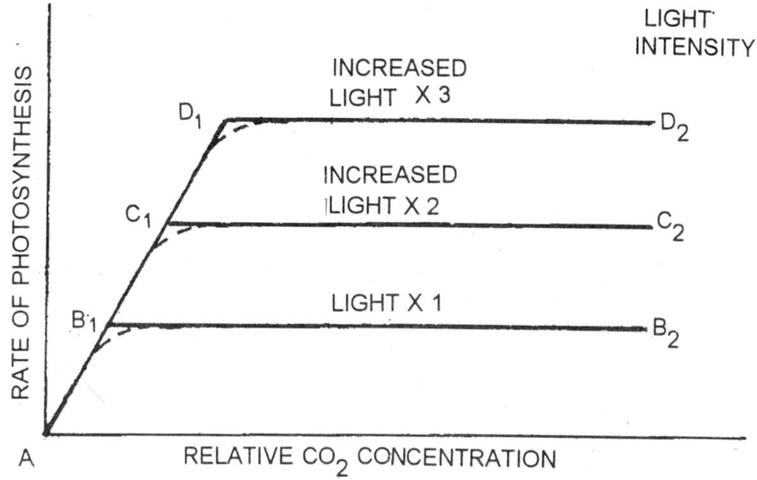

Fig. 16.27 Photosynthesis

Graph representing Law of Limiting Factors

Chromatography, in particular, paper chromatography is an ideal technique for the separation of compounds present in a small mixture. The extract (mixture) of the plant to be analysed is placed as a spot on paper and a solvent is allowed to run. This solvent due to capillary action carries with it the different substances in the extract. The rate of migration of the differnt substances depends upon their solubility and specific gravity.

The filter papter (Chromatogram) should be removed from the solvent before it reaches the edge of the paper and dried. By spraying a suitable reagent (For e.g. ninhydrin for amino acids), the compounds can be detected as several spots widely separated on the paper. The substances may be identified by their relative positions on the Chromatogram or by comparing with a standard Chromatogram. (In a standard Chromatogram, compounds whose identity are known are spotted, and their position ismarked after the solvent run).

Radioautography is a technique to find out, if a particular source consists of a particular type of isotope. Different radioactive isotopes emit different types of radiation. Chiefly, there are 3 types. These are : (a) Alpha rays (b) Beta rays (c) Gamma rays.

When we use a particular we already know what is the type of emission. For instance, C_{14} emits β rays. When the rays are directed to a photographic film they from characteristic markings, and by these one can identify the type of radiation and thereby the isotope.

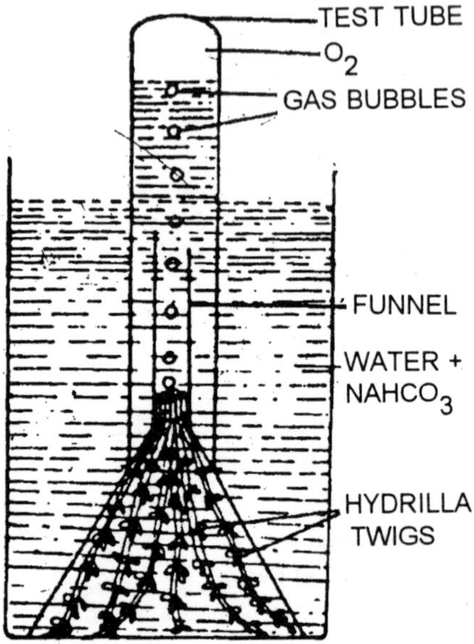

Fig. 16.28 Photosynthesis
Experiment to demonstrate the release of O_2 during Photosynthesis

In the present expirement, Melvin Calvin and associates allowed the algae to photosynthesise for varying periods (using C_{14}), after which they (plants) were killed in boiling methanol. The plant extract was then spotted on the Chromatogram to identify the compounds. In order to make sure that these contain C_{14}, the chromatogram was held against a photographic film of equal size. Only those compounds which had incorporated C_{14} leave markings on the film and these were taken into consideration (as produced during experimentation). In order to vary the time of photosynthesis, flashes of light from 1/16 sec to 5 sec were given to identify the earliest compound formed. In this way, Calvin and his associates found out all the intermediaries in the CO_2 reduction cycle.

The first stable compound formed after the initial entry of CO_2 is a triose (PGA). The triose cannot be the acceptor molecule for it has to initially, have a 2C compound to accept one CO_2 molecule to become triose. Benson, however, soon found out that a 5 carbon compound is the first acceptor molecule (Ribulose 1.5 diphosphate). After accepting a molecule of CO_2, an unstable 6C compound is formed, which immediately breaks up into 2 molecules of PGA (triose), the first stable compound.

The essentials of the Calvin cycle are as follows. The cycle should be self exclusive in that it should remain unaltered, but provide for the continuous entry of 6 molecules of CO_2, formation of one molecule of glucose and regeneration of six molecules of Ribulose 1.5 diphosphate (RUDP) 1, the initial acceptor molecule. The important events of CO_2 cycle can be detailed as under.

(a) Fixation of CO_2

(b) Reduction of PGA
(c) Formation of sugars, and
(d) Regeneration of RUDP

Fixation of CO_2

CO_2 is first accepted by molecules of a pentose (5C) sugar Ribulose 1.5 diphosphate, and an unstable 6 carbon compound is formed. This compound (tentatively called carboxy dimsutase system) immediately breaks up into 2 molecules of a triose (3C) called phosphoglyceric acid (PGA). The reaction is catalyzed by the enzyme carboxy dismutase of Rubp carboxylase (Rubisco). For every molecule of CO_2, 2 mols of PGA are formed; hence 12 PGA moles for $6CO_2$ mols.

Reduction of PGA

The molecules of PGA are reduced to phosphoglyceraldehyde (PGAI) by NADPH + H. The reaction requires energy and hence. ATP participates. The enzyme triose phosphate dehydrogenase catalyses this reaction..

Rudp + CO_2 H_2O ——— PGA (2 molecules)

ATP

PGA + NADPH + H ——— PGAI

The above reaction takes place at two stages. In the first stage, PGA + ATP Phosphoglycerokinase 1,3 - diphosphoglyceric acid + ADP In the ssecond stage.

1.3 Diphosphoglyceric acid + NADPH + H —— 3 PGAI + NADP

Since 6 molecules of CO_2 are taken into the system (by 4 mols of Rudp), 12 PGAL mols are formed after reduction of the 12 molecules of PGA. Only two are utilized for the formation of sugars, while the rest are channeled back for the regeneration of Rudp molecules. Several intermediate compounds are formed in the process.

Formation of sugars

The PGA molecule (one out of two) undergoes isomerization to form a mol of dihydroxyacetone phosphate. The enzyme triose phosphate isomerase catalyses this reaction. Ultimately, sugars are formed as follows :

Triosephosphate

1. 3 PGAL ——————— Dihydroxyacetone phosphate

Isomerase

2. 3 PGAL + Dihydroxyacetonephosphate

Aldolase

——— Fructose 1.6-diphosphate

Fructose 1,6-diphosphate now gets converted into fructose 5 monophosphate due to the loss of one phosphate group under the influence of the enzyme phosphotase.

$$\text{3. Fructose 1.6-dip} \xrightarrow[+H_2O]{\text{Phosphatase}} \text{Fructose 6. Monophosphate (FMP).}$$

Fructose 6 monophosphate isomerises into glucose monophosphate under the influence of the enzyme isomerase.

4. EMP Isomerase Glucose 6 monophosphate (GM)

Now FMP or GMP may undergo dephosphorylation to form either fructose or glucose. These monosaccharides unite to form sucrose and various other carbohydrates.

Regenration of Rudp

In the above explanation of formation of sugars, it has been pointed out that Rudp accepts a molecule of CO_2 to ultimately form a molecule of sugar. But, the overall formula for photosynthesis accounts for the entry of 6 molecules of CO_2, hence, there must be 6 molecules of Rudp to accept the CO_2 molecules and for every turn of the cycle, this should result in the formation of 6 molecules but there is only one molecule of glucose. Then what happens to the remaining 6 molecules of hexose (5moles of 6C). Obviously they undergo transformation to regenerate 6 molecules of Rudp. This is necessary format Rudp to be ready to accept and be self-sufficient; and there must always be 6 molecules of Rudp ready, to accept 6 molecules of CO_2 and form 6 molecules of glucose, the cycle will cease to operate for the next turn of the cycle there will be no Rudp to accept CO_2 molecules. Thus for the entry of every 6 CO_2 molecules, the output is one mol of hexose and the rest go for the regeneration of Rudp.The balance sheet of the molecules of carbon from the entry of CO_2 to the generation of Rudp is given in Fig.In the following equations, regeneration of one mol of Rudp is explained (and this may be multiplied to obtain 6 molecules of Rudp).

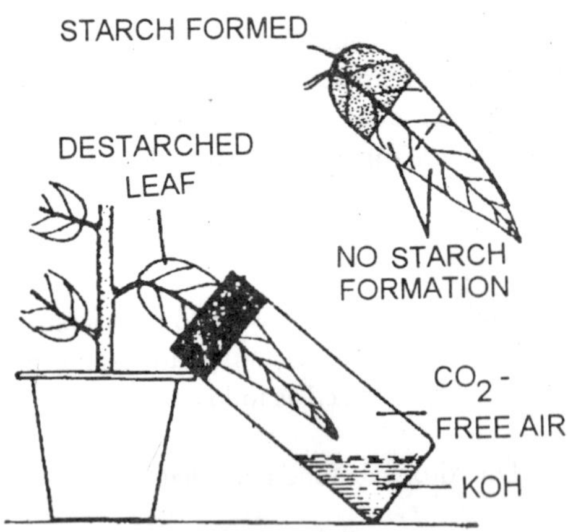

Fig.16 .29 Photosynthesis
Experiment to demonstrate the necessity of light for photosynthesis

1. Fructose 6 Phosphate + PGAL $\xrightarrow{\text{Transketolase}}$ Erythrose4, Phosphate+ Xylulose 5. Phosphate
2. Erythrose 4, Phosphate + Dihyrosxyacetone phosphate $\xrightarrow{\text{Aldolase}}$ sedophetulose 1.7. Diphosphate
3. Sedoheptulose 7. Phosphate + PGAL $\xrightarrow{\text{Phosphatase}}$ Xylulose 5 H_3PO_4.
4. Sedoheptulose 7. Phosphate + PGAL $\xrightarrow{\text{Transketolase}}$ Xylulose 5 Phosphate + Ribose 5 Phosphate

Xylulose 5 Phosphate and Ribose 5 phosphate get converted into Ribulose 5 phosphate under the influence of the enzyme phosphophentose isomerase

5. Ribulose 5 Phosphate $\xrightarrow[\text{-Kinase}]{\text{Phosphopento}}$ Rudp

Reductive carboxylic acid cycle

This has been demonstrated in green bacteria such as *Chlorobium thiosulfatophilum* and *Chromatium* spp (Evans et albugo, 1966). In this type of fixation, the first detectable photosynthetic products are amino acids. CO_2 is used to form pyruvate by means of pyruvate synthetase reaction. Reducing power is provided by the oxidation of H_2S. Reducing power obtained thus will reduce ferrosoxin (Fd). Acetyl CoA (of Krebs cycle), accepts CO_2 and is reduced by Ferrodoxin to yield pyruvate. The reactions take place as follows –

1. *Formation of pyruvate*

Acetyl CoA + CO_2 + Ferredoxin (reduced) —— pyruvate + CoA + Ferredoxin (oxidized)

2. *Formation of Oxaloacetate*

Pyruvate + ATP + CO_2 —— oxaloacetate + ADP + Pi

Oxaloacetate enters the reversed tricarboxylic acid cycle to yield succinyl CoA.

3. *Formation of α Ketoglutarate*

Succinyl CoA + CO_2 + Ferredoxin (reduced) —— α Ketoglutarate + CoA + Ferredoxin (Oxidized)

4. *Formation of Ctirate*

α Ketoglutarate is converted into citrate through oxalosuccinate. Citrate the, splits into oxaloacetic acid and acetate.

α Ketoglutarate CO_2 —— oxalosuccinate —— citrate —— oxaloacetic acid + Acetate

The above reactions are generally irreversible, but because of string reducing potential of the Ferrodoxin, the reaction can be reversed. As these bacteria can generate reduced ferrodoxin by means of photosynthesis, it is called reductive carboxylic acid cycle (Evans,

Buchanan, Arnon, 1966).

The net result for every turn of the cycle is that four molecules of CO_2 are incorporated (reductive fixatin), and on molecule of oxaloacetate is synthesized. Besides ferredoxin, the cycle requires all other cofactors of Krebs cycle in reduced form. Three molecules of ATP ar required as follows for this fixation - 1. Activation of acetate 2. Carboxylation of pyruvate and 3. Activation of succinate.

Organic carbon assimilation pathway

Certain types of bacteria use organic compounds to provide CO_2 for fixation (not atmospheric CO_2) Stainer (1961), opined that a part of the organic carbon is directly assimilated. Accordingly to her, photoassimilated organic substrates (from organic polymers) such as acetate, is directly converted into poly β hydroxybutyrate as follows :

According to Gaffron (1963), however, during photoassimilation CO_2 reduction occurs only when excess reductants are available. Under anaerobic conditions, photometabolism of organic acids is strictly light dependent, and may be dependent on light generated ATP.

COMPARISON BETWEEN BACTERIA AND GREEN PLANTS (in terms of photosynthesis)

In the photosynthetic activity, there are a number of similarities and a few differences between green plants and photosynthetic bacteria. It is believed that photosynthesis in green plants is evolved from that found in simple green and purple bacteria. There is difference between green plants and bacteria in the localization of the photosynthetic pigments; while well organized chloroplasts are there in the former, in the latter, pigments are localized in membranous vesicles. (There is infact an opinion that the chloroplasts with their own DNA can be regarded as some primitive phototrophic organisms which have been trapped by cells of plants in the course of evolution).

The photosynthetic pigments viz., the chlorophylls of bacteria are different from those of higher plants in terms of their absorption spectra. It should be pointed out, however, that the basic chemical framework of all chlorophylls is the same. In carotenoid composition, however, green plants and bacteria differ– open aliphatic chain Carotenoids are seen in the latter, while the former have bicyclic Carotenoids.

In terms of the photochemical processes, bacteria are *anoxygenic,* while green plants (Cyanobacteria also) are *oxygenic.* This is due to the fact that in bacteria, the reducing power comes from H_2S and other inorganic and organic compounds, while in green plants water gets reduced and oxygen is released in the process. It is an enigma as to why bacteria have not chosen H_2O as a source of reducing power.

The source of cellular carbon is always atmospheric CO_2 in green plants, while it may be CO_2 or other organic compounds in bacteria.

There are a number of similarities also between bacteria and green plants in pigment composition. *Chl a* is present in both, though bacteriochlorophyll has two more hydrogen than green plant chlorophyll. Both chlorophylls are formed in membrane bound particles in the cell. In both, chlorophyll systems reaction centres exist with antenna pigments channelising light energy for harvesting.

In both green plants and bacteria, the overall photosynthesis is an oxidation reduction reaction. A reductant H_2A gets oxidized and provides the reducing power to convert carbon to cellular carbon. Oxidized 'A' is released in the process. Cyclic and non-cyclic photophosphorylation is found in both. Electron transport chain leading to the formation of ATP and reduced NAD compound is found in both green plants and bacteria. Carbon assimilation is similar as well as different in both the groups with Calvin cycle being common.

There is some difference in the photoreductant compound. It is $NADPH_2$ in green plants and $NADPH_2$ in photosynthetic bacteria. But the chromatophures of bacteria have an enzyme transhydrogenase and can photoreduce NADP in the presence of NAD. The following are the summarized similarities and differences between bacterial photosynthesis and green plant photosynthesis.

Similarities

1. Presence of chlorophyll *a*
2. Arrangement of photosynthetic pigments in membranous particles in the cell.
3. Presence of a definite reaction centre for harvesting light energy.
4. Accessory pigments channelise light energy towards reaction centre.
5. Occurrence of photophosphorylation, cyclic as well as non-cyclic.
6. Electron carriers of the chain are almost similar.
7. Light energy produces the reducing power.
8. Overall photosynthesis is an oxidation–reduction process.

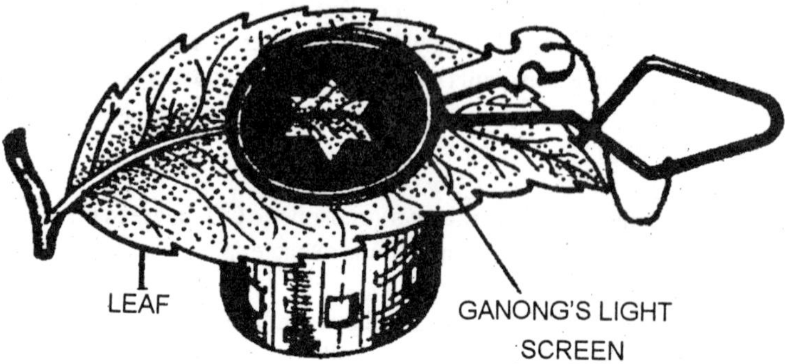

Fig. 16.30 Photosynthesis
Experiment to demonstrate the necessity of light for photosynthesis

Differences between photosynthesis found in bacteria and green plants

	Bacteria	Green Plants
1.	Bacteria have no distinct chloroplast	Chloroplast is well developed
2.	Bacteria absorb light of longer wavelength (800-900 nm or infra red)	These absorb light of relatively shorter wavelength (450-700 nm).
3.	P_{890} is the reaction centre	P_{680} and P_{700} are two reaction centres
4.	Chlorophyll *a* is absent and bacteriochlorophyll takes its functions	Chlorophyll *a* is present which converts radiant into chemical energy.
5.	The carotenoids are open chain alephatic type	Carotenoids are bicyclic
6.	Oxygen is not evolved during the product (anoxygenic photosynthesis)	Oxygen is evolved as by product(oxygenic photosynthesis).
7.	Water does not serve as a source of	Water serves as source of

	reducing power (electron donor)	reducing power.
8.	Bacteria can use CO_2 as well as organic compounds as a source of carbon	Only CO_2 is the source of carbon
9.	The photoreductant is $NADH_2$	The photoreductant is $NADPH_2$.
10.	The process occurs in the presence of light but in the absence of O_2	Both are present during the process
11.	Emerson effect is not found	Emerson effect is found dominant
12.	Cyclic photophosphorylation is dominant	Non-cyclic photophosphorylation isdominant
13.	Plastocyanin is absent	Plastocyanin is present
14.	Lycopene is found higher	Lycopene is not found in the chloroplast of plants.

17

RESPIRATION

The most basic requirement of all living organisms is *energy*. Without energy, no life activity is possible. The solar energy that is trapped by green plants are be stored in cells in the form of carbohydrates and other nutrients. These are oxidised and the energy so released utilized to carry on the vital processes of living beings.

The process of energy release by the (oxidation of) substrate material is called respiration. The term respiration, which has been in usage since a long time is derived from the latin word *respirare* (= breathe). Breathing as a phenomenon of life has been noticed in animals long before its relation to respiration and its universal occurrence in all forms of life.

Breathing and respiration

Breathing and respiration are often used as synonymous. But it is not correct. Breathing as a separate process is noticed only in few organisms (animals), while respiration occurs in all organisms. Further, breathing is a physical process, while respiration is a chemical process taking place in all the cells. Breathing may be regarded as the last step found only in aerobic respiration, where atmospheric oxygen participates. There can be respiration without breathing, but the reverse is not possible.

Respiration and combustion

Respiration is often compared to the process of combution, because in both the instances the substrate material gets burnt (oxidized) up. But the process of respiration differs from mere combustion in many respects. In combusion, the burning up or oxidation of the substrate is sudden, while in respiration it is stepwise. In combustion, a large amount of energy is wasted as heat, while in respiration it is channelised and stored up for further usage. Hence, respiration may be termed as an orderly combustion found in living organisms.

Cellular respiration

In a multicellular organism where there are thousands of cells, how does energy reach each cell? Each cell obtains its own energy by the oxidation of substrate material present within. Hence, respiration takes place in all the cells wherever they may be present in animals and plants. This is called cellular respiration.

Cellular respiration is a complex process which includes various aspects. These are the following.

(a) Absorption or oxygen (not in all)
(b) Degradation of substrates such as carbohydrates to CO_2 and H_2O (oxidation).
(c) Energy is released in the process of oxidation; some amount will also be lost in the form of heat.
(d) Formation of many intermediate products which play different roles in metabolism.
(e) There will be some loss of weight due to the process of oxidation. From the point of view of plant metabolism, cellular respiration is of paramount significance as it is the basic biological process that provides energy for all the vital activities. In short, the following aspects may be mentioned as indicating the significance of respiration.
    (i) Energy released in respiration is used for various metabolic processes. In fact, most of the physiological activities are driven by respiratory energy.
    (ii) Many intermediate compounds are formed which play a key role in the intermediate metabolism.
    (iii) Insoluble food is converted into soluble form.
    (iv) Potential energy is converted into kinetic energy.
    (v) CO_2, which is released during respiration is an important step in maintaining the carbon balance of nature.

The overall formula for respiration is as follows :

$$C_6H_{12}O_6 + H_2O + 6O_2 \; 6CO_2 + 12\,H_2O + 686 \text{ K cal energy.}$$

Respiratory substrate

The energy rich material (food substrate), that is used for the purpose of oxidation is called the *respiratory substrate*. In other words a respiratory substrate is defined as "any organic plant constituent oxidized partially or completely with the concomitant release of energy together with CO_2 and H_2O".

Plants use a variety of organic constitutents as their respiratory substrates. These may be a variety of carbohydrates (hexose sugars, disaccharides, polysaccharides), organic acids, fats etc. Under certain circumstances like starvation, even proteins may be used as respiratory substrates.

Among the carbohydrates, sucrose and starch are the principal respiratory substrates. When simple monosaccharides like glucose and fructose are available, they are directly utilized as respiratory substrates, while sucrose and starch are first hydrolysed to simple sugars before being channeled into respirtory metabolism.

In certain seeds like castor bean, fats are reserve food materials. They are first broken down into fatty acids and glycerol. As already mentioned, proteins will be channeled into respiratory metabolism under special circumstances such as starvation. Blackmann, who has studied respiration in detail, observed that in leaves kept in darkness, the respiratory rate which will be normal initially, declines, but does not stop altogether. Blackmann termed the normal respiration (when carbohydrates are utilized), as *floating respiration* and the slowed down rate of respiration as *protoplasmic respiration,* when the essential proteins are oxidised to keep the cells alive. In the long run, however this is harmful to the cells. When proteins are the respiratory substrates, they are first broken down into amino acids and these by deamination get converted into keto acids and enter the respiratory chain.

Respiratory Quotient (RQ)

RQ may be defined as *"the ratio between the volume of carbon dioxide given out and oxygen taken in simultaneously, by a given weight of the tissue in a given period of time, at standard temperature and pressure".*

RQ may be calculated as follows

$$RQ = \frac{\text{Volume of } CO_2 \text{ evolved}}{\text{Volume of } O_2 \text{ consumed}}$$

The value of RQ depends upon the nature of the respiratory substrate, amount of oxygen present (in the substrate), and the extent to which the substrate gets broken. Normally, the value of RQ should be unity (1), but many deviations are noticed due to the alterations in the levels of oxidation and reduction. The value of RQ also depends upon whether all the oxygen consumed is used up only for respiration or some of it is utilized for other purposes also. It is for these reasons that RQ may be unity, less than unity or even more than unity. RQ value also gives an idea of the respiratory substrate that is being used. The following are some RQ values when different respiratory substrates are involved.

RQ for Carbohydrates

Carbohydrates are the principal respiratory substances in a large majority of the organisms. These include a variety of monosaccharides (glucose, fructose etc), disaccharides (sucrose) and polysaccharides (starch, inulin etc). When simple sugars are available they are directly, channelised into the respiratory stream, but when poly and disaccharides are involved, they are first hydrolysed into monosaccharides.

When hexose sugars are the respiratory substrate, volume of CO_2 evolved equals volume of O_2 used, and hence the value will be unity. This will be illustrated by the following equation.

$$C_6H_{12}O_6 \rightarrow 6O_2 + H_2O$$

$$RQ = \frac{CO_2}{O_2} = \frac{6CO_2}{6O_2}$$

It has to be noted that not in all cases when sugar is the respiratory substrate, the RQ will be unity. There will be many deviations due to the following reasons.

(i) Incomplete oxidation of sugars (as in anaerobic respiration).
(ii) Involvement of many reductive events such as sulphate and nitrate reduction.
(iii) Oxidation and decarboxylation processes unrelated to respiration.
(iv) Retention of CO_2 within the cell fluids instead of releasing it to the exterior.
(v) Non photosynthetic carboxylation reactions resulting in accumulation of these products as seen in succulent plants.
(vi) Metabolic utilization of CO_2

RQ for Fats

Fats are not generally found in vegetative parts, but they are found as storage products in many of the angiosperm seeds like sunflower, castor, niger seed, groundnut etc. At the time of the germination of seeds, a major amount of fat is converted into carbohydrates, while the rest is used in respiration.

Fats are poor in oxygen content, and the ratio of oxygen to carbon is always less in fats when compared with carbohydrates. Fats are rich in hydrogen and as a result they require more O_2 for complete oxidation. In addition to this, fats are not directly oxidised. They are first broken into *fatty acids* and *glycerol,* which then act as respiratory substrates. Oxygen is required for the breakdown of fats into fatty acids and glycerol. As a result, for the overall respiratory process (when fats are involved), more oxygen is required. Hence, the value of RQ will be less than unity.

(a) $C_{18}H_{36}O_2 + 26O_2 \rightarrow 18\,CO_2 + 18H_2O$
(Stearic acid)

$$RQ = \frac{26\,O_2}{18CO_2} = 0.69$$

(b) $C_{57}H_{16}4O_6 + 80O_2 \rightarrow 57CO_2 + 52H_2O$
(Triolein fat)

$$RQ = \frac{57}{80} = 0.7$$

When simple fatty acids like acetic acids are involved, RQ will be unity as seen below.

(c) $CH_3COOHY + 2O_2 \rightarrow 2CO_2 + 2H_2O$
(Acetic acid)

$$RQ = \frac{20_2}{2\,CO_2} = 1.\ (\text{unity})$$

(d) $2C_{51}H_{98}O_6 + 145\,O_2 \rightarrow 102\,CO_2 + 98H_2O$
(Tripalmitin)

$$RQ = \frac{102}{145} = 0.7$$

Interestingly, fats liberate approximately three times more energy when compared with Carbohydrates. While a gram of carbohydrates yeilds about 3.8 K cal, the same amount of fat produces approximately 9.1 K cal of energy. This may be the reason as to why the seeds prefer to store reserve food material in the form of oil, more amount of energy can be obtained by less amount of food (it requires less space).

RQ for Proteins

Proteins and amino acids serve as respiratory substrates only under certain situations like starvation etc. Like fats, proteins also have less oxygen and the volume of oxygen to carbon is low. When proteins are hydrolysed, they require more oxygen for complete oxidation as a result of which RQ value will be less than unity. The RQ value for proteins and their derivatives varies between 0.5–0.8; very often it will be 0.79.

The value of RQ may be 0.99 when ammonia is released or it may be 0.79 when amides are formed.

RQ for Organic acids

The value of RQ is more than unity when organic acids are used as respiratory substrates. This is due to the fact that in organic acids, the proportion of oxygen in relation to carbon is always more, or less of O_2 is required for oxidation. The following are some of the examples.

(a) $C_4H6O_5 + 3O_2 \rightarrow 4CO2 + 3H2O$
(malic acid)

$$RQ = \frac{4CO_2}{3O_2} = 1.3$$

(b) $2(COOH)_2 + O2 \rightarrow 4CO_2 + 2H_2O$
(oxalic acid)

$$RQ = \frac{4}{1} = 4.0$$

(c) $2C_4H_6O_6 + 5O_2 \rightarrow 8CO_2 + H_2O$
(tartaric acid)

$$RQ = \frac{8}{5} = 1.6$$

(d) $2C6H8O7 + 902 \rightarrow 12CO2 + 8H2O$
(Citric acid)

$$RQ = \frac{12}{9} = 1.33$$

RQ for succulent plants

In succulent xerophytes such as *Opuntia,* and many other plants (*Bryophyllum, Sedum* etc), carbohydrates are incommpletely oxidized resulting in the formation of intermediate compounds, there being no evolution of CO_2. As only O_2 is taken in and there is no release of CO_2, RQ is zero.

In succulent plants, stomata are open during night, O_2 is taken in with the result carbohydrates are incompletely oxidized, while during day time, the organic acids are completely oxidized releasing CO_2. But this CO_2 is not released as it is used for photosynthesis. Hence, with no evolution of CO_2 RQ will be zero.

$$2C_6H_{12}O_6 + 3O_2 \rightarrow 3C_4H_6O_5 + 3H_2O$$

$$RQ = \frac{CO_2}{O_2} = \frac{O}{3} = 0.0$$

RQ when O_2 is used for other activities

O_2 is also utilized for various other processes like converon of fats to carbohydrates, synthesis of anthocyanins etc. In such cases, the amount of CO_2 evolved will not correspond to the amount of O_2 consumed. Here RQ falls below unity.

RQ for anaerobic respiration

Certain tissues respire anaerobically (no participation of atmospheric, O_2), in which case only CO_2 is evolved without the absorption of O_2. In such circumstances, RQ will be more than unity.

$$C6H12O6 \rightarrow 2C2H5OH2 + 2CO2$$

$$RQ = \frac{CO2}{O_2} = \frac{2}{O} = 2$$

The following is a list of RQ values for some of the substrates occuring in various plant parts.

No.	Plant part	Substrate	SQ value
1.	Leaves	Carbohydrates	1.0
2.	*Opunitia* shoots	Organic acids	0.03
3.	Seeds	Starch	1.0
4.	Linsed	Fat	0.64
5.	Buckwheat seed	Protein	0.5
6.	Pea Seeds	Carbohydrates	1.5-2.4

Measurement of RQ

RQ can be measured with the help of *Ganong's respiroscope.* The apparatus essentially consists of a graduated tube with a bulb at the upper end and a levelling tube. The two tubes are connected at their base with a long rubber tube which assumes a 'U' shape. The bulb in the graduated tube has a glass stopper fitted into its neck. Both the neck and the stopper have holes which can be brought into a straight line by twisting the stopper. The apparatus is fitted into an iron or wooden stand, and filled with mercury (instead of Hg, saline solution also can be used). Respirable material such as plant parts or germinating seeds are kept in the bulb and the stopper is rotated to allow the air enter into the bulb. Mercury level in both the tubes is brought at the same level. Now the stopper is rotated to cut off the communication.

The initial level of Hg is noted. If the respiratory substrate is carbohydrate, CO_2

released will be equal to O_2 observed, so that there is neither rise nor decrease in the level of Hg. If the substrate is fat or protein, CO_2 released will be less than O_2 consumed and therefore a vaccum will be created in the chamber. This results in the increase of the level of Hg.

If the CO_2 released is more than O_2 used up, the initial level of Hg will fall after the experiment. RQ in this case will be more than one. RQ may be calculated with the help of the following formulae.

$$RQ = \frac{v^2}{v^1 + v^2}$$

v^1 = Initial rise of Hg; v^2 = further rise in Hg level after introduction of caustic potash. This is understood to find out the excess release of CO_2 if any.

$$RQ = \frac{v^2}{v^2 --- v^2}$$

v^1 = Fall in the initial level of Hg. v^2 = Rise in level after introduction of caustic potash.

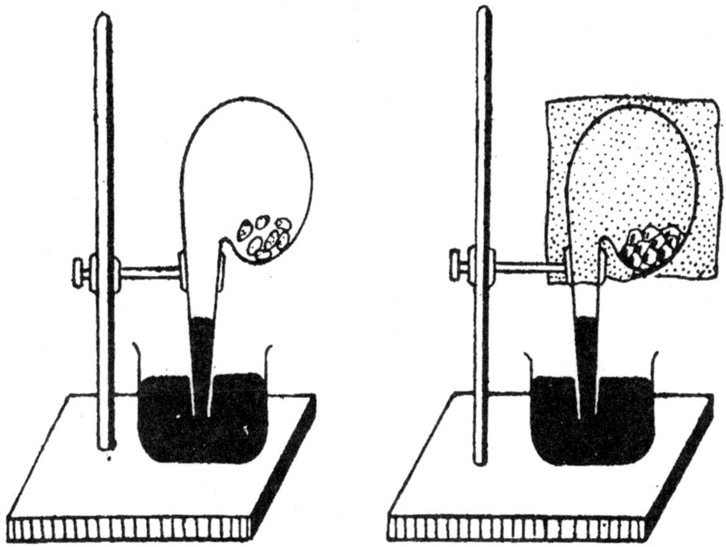

A B

Fig. 17.1 Respiration
Demonstration of value of RQ for carbohydrates. **A.** Initial leel, **B.** Ater 24 hours

TYPES OF RESPIRATION

The involvement of atmospheric oxygen during respiration constitutes the basis for its categorization into aerobic and anaerobic types.

Aerobic respiration

The respiratory substrate is completely broken down into CO_2 and H_2O with the participation of atmospheric O_2.

$$C_6H_{12}O_6 + 6O_2 \rightarrow 6CO_2 + 6H_2O + 686K\ Cal.$$

Anaerobic respiration

Here there is no participation of atmospheric oxygen. Food substances are incompletely oxidized. Hence, all the carbon in the substrate will not be released in the form of CO_2. The end products are not only CO_2, but intermediate products like ethyl alcohol, acids etc.

$$C_6H_{12}O \rightarrow 2C_2H_5OH + 2CO_2 + 56K\ Cal.$$

Besides the above, there are some more differences between aerobic and anaerobic respiration. They are listed below.

	Aerobic	Anaerobic
1.	O_2 necessary	O_2 not necessary
2.	Common to most of the organisms	Restricted to certain organisms, and certain tissues.
3.	Occurs always	Occurs temporarily in certain tissues, besides microorganisms
4.	Oxidation complete	Oxidation incomplete
5.	Large amount of energy is liberated	Less energy is liberated
6.	End Products are CO_2 and water	End products are ethyl alcohol, organic acids etc and CO_2
7.	The entire carbon content of the substrate comes out in the form of CO_2 (inorganicc	Due to incomplete oxidation of the substrate, carbon will still be left in the organic compounds (ethyl alcohol)
8.	Process non toxic to plants	Toxic to plants
9.	Respiratory enzymes present in mitochondria and cytoplasm.	Enzymes present only in cytoplasm
10.	Output in 38 molecules of ATP	Output is only 2 molecules of ATP

MECHANISM OF RESPIRATION

As has already been pointed out, during respiration the substrate material is broken down with release of energy, CO_2 and H_2O.

In the mechanism, we will be studying the various stages of the breakdown of the substrate material, enzymes involved etc., and account for the release of CO_2 and energy released in the form of molecules of ATP. In the foregoing paragraphs, various steps involved in the breakdown of carbohydrates are given.

Broadly categorizing, the following steps are identifiable in the oxidative breakdown of carbohydrates.

1. Hydrolysis 2. Glycolysis 3. Pyruvic acid oxidation 4. Krebs cycle or citric acid cycle and 5. Electron transport system or oxidative phosphorylation.

The mechanism of anaerobic of anaerobic respiration consists of pyruvic acid.

1. Glycolysis and 2. Anaerobic oxidation of pyruvic acid.

From the above amount it can be understood that the step glycolysis is common for both aerobic and anaerobic respiration.

HYDROLYSIS

When starch is the reserve food material, it should first be hydrolysed into simple sugars before it can enter into the respiratory cycle. The process of breakdown of starch into simple sugars with the addition of water is called hydrolysis.

GLYCOLYSIS

During glycolysis, one molecule of glucose will be broken into two molecules of pyruvic acid. Originally, the name glycolysis was meant for the breakdown of glycogen (*lysis* or breakdown of glycogen) found in muscle cells. Enzymes responsible for glycolysis are present in the cytoplasm. As will be seen later, no participation of atmospheric O_2 is required for the formation of pyruvic acid. Glycolysis is the common pathway for both aerobic and anaerobic respiration.

Glycolytic breakdown of glucose into pyruvic acid is also known as the **Embden - Meyerhof - Parnas** pathway (EMP pathway), after three scientists who discovered the various steps of glycolysis. EMP pathway is also called cytoplasmic respiration as the whole process of glycolysis takes place only in cytosol (cytoplasm minus organeloles). EMP pathway was first worked out in muscle cells and yeast, and later it was observed that it is common for all the organisms. The overall reaction of glycolysis may be summed up as follows.

$$C_6H_{12}O_6 \rightarrow 2CH_3COCOOH + 4H$$

The details of the above reaction are given below.

Phosphorylation

Any sugar that has to enter into the glycolysis pathway must first undergo phosphorylation (i.e., a phosphate group has to be added). In this process, ATP participates and donates a phosphate group (there will be utilization of a molecule of ATP).

(i) Glucose + ATP Glucose 6 Phosphate + ADP

The reaction is catalyzed by the enzyme *hexokinase* with Mg ++ as the cofactor. This reaction was first worked out by Meyerhof in 1927.

Glucose 6 phosphate may be produced by another means also, if starch is the original food material. It is as follows.

Starch + H3PO4 → Glucose 1 phosphate

The enzyme phosphorylase catalyses this reaction. In the next step

Glucose 1 phosphate → Glucose 6 phosphate

In the above reaction Glucose 1 phosphate changes into Glucose 6 phosphagte i.e., the phosphate group attached to the first carbon atom is shifted to the sixth carbon atom. The reaction is catalyzed by the enzyme *Phosphoglucomutase.*

Isomerization

In the step, glucose 6 phosphate isomerises into Fructose 6 phosphate under the influence of the enzyme *Phosphohexoisomerase.*

Glucose 6 phosphate → Fructose 6 phosphate

Fructose 6 phosphate may be produced from fructose directly when the latter is available.

$$\text{Fructose + ATP} \xrightarrow[\text{Mg ++}]{\textit{Hexokinase}} \text{Fructose 6 phosphate}$$

Formation of Fructose 1.6. diphosphate

This is a key reaction in the initial breakdown of glucose. Fructose 6 phosphate utilizes a molecule of ATP to get converted to Fructose 1.6 diphosphate under the influence of the enzyme *Phosphohexokinase.*

Fructose 6 phosphate + ATP → Fructose 1.6 Diphosphate + ADP

The reactions of phosphorylation leading to the formation of Fructose 1.6 diphosphate may be summarized as follows.

(i) $\text{Glucose + ATP} \xrightarrow[\text{Mg ++}]{\textit{Hexokinase}} \text{Glucose 6. Phosphate + ADP}$

(ii) Glucose 6 Phosphate —— *Phosphohexoisomerase* ——→ Fructose 6 Phosphate

(iii) Fructose 6 Phosphate + ATP —— *Phosphohexokinase* (Mg ++) ——→ Fructose 1.6 Diphosphate + ADP

Cleavage of Fructose 1.6 dip

Further reactions in glycolysis begin with the breaking up of Fructose 1.6 diphosphate into two molecules of triose under the influence of the enzyme *Aldolase.* The presence of this enzyme in plant tissues was first discovered by Tewfik and Stumf (1949).

1. Fructose 1.6 diphosphate ______ Glyceraldehyde 3 phosphate + Dihydroxyacetone Phosphate.

97% of Fructose 1.6 diphosphate gets converted into glyceraldehyde 3 phosphate, while about 3% will change into dihydroxyacetone phosphate. Dihydroxyacetone phosphate isomerises into glyceraldehyde 3 phosphate under the influence of the enzyme *phosphotriose isomerase.* The reaction may be represented as follows.

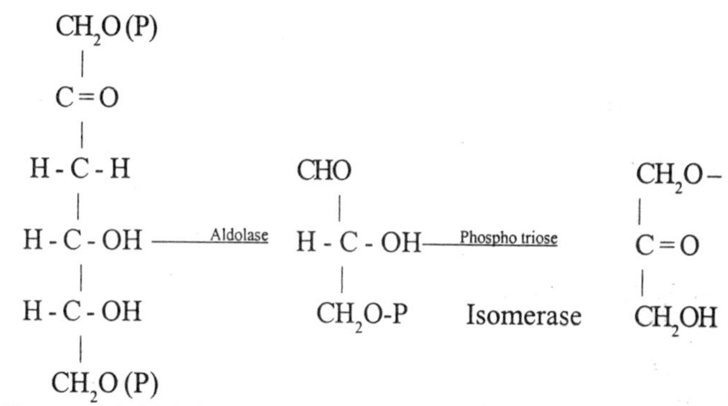

Fructose 1.6 Diphosphate   Glyceraldehyde 3 - Phosphate   Dihydroxyacetone Phosphate

To sum up, Fructose 1.6 diphosphate (6C) breaks up into 2 molecules of glyceraldehyde 3 phosphate (3C). From now on the reactions are explained for a molecule of triose (glyceraldehyde 3 phosphate). While accounting for hexose, each reaction has to be doubled. In the next step, glyceraldehyde 3 phosphate

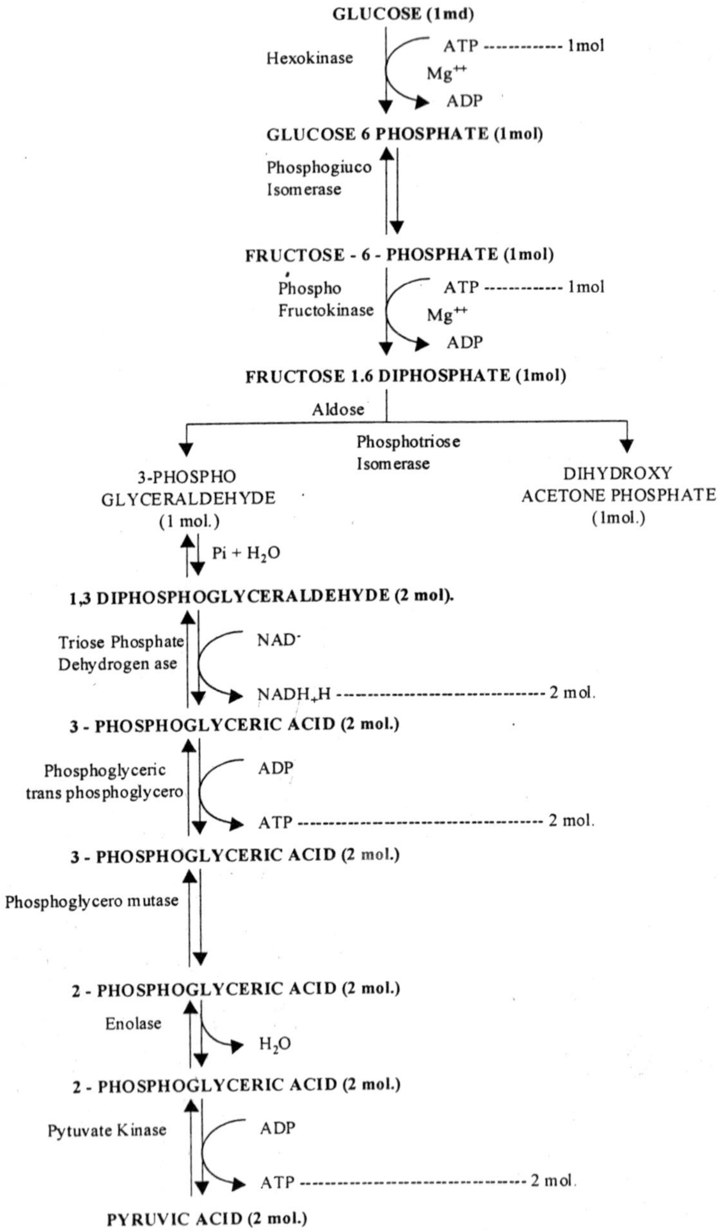

Fig.17.2 Respiration

Glycolysis - Various reactions leading to the formation of Pyruvic acid is oxidized (hydrogen removed) with the help of the enzyme *phosphoglyceraldenhyde dehydrogenase.* Inorganic phosphate and NAD participate in this reaction. 2 electrons and two protons (H+) are released from glyceraldehyde 3. Phosphate and the energy is utilized to link the inorganic phosphate to the oxidized glyceraldehyde 3 phosphate to form 1,3 - diphosphoglyceric acid.

Glyceradehyde 3 phosphate + H_3PO_4 + NAD

———————— 1.3 Diphosphoglyceric acid +
NADH + H +

As may be noticed, while the electrons provide the energy necessary for the linking of inorganic phosphate, the H + goes for the reduction of NAD to NADH +H^+

CHO
|
H - C - OH + H_2PO_4 + NAD
|
CH_2OP

O
||
C - O - P
|
H - C - OH + NADH + H^+
|
CH_2OP

lyceraldehyde - 3 Phosphate 1.3 - Diphosphoglyceric acid

3. 1,3 - Diphosphoglyceric acid is transformed into 3 phospoglyceric acid when one of the phosphate groups is released, which is accepted by ADP to synthesize a molecule of ATP. *Phosphoglyceric kinase* enzyme catalyzes this reaction.

The enzyme has Mg^{++} as the cofactor

1.3 Diphosphoglyceric acid + ADP 3 Phosphoglyceric acid + ATP

O
| |
C - O - P
|
H - C -- OH + ADP
CH_2O - P

Phosphogyceric
————————
Kinase + Mg^{++}

COOH
|
H - C - OH + ATP
CH_2O - P

4. 3 Phosphoglyceric acid gets transformed into 2 phosphologlyceric acid (phosphate group is shifted from the 3rd carbon position to the 2nd carbon position), under the influence of the enzyme *Phosphoglyceromutase.*

COOH
|

Phosphoglyceromutase

COOH
|

H - C - OH ———————— H - C - O - P

|

CH_2O - P CH_2OH

3. Phosphoglyceric acid

5. 2 - Phosphoglyceric acid (2PGA) gets converted into phosphoenolpyruvic acid (PEP) under the influence of the enzyme, *enolase* with Mg^{++} ions as cofactor. A molecule of water is eliminated from 2PGA in the process with the result an energy rich (-P) centre is created.

COOH COOH

| *Enolase* |

H - C - O - P ———————— H - C - O - P + H_2O

| Mg^{++} |

CH_2OH CH_2

(2 PGA) (2 PEP)

6. In the molecule of PEP, much of the energy is centered round the phosphate group. This energy rich phosphate group is removed from PEP, which is accepted by ADP to produce a molecule of ATP. In the process, the dephosphorylated PEP becomes converted to pyruvic acid. This reaction is catalyzed by the enzyme *pyruvate kinase* with Mg++ and K+ ions as cofactors.

COOH COOH

| *pyruvate kinase* |

H - C - O - P + ADP ———————— C = O + ATP

| $Mg^{++}K^{++}$ |

CH_2 CH_3

(Pyruvic acid)

With the formation of pyruvic acid, the reactions of glycolysis come to an end. One molecule of pyruvic acid is produced for every triose, hence for hexose, it should be two molecules of pyruvate.

ATP account during glycolysis

The production of pyruvic acid terminates glycolysis. When we briefly glance through the glycolytic reactions we realize that there are two important stages – (i) phosphorylation and (ii) oxidation of Fructose 1.6 diphosphate to pyruvic acid.

During phosphorylation of glucose to form Fructose 1.6 diphosphate, two molecules of ATP are used up. Thus during phosphorylation of glucose, 2 molecules of ATP are used.

In the second stage, at two stages ATP molecules are synthesized–once during the conversion of 1,3 - diphosphoglyceric acid to 3 phosphoglyceric acid (2 molecules - i.3., one for every triose and two for every hexose), and again during the conversion of phosphoenol pyruvic acid to pyruvic acid (2 molecules of ATP, reactions are to be doubled while accounting for hexose). Thus in all, towards the end of glycolysis 4 molecules of ATP are produced and two molecules of ATP are consumed (during phosphorlation). Hence, the net gain of ATP during glycolysis is only two molecules.

But the statement that the net gain of ATP during glycolysis is only two molecules is not completely correct, because during the oxidation of glyceraldehyde 3 phosphate to phosphoglyceric acid, 2 molecules of NADH + H + are produced. These two molecules pass through the electron transport chain producing a total of 6 (3 + 3) ATP molecules (Details explained later). Hence, the net gain of ATP during aerobic glycolysis is, 6 + 2 = 8 molecules of ATP.

$$\underset{\text{Pyruvic acid}}{CH_3COCOOH} + NADH_2 \xrightarrow{\text{Dehydrogenase}} \underset{\text{Lactic acid}}{CH_3CHOH\,COOH} + NAD$$

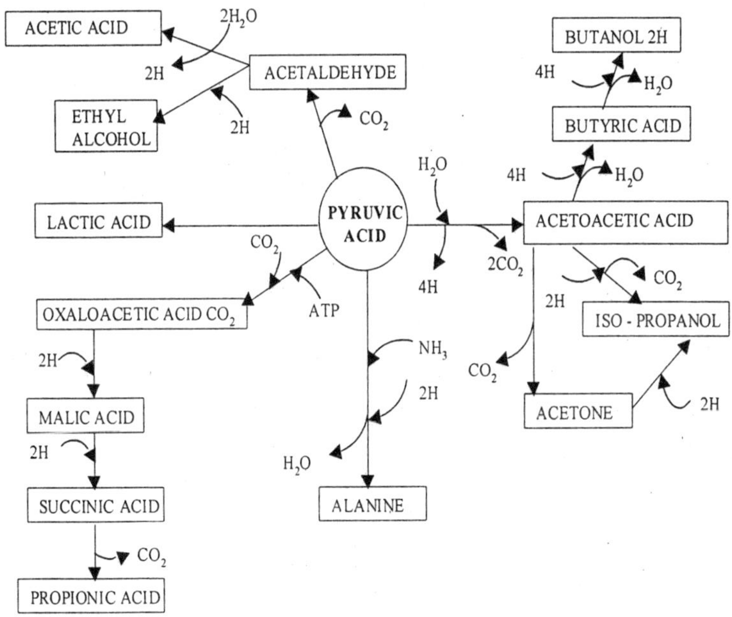

Fig.17.3 Respiration

Anaerobic degradation of Pyruvic acid

Breakdown of pyruvic acid

Further breakdown of pyruvic acid depends upon the nature of the organism, type of respiratory pathway and the presence or absence of O_2. Broadly these breakdown are –

(i) Alcoholic fermentation
(ii) Lactic acid fermentation
(iii) Amino acid synthesis and
(iv) Aerobic breakdown to produce CO_2 and H_2O

Options (i) and (ii) can be carried out in the abscence of O_2 and (iv) is possible in the presence of O_2, which happens to be the major pathway. Option (iii) leads to the biosynthesis of alanine by transaminase reaction with glutamate. This is not discussed here as it is not relevant to the respiratory pathway.

FERMENTATION

Fermentation is essentially anaerobic respiration found in microorganisms like fungi and bacteria. In fermentation, the substrate for oxidation is extracellular (not present within the cell), since the organisms are basically saprophytic.

Fermentation as a process to produce alcohol was known to human beings since time immemorial, eventhough its scientific understanding dates back to only the middle of the 19th century. Experiments conducted by ***Louis Pasteur*** established that fermentation is carried out with the participation of living cells.

Buchner in 1897 found out that the extract of yeast cells (which does not contain living cells) can bring about fermentation of sugar solution. Subsequently, it was discovered that the enzyme *Zymase* present in the yeast extract is capable of fermenting activity (breakdown of carbohydrates together with the formation of alcohol and CO_2).

The enzyme *Zymase,* will be active only in the presence of phosphate. *Sucrase* and *Maltase* are the other carbohydrate fermenting enzymes secreted by yeasts.

In fermentation also, as in anaerobic respiration, atmospheric oxygen does not participate. Oxidation is accomplished by the intermolecular atomic shifts, with the result the product will have less energy than the original substrate. Much of the energy released during fermentation is wasted as heat, while the remainder is used for the conduct of metabolic activities.

The end product of fermentation need not always be alcohol, it may be various types of organic acids such as butyric acid, lactic acid etc. Depending on the end product, fermentation may be termed as alcoholic fermentation, lactic acid fermentation etc. The following are formulae for these fermentations.

(i) $C_6H_{12}O_6$ → $2C_2H_5OH + 2CO_2$ (ethyl alcohol)
(ii) $C_6H_{12}O_6$ → $C_4H_8O_2 + 2H_2 + 2CO_2$ (butyric acid)
(iii) $C_6H_{12}O_6$ → $2_3H_6O_3$ (Lactic acid)
(iv) $C_2H_5OH + O_2$ → CH_3 $COOH + H_2O$ + energy (acetic acid)

In the formula (iv) given above, for the formation of acetic acid from ethyl alcohol, O_2 is necessary. Hence acetic acid fermentation is different from other types of fermentation.

Fermentation is carried out by microorganisms, a point which has been emphasized earlier. The following is a list of organism which carry out the various types of acid fermentation.

	Organism	Type of fermentation (based on end product)
1.	*Bacillus butyricus and Clostridium butyricum*	Butyric acid
2.	*Bactyerium lactic acid*	Lactic acid
3.	*Acetobacter aceti*	Acetic acid

The brief description of fermentation given above may now be followed by certain details of a few specific types of fermentation and a mechanism of regeneration of ATP.

ATP Generation by fermentation

During the fermentation of the glucose by microorganisms about 1–4 molecules of ATP may be produced. While there are large numbers of end products of fermentation starting from glucose, it is surprising that there are only a few reactions concerned with energy conversions of substrate level phosphorylation. Of these, the three most important reactions are discussed below.

1.1m3 - Bisphosphoglycerate + ADP ———— 3 - Phosphoglycerate + ATP

2. Phosphoglycerate + ADP ———— Pyruvate + ATP
3. Acetyphosphateor Butyrylphosphate + ADP ———— Acetate or Butyrate + ATP

Most of the fermenting microorganisms use only reaction 1 and 2 above, and use pyruvate compounds synthesised from acetyl - Co - A as the necessary hydrogen acceptors. In the above, reaction 1 is catalysed by phosphoglycerate kinase and reaction 2 is catalysed by pyruvate kinase, during the fermentation of one molecule of glycose only two or a maximum of 4 molecules of ATP are generated and the following products may be formed: lactate, ehtanol, acetone, butyrate, n-butanol, 2-propoanol, 2,3 - butanediol, caproate, acetate, carbon dioxide, and molecular hydrogen.

In reaction 3 given, above the enzyme involved is acetate kinase, and it offers the advantage of additional ATP regeneration. Acteyl phosphate is produced from acetyl - Co - A by phosphotransacetylase.

acetyl - Co - A + P1 ——— acetylphosphate + Co - A

Acetyl phosphate may also be produced from sugar phosphates such as fructose - 6 - phosphate via reactions catalysed by phospoketolase.

The ability of microbes to utilise acetate kinase depends on whether they can evolve molecular hydrogen. No hydrogen acceptors need to be synthesised for reducing equavalents that may be liberated as molecular hydrogen. Fermentations may be basically classified into two types :

1. Alcoholic fermentation,
2. Acid fementation

1. Alcoholic fermentation

When respiratory substrates are sugars, during anaerobic respiration many of the microbes ferment them and produce ethanol (ethyl alcohol). The wide variety of microorganisms are involved in ethanolic fermentation. The chief ethanolic producers are yeasts (particularly strains of *Saccharomyces cerevisiae)* and fungi. Yeasts also respire aerobically but in the absence of air, they ferment carbohydrates into ethanol and carbon dioxide. Among bacteria also a number of anaerobes or facultative aerobes produce ethanol as the main product during fermentation of sugars. Unusuall accumulation of ethanol can occuı even in plants under anaerobic conditions.

Ethanol formation of yeasts

The fermentation process was infact studied in great detail in yeasts. The conversion of glucose to ethanol was formulated in 1815 by Guy – Lussac as follows.

$$C_6H_{12}O_6 — 2CO_2 + 2C_2H_5OH$$

The normal fermentation of glucose by yeasts

The breakdown of glucose to ethanol and carbon dioxide by yeasts proceeds via the usual glycolytic pathway. From their onwards, the conversion of pyruvate to ethanol comprises few steps. In the first step, pyruvate is decarboxylated by pyruvate decarboxylase with the participation of thiamine pyrophosphate. As a result of this, a molecule of acetaldehyde is formed which is then reduced to ethanol by alcohol dehydrogenase with $NADH_2$ as the hydrogen donor. In this hydrogen transfer, the hydrogen obtained during the dehydrogenation of triphosphate is utilised. Thus the oxidation reduction balance in maintained.

History of yeast fermentation

Louis Pasteur was perhaps the first to have carried out experiments on fermentation by microorganims. It was Buchmer and Hahn (1896), however, who connected the fermenting activity with yeast. They observed that the extract obtained from Brewer's yeast by grinding with kieselguhr (diatomaceous earth), and sand produced gas bubbles after the addition of sugar. This was perhaps a first example of biochemical reaction *in vitro.* Harden and Young (1906), made the discovery that inorganic phosphate is a requirement for the fermentation of glucose by yeast extract and it is incorporated in fructose - 1,6 - bisphosphate. According to Harden and Young equation, fermentation of glucose by yeast is as follows.

$$2C_6H_{12}O_6 + 2Pi — 2CO_2 + 2C_2H_5OH + H_2O$$

+ fructose - 1,6 - bisphosphate

Fermentation formulae by Neuberg

The studies of Neuberg on fermentation are of great significance. He showed that yeast can ferment both glucose as well as pyruvate. The formation of intermediate compounds such as acetyldehyde could by demonstrated by trapping it with hydrogen sulphite. On addition of hydrogen sulphite to glucose fermenting yeast, acetyldehyde is precipitated as the addition product.

$$CH_3CHO + NaHSO_4 \text{———} CH_2 - CHOH - SO_3Na$$

Under the above conditions, glycerol is formed as a new product while the yield of ethanol and carbon dioxide gets reduced. This kind of fermentation has been exploited industrially for the production of glycerol. By trapping the acetyldehyde (using hydrogen sulphite), it is not allowed to function as hydrogen acceptor. As a result, dihydroxycetone phosphate becomes the hydrogen acceptor and is reduced to glycerol - 3 - phosphate which is then dephospharylated glucose + hydrogen sulphite ——— glycerol + acetaldehyde - sulphite + CO_2

The above fermentation which is a modification of the normal ethanolic fermentation by yeasts is known as Neuberg's second fermentation. The principle of trapping a metabolite along the pathway has now become known as the trapping method and has been extensively used in biochemical research.

Addition of alkali to fermenting yeast extract, also leads to the formation of glycerol because acetyldehyde undergoes dismutation to ethanol and acetate, and hence cannot function as a hydrogen acceptor. This reaction is known as Neuberg third fermentation and is as follows:

$$2\ \text{GLUCOSE} + H_2O \longrightarrow \text{ethanol} + \text{acetate} + 2\ \text{glycerol} + 2CO_2$$

The normal ethanolic fermentation is referred to as Neuberg's first fermentation and the reaction is as follows.

$$C_6H_{12}O_6 \longrightarrow 2C_2H_5OH + 2CO_2$$

Relationship of yeast to oxygen

Under anaerobic conditions, yeasts ferment very actively but they can hardly grow. This is because yeasts are aerobes. On admission of air, fermentation decreases in favour of aerobic respiration. As shown in Pasteur effect, fermentation can be completely supressed by vigrous aeration. Pasteur discovered this effect, more than 100 years ago during his investigations on wine making. Pasteur effect, however is not restricted only to yeast but has been found in all anaerobic cells. Aeration of yeasts supports growth, but leads to a decline in glucose consumption and in the reduction of ethanol and carbon dioxide. This observation appears to be sensible from the stand point of energy conversion and suggests an ingenious regulatory mechanism. One molecule of glucose anaerobically yields only two molecules of ATP in contrast to the aerobic yield of 38 molecules of ATP per glucose molecule. Thus, the cell adopts by regulation of substrate consumption to the energy gain obtainable under two conditions.

The Pasteur effect involves several regulatory mechanisms working tegether. One of these perhaps applies at the level of phosphorylation and can be explained by competition for ADP and P.

A second regulatory mechanism perhaps is also responsible for the Pasteur effect. The enzyme phosphofructokinase is allosterically inhibited by ATP.

Harden and Young's explanation of fermentation balance

According to Harden and Young, fructose, 1,6 biphosphate accumulates due to lack to ATP utilisation for energy requiring reactions in the cell - free system. This leads to an excess of ATP balance. The yeast extract does not have any phosphates activity so that ADP must be constantly regenerated by phospharylation of excess glucose or fructose - 6 phosphate.

Industrial uses of yeast

For industrial purposes, different strains of yeast are selected. These are of two types.

Bottom yeast and *Top yeast.* Bottom yeasts are weekly respiring and actively, fermenting ones, and hence are used in the brewing of beer. Top yeasts on the other hand are used for production of ethanol and wine making.

Baker's yeast (*Saccharomyces cerevisiae),* are mostly top yeasts and are mainly used in baker's to produce carbon dioxide to raise the dough, i.e, these are grown in tanks with good aeration. In this reaction, ethanol is produced as a byproduct.

Brewer's yeasts are mostly bottom yeasts and are mainly used in brewing of beer from barley.

Industrial alcohol is produced from the residue of cane sugar refining called molasses. Occassionally, potatoes also may be used as rawmaterial for fermentation. Alcohol may also be obtained by hydrolysed wood chips. Microbial species used for these fermentations are species of *Candida* and *Endomyces*.

Ethanol formation by bacteria

Not many bacteria have been studied and exploited for ethanol production as has been done in the case of yeast. Among all the bacteria examined so far, only in the case of *Sacrcina ventriculi,* the pathway of ethanol formation is similar to that of yeast. Another bacterium that seems to produce ethanol has been isolated from the juice of the Agave plant. This bacterium (*Zymomonas mobilist*) breaksdown glucose via 2-keto-3-deoxy-6-phosphogluconate pathway and splits the pyruvate to acetyldehyde and carbondioxide under the influence of the enzyme pyruvate decarboxylase. The acetyldehyde is further reduced to ethanol along with carbondioxide and some quantity of lactic acid. It has further been discovered that the agave alcohol contains carbon atoms 2,3,5 and 6 of glucose, whereas a yeast ethanol contains carbon atoms 1,2,5 and 6 of glucose.

Some of the bacteria belonging to Enterobacteriaceae and Clostridia are also known to produce ethanol as a product of secondary fermentation. In these microorganisms, however, the precurosor of ethanol namely acetyldehyde is not produced from pyruvate by pyruvate decarboxylase but originates by the reduction of acetal Co-A.

Some of the acid fermenting bacteria like *Leuconostoc mesenteriodes,* produce alcohol by an entirely different pathway. Glucose is first broken down by the early steps in the pentose phosphates cycle to produce pentose phosphate. This pentose phosphate (xylulose -5 - phosphate) is then acted upon by the

enzyme phosphoketolase to produce acteyl phosphate. This is reduced to ethanol by the enzymes acetaldehyde dehydrogenase and alcoholdehydrogenase. The remaining product of breakdown glyceroldehyde - 3 - phosphate is converted into pyruvate xylulose - 5 - phosphate + Pi — acetylphosphate
+ glyceraldehyde -3 - phosphate.

Acid fermentation

Anaerobic breakdown of sugars by certain microorganisms results in the formation of various types of acids such as lactic acid, acetic acid, butyric acid, propionic acid, formic acid etc. Some of these fermentation are discussed below in some detail.

Lactic acid fermentation

The lactic acid bacteria belong to the family *lactobacteriaceae* and includes a wide variety of rods and cocci. All the members of lactobacteriaceae whether it be the physical form have a characteristic physiology, are gram positive and non-sporing (with few exceptions). They are usually non-motile and have carbohydrate as the main energy source and produce lactic acid. Another family of bacteria which produce lactic acid is Enterobacteriaceae. The lactobacteriaceae, however, are obligate fermenters. Basically, lactobacteria are anaerobic but aerotolerant.

Generally, the members of lactobacteriaceae require many accessory factors in the culture medium in addition to the basic nutrients. Addition of vitamins, amino acids, purines and pyrimidines is necessary to cultivate lactobacteria. The culture medium is usually complex one containing yeast extract, tomato extract or even blood.

Generally, the metabolic machinery of lactobacteria appears to be quite defective as it cannot produce a number of compounds which are required for its growth and consequently they have to be added to the medium. Many microbiologists regard the lactobacteriaceae as "metabolic cripples", as they have lost the ability to synthesise a number of metabolants. This may be due to the fact that in the course of evolution they seem to have acquired an affinity to milk and other nutrient rich media. This has lead to a special metabolic mechanism in them to utilise lactose, an ability that most microorganisms do not possess. The ability to use lactose is also possessed by many intestinal bacteria of the coliform type.

Milk as such is produced as a nutrient factor by mammals and consists of lactose. It is not found in the plant kingdom. The ability of some microorganisms to utilise lactose may be regarded as an adaption to the environmental conditions prevailing in the mammalian digestive tract.

Lactose is a disaccharide which must be hydrolysed before it can enter the catabolic pathway to produce Hexoses. This reaction is catalysed by the enzyme β-galactosidase. This enzyme however, occurs only in, few bacteria. The product of hydrolysis mainly ngalactose is converted to glucose - 1 - phosphate after phosphorylation.

Occurrence of lactobacteria

Lactic acid bacteria are largely distributed in places where there is a generous amount of lactose, as such, they are hardly found in soil or water. The natural habitats of natural bacteria (Schlegel, 1993) are –

1. Milk and the places where milk is produced and processed (*Lactobacillus lactis, L. bulgaricus, L.Helveticus, L.casei, L.fermenttium, L.brevis, Lactococcus lactis, L.diacetilactis*).

2. Intact and rotting plants (*Lactobacillus plantarum, L.delbruackii, L.fermentum, L.brevis, Lactococcus lactis, Leuoconostoc mesenteriodes*)

3. Intestinal tracts and mucous membranes of animals and humans (*Lactobacillus acidophilus, Bifidobacterium, Enterococcus facealis, Streptococcus salivarius, S.bovis, S.oyogenes, S.pneumoniae*).

Classification of lactobacteria

They are classified into two basic categories based on their ability to ferment glucose into solely lactate or to other additional products. If the fermentation produce is only lactat the bacteria are said to be **homofermentative** and if the fermentation product is many other compounds in addition to lactate they are said to be **heterofermentative.** The following table gives a list of heterofermenting and homofermenting bacteria.

Cocci	Rods
Homodermentative : $C_6H_{12}O_6$ —— $2CH_3$ - CHOH - COOH	
Streptococci	Lactobacilli
Lactococcus lactis subsp. Lactis	Thermobacteria (temp. Opt 40°C,do not grow at 15°C)
Lactococcus lactis subsp. Lactis	subsp, Delbrueckii
var.diacetilactis	Lactobacillus Delbrueckii
subsp.cremoris	Subsp. Lactis
Enterococcus faecalis	Lactobacillus delbrueckii

Streptococcus salivarius subsp.salivarius
Streptococcus salivarius subsp.thermophilus
Streptococcus pyogenes

subsp.bulgaricus
Lactobacillus helveticus
Lactobacillus acidophilus
Lactobacillus salivarius
Streptobacteria (temp.opt. 30-37°C, alwaysgrow at 15°C)
Lactobacillus casei
Lactobacillus alimentaris
Lactobacillus coryniformis
Lactobacillus plantarum

Heterofermentative : $C_6H_{12}O_6$ — CH_3 - CHOH - COOH + CH_3 - CH_2OH + CO_2 (or CH3 - COOH)

Streptococci
Leuconostoc mesenteriodes subsp. mesenteriodes
Leuconostoc mesenteriodes subsp.dextranicum
Leuconostoc mesenteriodes subsp.cremoris
Leuconosoc lactis

Lactobacilli
Lactobacillus bifermentans
Lactobacillus brevis
Lactobacillus fermentum
Lactobacillus kandleri
Lactobacillus viredescens

Homolactic fermentation

Homofermentative lactobacteria produce as the end product of metabolism almost pure lactate. They breakdown glucose via the fructose biphosphate pathway as they posses all the required enzymes. Aldolase, which is a key enzyme helps in the anaerobic breakdown of glucose to lactate. From glucose pyruvate is produced first, which is then reduced to lactate. The hydrogen necessary for reduction of pyruvate comes from the dehydrogenation of glyceraldehyde - 3 - phosphate. The type of lactate formed whether it is D, L or DL depends on the stereospecificity of the enzyme lactate dehydrogenase. Only the small fraction of the pyruvate is decarboxylated and converted to acetate, ethanol, carbon dioxide or possibly acetation. The extent of the formation of subsidiary products besides lactose depends on the availability of the oxygen.

Hetero lactic fermentation

The hetero fermentive bacteria lack some key enzymes that are necessary for the breakdown of fructose biphosphate. These enzymes are aldolase and triosephosphate isomerase. The initial breakdown of glucose therefore is entirely along the pentose phosphate pathway i.e., via glucose - 6 - phosphate —— 6 - phosphoclugonate —— ribulose - 5 - phosphate. Ribulose - 5 - phosphate is further converted into xylulose - 5 - phosphate by the enzyme epimerase. This pentose sugar is cleaved by a TPP dependent reaction catalysed by the enzyme phosophoketolase resulting in the formation of glyceroldehyde - 3 phosphate and acetyl phosphate.

Lactobacteria such as *Leuconostoc mesenteroides* ferment glucose to yield lactate, ethanol and carbondioxide. Obviously, they reduce the acetyl phosphate via acetyl - Co - A and acetyldehyde to ethanol. Many other heterofermentive bacteria convert the acetyphosphate either partially or completely to acetate, transfering the energy rich phosphate bond to ADP and thus producing ATP. The excess of hydrogen in this instance is transferred to glucose to give mannitol. The other compound glyceraldehyde - 3- phosphate is converted to lactate via pyruvate.

Fructose is also fermented by heterofermentive bacteria resulting in the formation of lactate, acetate, carbon dioxide and mannitol.

$$3 \text{ fructose} \longrightarrow \text{lactate} + \text{acetate} + CO_2 + 2 \text{ mannitol}$$

Another bacterium, *Lactobacillus plantarum* can ferment glucose homofermentatively, but breaks pentoses by the enzyme phosphoketalase to lactate and acetate. There are some bacteria such as *Lactobacillus casei,* which are homofermentive on glucose; become heterofermentive when the source substrate is ribose, and produce acetate and lactate.

Another important example of heterofermentive lactic acid bacteria is *Bifidobacterium bifidium,* which owes its name to the shape of its cells. This is predominantly present in the intestinal tract of breast fed babies. These bacteria depend heavily on sugars that contain N - acetylglucosamine, which is found only in human milk and not in any other milk. The species of *Bifidobacterium,* are strict anaerobes and require an atmosphere containing atleast 10% carbon dioxide. The metabolism of glucose by *Bifidobacterium* is as follows.

Glucose fermented via the phosphoketolase pathway. The microbe lacks aldolase and glucose - 6 - phosphagte dehydroganse, but has phosphoketolase. This enzyme can split fructose phosphate and xylulose - 6 - phosphate to acetylphosphate and erythrose - 4 - phosphate or glyceraldehyde - 3 - phosphate respectively.

$$2C_6H_{12}O_6 \longrightarrow 2CH_3 - CHOH - COOH + 3CH_3COOH$$

Propionic acid fermentation

The bacteria that breakdown glucose to propionic acid are called *Propionibacteria.* Generally, these bacteria are present in the intestine of ruminants such as cattle and sheep where they play an important part in the formation of fatty acids particularly propionic and acitic acids. These bacteria are responsible for the conversion of lactate to propionate. *Propionibacteria* rarely occur in milk nor are they found in soil or water. They can be isolated by enrichment cultures under anaerobic conditions in a lactate yeast extract medium inoculated with Swiss cheese.

The following is a list of propionic acid fermenting bacteria.

Propionibacterium freudenreichii
P.acidi - propionici
P.acnes
Veillonella alcalescens (Micrococcus lactilyticus)
Clostridium propionoicum
Selenomonas
Micromonospora

Character of Propionibacterium

Cells are gram positive, non-motile rods and also non-spore forming. Propioni bacteria are strict anaerobes and are intolerant to atmospheric oxygen. Because of its intolerance, they are long regarded as obligatory fermentors. Some recent cytochromes and catalase. This suggests that under conditions of regulated aeration the cell growth perhaps will be better than under anaerobic conditions.

Hence propionibacteria are regarded as micro aerotolerant.

Under anaerobic conditions, propionibacteria ferment a variety of sugars such as glucose, sucrose, lactose and pentoses to porpionic acid. Even glycerol can be converted to propionic acid.

Wood and Werkman (1936), who have made a detailed study of glycerol fermentation by Propionibacterium acidi - propionici found out that the fermentation involves the net fixation of carbon dioxide. The fixed carbon dioxide was found in the excrete succinate. Carboxylation of pyruvate here is called, the Wood Werkman reaction.

$$3CH_3 - CHOH - COOH — 2CH_3 - CH_2 - COOH + CH_3 - COOH + CO_2 + H_2O$$

Reduction of lactate or pyruvate to propionate follows a pathway called methylmalonyl - Co - A pathway because of the characteristic intermediate compound methylmalonyl - Co - A. Pyruvate is carboxylated to oxalocetate by the enzyme methylmalonyl - Co - A. carboxytransferase. Oxalocetate is then reduced to succinate, this step is linked to an electron transport chain. The succinate gets converted to succinyl - Co - A and then gets transformed into methylmalonyl - Co - A by the enzyme metholmalonyl - Co - A mutase. This intermediate compound undergoes dcecarboxylation to yield propionyl - Co - A. Propionate is liberated from propional Co - A to succinate. Most of the propioni bacteria form propionate by the above mentioned pathway. The same pathway is used by Veillonella alcalescens and Selenomonas ruminatium also.

The methylmenloly - Co - A pathway is reversible and can be used to produce succinly - Co-.A.

Acryloly - Co - A pathway

Bacteria such as Clostridium propionicium, Bacteroides ruminicola and Megasphera elsedeni produce propionic acid by a simple route with derivatives such as acrylic acid, acryloyl - Co -A acting as the intermediate compound. This results in the formation of propionil - Co -A leads to the formation of propionate.

Formic acid fermentation

Some acid producing fermenting bacteria are collected together in a physiological grouping whose characteristic feature is the production of formic acid. In these bacteria, besides formic acid several other acids are also formed. Systematically, most of these bacteria are included in the family enterobacteriaceae. Formic acid producing interbacteria are gram negative, peritrichous, high motile and non spore forming rods. Physiologically they are facultatively aerobic and can obtain the energy either by aerobic, respiration or by fermentation.

Some of the common enterbacteriacease which produce formic acid are *Escherichia coli, Proteus, Enterobacter, Bifidiobacterium etc.*

Fermentation products

It has been pointed out early, that the fermic acid producing bacteria are also called mixed acid producing bacteria as they produce several acids in addition to formic acid. Many of the enterobacteriaceae bacillus species produces a number of organic products such as acetate, formate, succinate, lactate, ethanol, glycerol, acetoin etc. Sugars are largely metabolized via the fructose 1.6 diphosphate route and to a lesser extent by the pentose phosphate pathway.

Fermentation of glucose by Escherichia coli is characterized by the following reactions (Schlegel, 1986).

1. Conversion of pyruvate to acety Co - A and formate
2. Cleavage of formate to carbon dioxide and hydrogen
3. Reduction of acety Co - A to ethanol

The cleavage of pyruvate to acetyl Co - A and formate occurs only under anaerobic conditions and is catalysed by the enzyme pyruvate, formatelyase. This enzyme is very sensitive to oxygen. It is maintained in the reduced state by flavodoxin and requires for its activation, S - adenysyl - L - methionine.

Formate is split into carbon dioxide and hydrogen by most strains of C.coli as well as by other gas forming members of enterbacteriaceae. This reaction is catalyzed by the enzyme system formate: hydrogen lyase. The reaction is as follows :

$HCOOH \text{——} H_2 + CO_2$

The above reaction most probably involves the combined effects of a formate dehydrogenase and a hydrogenase

1. $HCOOH + X \text{—} CO_2 + XH_2$
2. $XH_2 \text{——} X + H_2$

Butyric acid - Butanol fermentation

Butyric acid, butanol, 2 - propanol and few other organic acids are produced as end products of carbohydrate fermentation by anaerobic bacteria belonging to the genus clostridium.

The genus clostridium belongs to the family Bacillaceae. The members are gram positive, motile (petritrichous), spore forming rods. Physiologically, clostridia are characterized by their intense fermentive metabolism. Species of clostridium, however, range from strict anaerobes (C.pasteurinum) to aerotolerants (C.acetobutylicum).

For their fermenting activity, clostridia use a variety of substrates such as polysaccharides, nuclei acids, proteins, purines, pyrimidnes etc.

Biochemistry of fermentation

Clostridal fermentation produces varying amounts of acids such as butryates, acetate, lactate etc. as well as alcohols (butanol, ethanol etc). In some instances, the end product of metabolism may even be acetone and gas (CO_2 or H_2.)

Clostridia metabolise glucose via the glycolytic pathway. Hydrogen removed from glyceraldehyde 3. Phosphate is added to organic acids synthesized from pyruvate and acetyl Co - A. Glucose fermentation by C. Butyrium and C. Acetobutylicum represents fermentation typical of the genus.

Butyrate is the product of condensation of two molecules of acetyl Co - A to acetoacetyl Co - A and subsequence reduction. Condensation of molecules is catalysed by the enzyme thiolase. Acetoacetyl Co - A is reduced by β-hydroxy butyryl - Co - A dehydrogenase and $NADH_2$ β-hydroxy butyryl Co - A. The enzyme corotonase removes water from this to give crotonyl CoA. This is then reduced by butyryl CoA dehydrogenase to butyryl CoA. The CoA can be transfered to acetate by CoA transferase to produce butyrate.

In a purely butyric acid fermentation, the hydrogen derived from the oxidation of pyruvate is liberated in gaseous form. When glucose is fermented to butyrate, CO_2 and H_2, a hydrogen balance is established and 3 mol ATP is regenerated per mol of glucose.

The end products of fermentation in Clostridium acetobutylicum are, butanol, butyrate, acetone and 2 - propanol. Initially, butyrate is excreted; with the increasing acidity, however acetoacelate decarboxylase enzyme becomes induced whose activity results in the formation of acetone and butanol. To be reduced to butano, butyrate must first get converted to butyryl CoA.

Fermentation of glutamic acid

Fermentation of glutamic acid attracted the attention of microbiologists, mainly because of the use of monosodium glutamate as a flavour increasing agent in food industry.

Japanese scientists devised a single stage fermentation to produce glutamic acid from strach (potato), using Micrococcus and Brevibacterium.

Presently, however, commercial production of glutamic acid is a two stage operation. In the first stage α- Ketoglutaric acid is produced by carbohydrate metabolism. A wide variety of bacteria produce α-Ketoglutaric acid (Pseudomonas fluroscens, Bacterium ketoglutarium etc). α Ketoglutaric acid is then converted to glutamic acid by the process of transamination.
E.coli and B. Ketoglutaricum produce glutamic acid with alanine or aspartic acid acting as amino group donor. The reactions take place as follows :
α Ketoglutaric acid + Aspartic acid — glutamic acid + Oxalo acetic acid
Homoacetate fermentation

Bacterial activity that metabolizes substrate and ultimately proudces predominantly acetate is called homoacetate fermentation. Many of the species of Clostridia (C.formicoaceticum, C.thermoaceticum, C.acidicum etc), are able to transfer the hydrogen equivalents liberated in the initial oxidation of substrate only to carbon dioxide producing acetate.

$$8\,[H] + 2CO_2 \text{ —— } CH_3 \text{ — } COOH + 2H_2O$$

C. Thermoaceticum and C.Formicoaceticum ferment glucose along the glycolytic pathway and produce almost 3 mol acetate per mol of glucose. Carbon dioxide released due to the decarboxylation of pyruvate gets refixed, and serves as H_2 acceptor. Pyruvate gives rise to acetate, CO_2, FdH_2 and ATP by the enzymes ferrodoxin oxidioreductase, phsophotransacetylase and acetate kinase.

MICROBIAL METABOLISM AEROBIC RESPIRATION

In the previous pages, anaerobic breakdown of carbohydrates has been studied ultimately leading to the formation of incompletely oxidised carbon compounds along with some amount of energy. In this chapter, we will study the aerobic breakdown of pyruvic acid leading to the complete oxidation of organic carbon and release of a large amount of energy. The essential feature of aerobic respiration is the participation of atmospheric oxygen.

Aerobic oxidation of pyruvic acid

This is the most important phase of aerobic respiration in which pyruvic acid gets converted into a two carbon compound called acetaldehyde, before entering the Krebs cycle for complete decarboxylation and oxidation. The reaction is mediated by a complex enzyme with atleast five essential cofactors (Gunsalnus 1954). The acetaldehyde combines with co enzyme A to form acetyl CoA which enters the Krebs cycle.

Formation of Acetyl Co A

As already pointed out, the enzyme which catalyzes the decarboxylation of pyruvic acids a complex one which includes 3 enzymes - *pyruvic acid decarboxylase, dihydroxylipoyl, transacetylase* and *dihydrolipoly dehydrogenase.* The enzyme complex also requires five cofactors – Thiamine pyrophosphate (TTP), Mg ions, NAD +, Coenzyme A (CoA) and lipoic acid.

The first step includes the formation of a complex between TTP and pyruvate, followed by the decarboxylation of the latter. In the second step, the acetaldehyde fragment reacts with the cofactor lipoic acid to form an *acetyl lipoic acid complex.* In this reaction, acetaldehyde is oxidized to the acid and lipoic acid is reduced. In the third step, acetyl group is removed from lipoic acid and added on to CoA, resulting in the formation of Acetyl CoA and reduced lipoic acid. In the final step, oxidized lipoic acid is regenerated by transforming the electrons from reduced lipoic acid to NAD +. This is necessary because, there should be a continuous supply of oxidized lipoic acid to accept pyruvate and produce acetyl CoA. The NAD which gets reduced to foom NADH + H^+ eventually enters into the electron transport system to produce 3 molecules of ATP. The following reactions summarize the formation of Acetyl CoA from pyruvate.

(i) Pyruvate + TPP $\xrightarrow{Mg^{++}}$ TPP Complex + Co_2

(ii) TPP Complex + Lipoic acid ——— Acetyl lipoic acid

(iii) Acetyl lipoic acid + CoA ——— Acetyl lipoic acid
(oxidized) complex; Complex + TPP; (reduced form)

(iv) Lipoic acid + NAD ——— Lipoic acid + NADH + H^+
(reduced form); (oxidized form)

The net result may be given by the following equation.

Pyruvate + CoA + NAD Acetyl CoA + CO_2 + NADH + H^+

Finally, the two carbon compound Acetyl CoA formed in cytoplasm enters the mitochondria to react with the acids of the 'Kreb's cycle for the removal of remaining 2 carbon and to produce energy to be stored in the molecules of ATP.

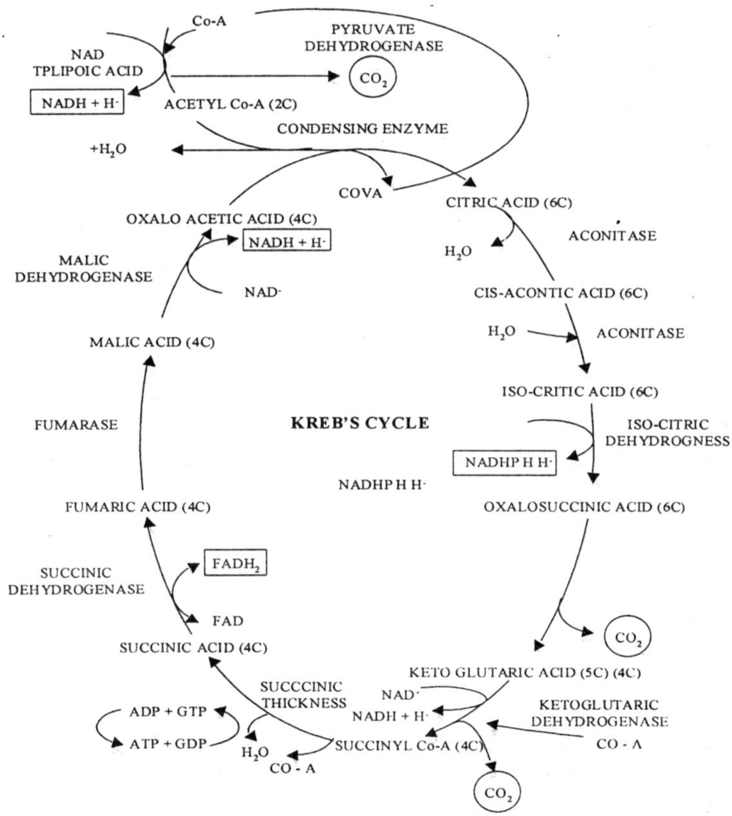

Fig. 17.4 Respiration
Reactions of Kreb's cycle

KREBS CYCLE

Acetyl CoA, the two carbon compound undergoes further decarboxylation when it enters a cycle of interconverting carboxylic acids present in mitochondria. Together with decarboxylation, there is oxidation also (dehydrogenation), resulting in the formation of molecules of ATP (to trap the released energy).

Krebs cycle (named after its discoverer), was actually worked out by Wood et al (1942), and Krebs (1943). The cycle has several names – Tricarboxylic XCycle (TCA), Citric acid Cycle, Mitochondrial respiration etc. Basically, the cycle consists of several 6 carbon, 5 carbon and 4 carbon acids. When these are getting interconverted, Acetyl CoA enters facilitating the interconversion, and in the process undergoes decarboxylation at two steps. The details of individual reactions are given below.

1. Formation of citric acid

Acetyl CoA, the two carbon compound reacts with a 4 carbon carboxylic acid - *oxaloacetic acid* in the presence of water to form a six carbon carboxylic acid, *citric acid.* This reaction is mediated by the enzyme *citrate synthetase.* In the process CoA is released which goes back to form some more Acetyl CoA.

CO_2COOH
|
$CH_2COOH + CH_3COS.CoA + H_2O$ —— *Citrate Synthetase* —— CH_2COOH
|
$C(OH)COOH$ + CoA + HS
|
CH_2COOH

Oxalo acetic acid   Acetyl CoA   Citric Acid

2. Citric acid loses a molecule of water to form cis *aconitic acid* under the influence of the enzyme *Aconitase* with Fe^{++} as cofactor.

CH_2COOH | $CH COOH + NAD$ | CH_3COOH (Citric acid) —— *Isocitric*, Fe^{++} —— CH_2COOH | $CH COOH + H_2O$ | $CH COOH$ (Cis aconitic acid)

3. Cis aconitic acid now gets converted into *Isocitric acid* (isomer of citric acid) by the addition of a molecule of water. The enzyme *Aconitase* catalyzes this reaction also

CH_2COOH
| *Aconitase* CH_2COOH
|

Cis aconitic acid		Isocitric acid
C COOH + H_2O | CH COOH	—————— Fe^{++}	CH COOH | CHOH COOH

The addition of water molecule to cis aconitic does not produce Citric acid, but its isomer isocitric because H_2O combines in a different position in the molecule of a cis aconitic acid.

4. Isocitric acid undergoes oxidation (removal of hydrogen), under the influence of the enzyme Isocitric dehydrogenase to form *oxalosuccinic acid.* The released H atoms are accepted by NAD to become NADH + H +
The NADH + H + gets reoxidised after entering into the electron transport chain (see later).

Isocitric acid		oxalosuccinic acid
CH2COOH | CH COOH + NAD | CHOH COOH	*Isocitric* —————— *dehydrogenase*	CH2COOH | CH COOH + NADH + H + | CO COOH

5. Oxalo succinic undergoes decarboxylation to form a 5 carbon acid, α *ketoglutaric acid.* The reaction is catalysed by the enzyme *decarboxylase* with Mn^{++} as the cofactor.

CH_2COOH | CH COOH | CO COOH	*Decarboxylase* —————— Mn^{++}	CH_2COOH | CH_2CO COOH + CO_2

6. The next step is the oxidation of a α ketoglutaric acid which is achieved by the removal of H. Simultaneously, it also undergoes decarbosylation leading to the formation of Succinyl CoA. The enzymes involved are α *ketoglutaric dehydrogenase* and α *ketodecarboxylase,* together with coenzymes NAD and CoA. NAD gets reduced to NADH + H^+, while a molecule of CO_2 is liberated. Together with CoA, TPP, Mg^{++}, FAD and lipoic acid also participate in the reaction.

α Ketoglutaric acid		Succinyl CoA
CH_2COOH | + HS.Coa + NAD CH_2 CO COOH	*Dehydrogenase* —————— Mg^{++} TPP, Lipoic acid, FAD	CH_2COOH | CH_2COS CoA + NADH + H^+ + CO_2

7. Succinly CoA is hydrolysed with a molecule of H_2O to *succinic acid*, regenerating CoA for continued participation in the cycle. Inorganic phosphate and a phosphorylating enzyme are involved in the formation of succinic acid. The release of CoA accompanies the release of energy, which will be used to synthesize a molecule of Guanosine triphosphate (GTP) by the addition of energy rich phosphate to Guanosine diphosphate (GDP). This is the only place where GTP instead of ATP is synthesized. This type of phosphorylation is called **substrate level phosphorylation.** The molecule of GTP subsequently reacts enzymatically with ADP to produce ATP and GDP.

(i) Succinyl Co.A + ip Sukccinyl-P + Co.A
(ii) Sukccinyl 1-P + GDP Succinic acid - GTP
(iii) GTP + ADP GDP + ATP

The enzyme succinate thiokinase catalyzes the reaction of conversion of succinyl CoA to succinic acid. From this reaction a molecule of ATP is synthesized.

$$\begin{array}{l} CH_2COOH \\ |+ GDP + Pi \\ CH_2CO - S.CoA \end{array} \xrightarrow[\textit{Dehydrogenase}]{\textit{Succinate}} \begin{array}{l} CH_2COOH \\ | \\ CH_2COOH \end{array} + GTP + HS.CoA$$

Succinyl CoA Succinic acid.

8. Succinic acid undergoes dehydrogenation in the presence of a flavoprotein enzyme – succinic acid dehydrogenase, to form fumaric acid. The pair of hydrogen atoms released are accepted by FAD (this is the only reaction in Krebs cycle where hydrogen is not accepted by NAD) and it gets reduced to $FADH_2$.s

$$\begin{array}{l} CH_2COOH \\ | \\ CH_2COOH \end{array} + FAD \xrightarrow[\textit{Dehydrogenase}]{\textit{Succinic acid}} \begin{array}{l} CH\,COOH \\ | \\ CH\,COOH \end{array} + FADH_2$$

Succinic acid Fumaric acid

9. One molecule of water is added to Fumaric acid enzymatically (Fumarase) to produce *malic acid.*

$$\begin{array}{l} CH\,COOH \\ | \\ CH\,COOH \end{array} + GDP + Pi \xrightarrow{\textit{Fumarase}} \begin{array}{l} CHOH\,COOH \\ | \\ CH_2COOH \end{array}$$

Fumaric acid Malic acid

10. In the last oxidation reaction of Krebs cycle malic acid undergoes dehydrogenation under the influence of *malate dehydrogenase* to prodeuce *oxalo acetic acid,* which combines with acetyl CoA to produce citric acid and continues the cycle. The hydrogen released from malic acid is accepted by NAD to produce NADH + H^+

CHOH COOH		*Malate*	CH COOH
|	+ GDP + Pi	————————	|
CH_2COOH		*Dehydrogenase*	CH_2COOH
Malic acid			Oxaloacetic acid.

Summary of Krebs Cycle

In the Krebs cycle, there are four 6 carbon acids – *citric, cisaconitic, isocitric* and *oxalosuccinic* acids, one 5 carbon acid α *Ketoglutaric acid* and four 4 carbon acids *succinic, fumaric, malic* and *oxaloacetic* acids. When the 2 carbon acetyl CoA enters the cycle decarboxylation takes place at two stages – once to form the α Ketoglutaric acid and again to form succinic acid. With this, all the carbon atoms of the original glucose molecule have been removed.

The reactions of the Krebs cycle are for one pyruvate; since two molecules of pyruvate (3C) are formed at the end of glycolysis, the reactions of the Krebs cycle are to be doubled to account for complete glucose. The removal of CO_2 from glucose may be summarized as follows

Glycolysis

$CH_6H_{12}O$ ———— $2CH_3CO\,COOH$

Glucose — Pyruvic acid

In the above, there is no decarboxylation as all six carbon atoms are found in the organic molecule of pyruvate.

2CH3CO COOH — 2CH3CHO + 2CO2

(Pyruvic acid) — (Acetaldehyde)

In the above, there is removal of two carbon atoms. Four more carbon atoms are left. They are removed as follows in the Krebs Cycle.

(i) 2 Molecules of Oxalosuccinic acid (6C) ——— 2 molecules of α Ketoglutaric acid + $2CO_2$ (5C)

(ii) 2 Molecules of α Ketoglutaric acid (4C) ——— 2 molecules of succinic acid (4C) + $2CO_2$

Thus in the aerobic respiration which involves Krebs Cycle, there is complete removal of all the organic carbon atoms to inorganic CO_2.

Another important aspect of Krebs Cycle is the dehydrogenation. The energy rich electrons released from the carboxylic acids at the following stages enter into the electron transport system (ETS) to produce molecules of ATP.

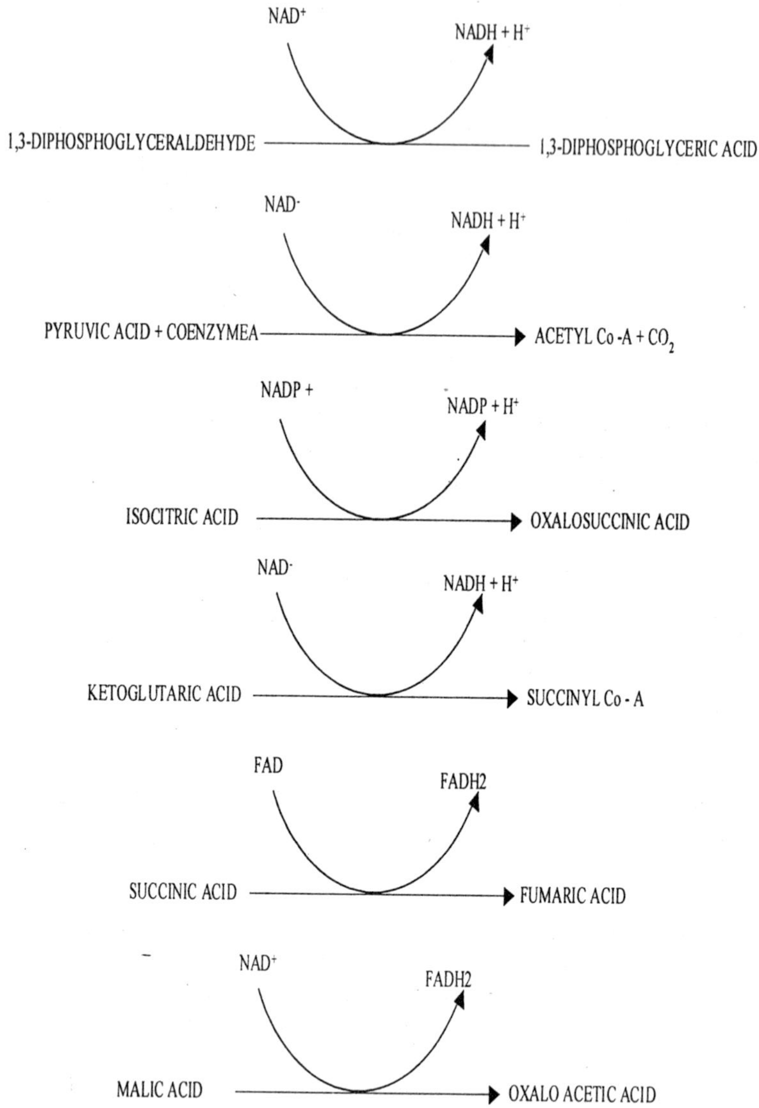

Fig. 17.5 Respiration

Dehydrogenation reactions in aerobic respirations (various stages)

Stages of Dehydrogenation in the Kreb's Cycle

(i) Conversion of Isocitric acid to oxalosuccinic acid
(ii) Conversion of α Ketoglutaric acid to succinyl CoA
(iii) Conversion of succinic acid to Fumaric acid
(iv) Conversion of malic acid to oxaloacetic acid

These are the stages at which the energy is trapped for the synthesis of ATP molecules. Besides the above, during the conversion of succinyl CoA to succinic acid there is substrate level phosphorylation (not through the ETS). All these are explained below in detail.

ELECTRON TRANSPORT SYSTEM (ETS)

Electron transport system or oxidative phosphorylation is the final stage in aerobic respiration. Oxidative phosphorylation takes place in the F particle or oxysomes of the cristae in mitochondria.

In various steps in the respiratory process, the substrates are dehydrogenated (oxidised), and a pair of hydrogen atoms are removed. These H atoms are accepted by a series of acceptors which in turn get reduced and oxidised alternately. As hydrogen is shuttled between acceptors it ultimately reaches the last acceptor and then combines with 1/2 molecule of O_2 (O) to form water. The series of acceptors constitute the ETS.

The bonded hydrogen cannot combine with atmospheric oxygen to form water. In ETS hydrogen is released from its bonded state. The H atoms get disassociated into two protons and two electrons ($2H^+ + 2e^-$). The two protons released in the medium as H^+ ions are available for combination with oxygen to form water. At each dehydrogenation, two atoms of hydrogen are released with the result at every step, after disassociation four protons ($4H^+$), and four electrons ($4e^-$) are released. While the H_+ ions are used for the formation of water, electrons, more specificaly the energy in them, is utilized for the formation of molecules of ATP.

Components of ETS

The paris of hydrogen ($2H^+ + 2e^-$) removed from the substrate at various steps are eventually transported to the oxygen through an assembly of enzymes which constitute the respiratory chain. The enzymes of the respiratory chain mediate in the transport of the hydrogen through their prosthetic groups or coenzymes. The prosthetic group of an enzyme in the respiratory chain undergoes reduction when it accepts hydrogen or electrons from the preceeding enzyme, and passes it on to the next enzyme in the chain getting oxidized (coming back to normal, in the process.

It is believed that most of the enzymes which are involved in dehydrogenating the substrate are loosely bound to the mitochondrial inner membrane, while the others (enzymes) constitute the integral part of the lipoprotein matrix of the inner membrane.

The ETS has the following enzyme complex and two mobile carriers.

Complex I
This consists of a flavoprotein with attached $NADH_2$ dehydrogenase enzyme having FMN (Flavin mononucleotide) as the prosthetic group. 2–4 atoms of nonheme iron with iron-sulphur (Fe-S centre, is found in the enzyme.

Complex II
This also consists of a flavoprotein called *succinic dehydrogenase* with FAD (Flavin adenise (dinucleotide) as its prosthetic group. In addition, there are two Fe-S centres an i-e- S protein. Between Complex I and II, is present a mobile carrier ubiquinone (CoQ). CoQ is reduced by $NADH_2$ as well as succinic hydrogenase.

Complex III
This consists of Cyhtochrome *B* and C_1 with a non heme iron protein with only one Fe-S centre. However, the presence of Cytochrome C_1, is doubtful in plant ETS.

Complex IV
This is also called cytochrome oxidase or terminal oxidase. The complex consists of cytochromes *a* and a_3, having one or two atoms of firmly attached copper. Another mobile carrier is found between Complex III and IV which has a prosthetic group made up of iron prophyrin IX. This is identical with the heme of hemoglobin.

Complex V
This is the enzyme system called ATP ase and is mainly concerned with the synthesis of ATP from ADP and inorganic phosphate.

The plant electron transport system consists of these complexes and their location and sequence may be decided by the evidence provided by redox potential and respiratory inhibitors.

Evidence from Redox potential
In the biological system, certain compounds more readily get oxidized (give out electrons), while others get reduced (accept electrons). This phenomenon of release or acceptance of electrons is called **Redox potential** (E'ϕ). It is usually

defined as the ratio between the reduced and oxidized forms of a compound. The standard electrode potential (Eo) may be defined as the potential at 50% oxidation for an oxidation reduction system at pH7. The standard electrode potential measurement (measured in milivolts) of the various components of the ETS gives the following sequence (in the order of decreasing potential), with exception of CoQ. The sequence is NAD, FMN, FAD, *Cytb, Cytc, Cyte, Cyta,* and *Cyta3*. The standard redox potential of NADH2 is 0.32 volts while that of water is + 0.82 volts. It is for this reason that electrons from hydrogen to oxygen will yield energy.

Evidence from respiratory inhibitors

The correct sequence of the components of ETS can also be ascertained by the use of specific inhibitors, which block the function of one or the other components. For instance, when dilute cyanides are used which block the functioning of *Cyta* and *a*3, all other components get reduced i.e., function normally indicating *Cyta* and *a*3 are at the end. If, on the other hand antimycin A is used, it inhibits the functioning of NAD, all other components remain in the oxidized state (will not get reduced), thus indicating that NAD must be the first electron acceptor.

Stages of dehydrogenation

Total oxidation of the substrate material (glucose) into CO_2 and water involves several dehydrogenations at various steps (in glycolysis and Krebs cycle). The released hydrogen is picked up by the NAD or FAD and passed on to the next electron acceptor.

1. 2H is removed in glycolysis when a molecule of glyceraldehyde 3. Phosphate is converted into 1.3 Diphosphoglyceric acid. NAD picks up the hydrogen and gets reduced to $NADH_2$.
2. During the decarboxylation and dehydrogenation of pyruvic acid into acetyl CoA. 2H is removed. Hydrogen is accepted by NAD to become NADH2.
3. 2H are removed from isocitric acid to form oxalosuccinic acid. NAD accepts hydrogen and is reduced to $NADH_2$.
4. During the oxidation of α Ketoglutaric acid to succinic and 2H are removed. This is again accepted by NAD to become $NADH_2$.
5. During the formation of fumaric acid from succinic acid, dehydrogenation occurs which reduces FAD to $FADH_2$ (Note : This is the only instance where FAD, and not NAD is the first electron acceptor from the substrate).
6. During the dehydrogenation of malic acid to oxaloacetic acid, 2H atoms are removed which are picked up by NAD to get reduced to $NADH_2$.

Thus in the above 6 steps (one in glycolysis, one in pyruvate and four in Krebs cycle), a total of 12 hydrogen atoms are removed. Since this number is for triose only, it has to be doubled for a hexose and it gives a number of 24 hydrogen atoms. These hydrogen atoms travel down their electrical gradient along the various electron acceptors, which alternately get reduced and oxidized as the electron in shuttled through. At specific stages, the energy in the electron in made use of for the synthesis of ATP (oxidative phosphorylation).

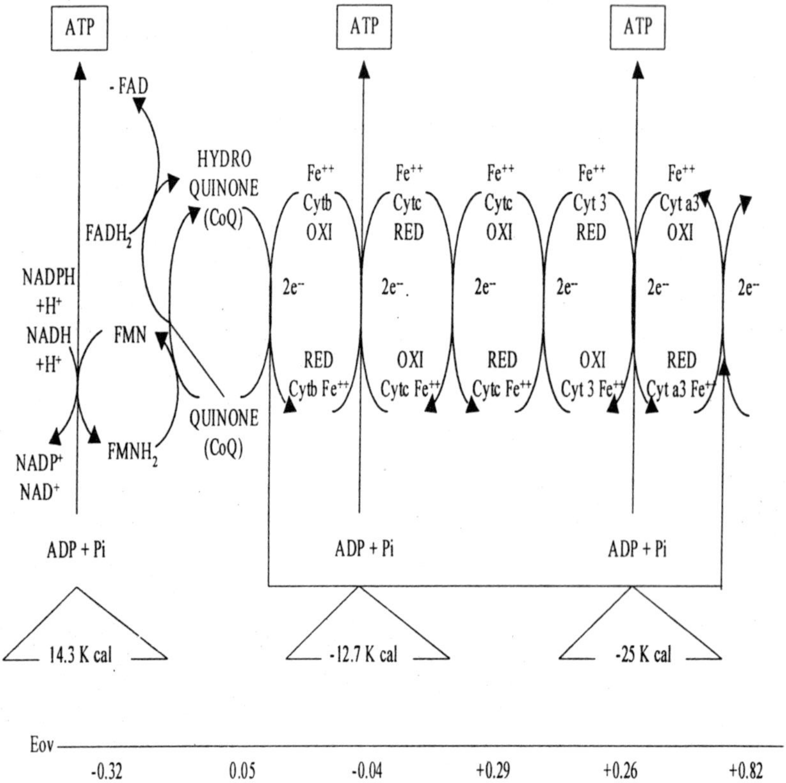

Fig. 17.6 Respiration

Schematic representation of electron transport system in oxidative phosphorylation

The Scheme of Electron Movement

The ETS begins with the acceptance of 2 atoms of hydrogen from the substrate by NAD which gets reduced to $NADH_2$ $NADH_2$ is oxidized when the 2H are transfered to the prosthetic group os lipoproteins of the FMN which in turn gets reduced to $FMNH_2$.

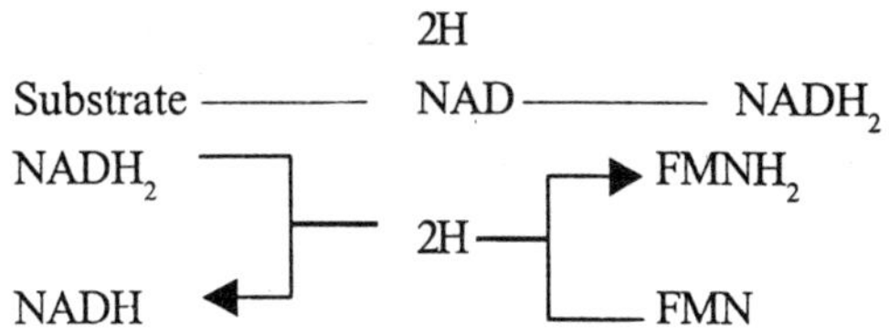

The 2H atoms are now transferred from $FMNH_2$ to the mobile carrier CoQ (Coenzyme Q or uniquinone). CoQ is chemically related to vitamin K and E and similar to plastoquinone of chloroplasts. This CoQ links the flavoproteins of the FAD to Cytochrome.

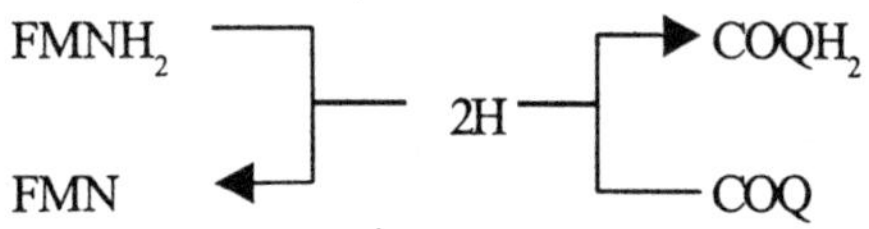

For hydrogen removed in all instances, NAD is the initial acceptor except in the case of succinic acid (Krebs Cycle). Here, the hydrogen is picked up by FAD (Flavin adenine dinucleotide), which is reduced as follows.

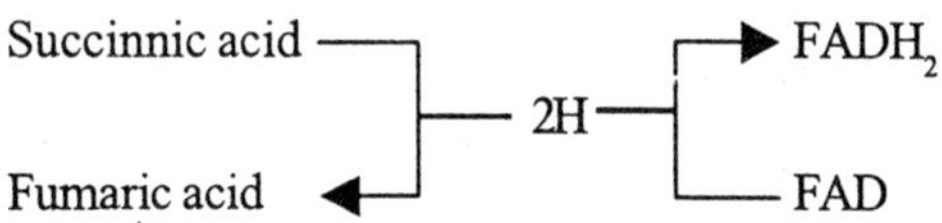

The reduced FAD ($FADH_2$), then passes on its hydrogen to CoQ or *Cytb*. The actual point of hydrogen entry (either to CoQ or *Cytb*) is not yet difinitely established.

Some of the current researches have indicated that at two places (pyruvate and α Ketoglutrate), hydrogen released from the substrate does not go direct to NAD but through lipoate and FAD and then to NAD. This is, however, subject to further proof.

According to Kligenberg and Kroger (1967), CoQ not only picks up hydrogen from succinate *via* FAD but also from α glycerophosphate and Acetyl CoA. The FAD of acetyl CoA, an additional flavoprotein ETF (electron transporting flavoprotein) is present.

The reduced CoQ is oxidized by the released of 2 hydrogen atoms. Soon after release, hydrogen undergoes disassociation into two, protons (2 H^+) are released into cytoplasm, while only the electrons are further transported through the chain (i.e., through *Cytb, Cyte, Cyta* and *Cyta* 3.) The two electrons finally combine with oxygen which in turn combines with protons (H+) to form water as follows.

$$O_2 + 4e^- \text{———} 2(O_2)$$
$$2(O_2) + 4H^+ \text{———} 2H_2O$$

(In the above equation, $4e^-$ and $4H^+$ are shown instead of $2H^+$ and $2e^-$. We have to remember that the reaction sequences of glycolysis (from 3 PGA onwards), and Krebs Cycle are for only half of glucose (one triose), hence they have to be doubled while accounting for a complete glucose molecule).

The passage of electrons are two at a time from NAD upto CoQ. But when electrons are transported from CoQ to *Cytb,* each molecule of *Cytb* can accept only one electron. But if we account for the passage of only one electron at a time, it will not be sufficient to reduce molecular oxygen. How the two electron chain is changed to a single electron chain (at the same time delivering two electrons towards the end), is as yet completely not known. The following are some of the views.

(a) Some unknown intermediate compound transfers the electrons one by one into cytochromes.
(b) The electron chain is forked at the point of CoQ so that two electrons can pass through the two branches (one in each branch). Each branch has the same enzyme system as the other. Finally, the two electrons released from the branches combine with molecular oxygen.

(c) The electron chain is formed at the point of CoQ so that the two electrons can pass through the two branches (one in each branch). Each branch has the same enzyme system as the other. Finally, the two electrons released from the branches combine with molecular oxygen.

The third view seems to be possible as it satisfactorily accounts for the smooth flow of both the electrons in separate but parallel channels.

The electrons give out by CoQ are first accepted by Cytochrome *b* which gets reduced. *Cytb* is reoxidised by transferring its electron to *Cyt c*. From these, the electron passes to *Cyt c, Cyt a,* and finally to *Cyt a*3. *Cyt a*3 gets oxidized by transferring electron to the atmospheric oxygen. How is it released? Each enzyme consists of a heme prosthetic group called *protoheme.* This iron alternately gets reduced and oxidized by accepting and donating an electron at a time as follows.

Cyt b CFe^{2+} Cyt*c* (Fe^{2+})

e

Cyt b (Fe^{3+}) Cytc$_1$ (Fe^{3+})

Some physiologists argue that *Cyt* cl is not a component of plant ETS. Hence, they believe that *Cyt b* transfers the electrons to *Cytc.*

ATP synthesis (oxidative phosphorylation)

All biological reactions require energy and this depends on the formation and utilization of, energy rich' phosphate bonds. A high energy bond does not actually mean that the bond is strong, on the contrary the bond is labile and when it is broken in a reaction, a large amount of energy is released. The energy requiring reactions always obtain energy by using certain compounds which readily release energy. These are called energy currencies. Similarly, during the storing of energy that is released in a reaction, these intermediate molecules are synthesized. Several molecules which act as energy intermediaries are known in biological systems. Of these, the most important are ATP (Adenosine Triphosphate), and ADP (Adenosine Diphosphate). Both ATP and ADP are derived from AMP (Adenosine Monophosphate). AMP consists of a purine (Adenine), a pentose sug (ribose), and phosphoric acid. ADP and ATP are similar to AMP except in the number of phosphate groups. While ADP has two phosphate groups, ATP has three. The second and third phosphate bonds are energy rich in ATP. On hydrolysis of the phosphate bond, energy equivalent to 7600 cal per mol is released.

To sum up, whereever energy is released, ATP molecules are synthesized and whenever energy is required ATP molecules are utilized. This may be illustrated by the following equations.

Synthesis

$$\text{ADP} + \text{Pi} \xrightarrow{\text{energy}} \text{ATP}$$

(inorganic phosphate)

Utilization

$$\text{ATP} + \text{X} \longrightarrow \text{X-(P)} + \text{ADP}$$

(X = Substrate that requires energy)

ATP molecules are synthesized in living systems by two methods.

(i) Photophosphorylation and

(ii) Oxidative phosphorylation.

In photophosphorylation (seen in green plants), radiant energy is used to add Pi to ADP to produce ATP. In oxidative phosphorylation, ATP is synthesized using the energy of oxidation.

Oxidative phosphorylation is the most significant functions of ETS. The number of ATPs produced may be explained on the basis of P/O index (phosphorylation/ Oxygen). P/O index is the ratio of ATP produced for each atom of oxygen used.

Observations of Lehninger point out that the P/O ratio in the case of NAD linked ETS is three *i.e.,* at three sites (between NAD and molecular oxygen), question as to why there are only three energy producing sites, the answer may be provided as follows.

When the electron passes through the chain from one acceptor to the other, at every step some energy is released. The energy so released wil be equivalent to the difference in energy level between the electron donor and acceptor. At certain steps, the amount of energy released is sufficient to bind Pi to ADP, and hence a molecule of ATP is synthesized. As per some calculations, it has been noted that 0.28 volts of energy is sufficient to synthesize a molecule of ATP from ADP and Pi, and in the ETS, whereever the energy difference between the donor and acceptor is 0.28 volts or more, a molecule of ATP is synthesized. As has already been pointed out, in substrate systems linked to NAD there are three sites, where a molecule of ATP is synthesized.

These are

(i) NADH	FAD
(ii) *cyt b*	*Cyt c* or *c,* and
(iii) *Cyt a*	$Cyta_3$

In the oxidation of succinnic acid which is linked to FAD directly, obviously there are only two sites for the synthesis of ATP, hence only two molecules of ATP are synthesized.

The scheme of ETS and the sites of ATP molecule synthesis proposed by Bidwell (1979), are slightly different, though he also agrees that in NAD linked substrate systems there are three sites of ATP synthesis and in FAD linked systems there are only two. But their places the CoQ (Ubiqinone) between *Cyt b, Cyt c* and not earlier to the cytochrome systems (between FAD and Cytochrome b). This poses a problem for the movement of electrons as ubiquione requires 2 H^+ atoms for reduction and not just electrons. But before the 2H reaches the Cytochrome b, it would have undergone disassociation with two protons ($2H^+$) getting released into cytoplasm. Hence, Bidwelo (1979), believes that once again (after ionization before they reach Cyt b) $2H^+$ and $2e^-$ unite to form H, and this reduces the ubiquinone. From ubiquinone when 2 H has to reach *Cyt c,* there is ionization, $2H^+$ is released into water while $2e^-$ continues in the ETS. Hence in this chain, there is ionization of hydrogen twice and not once as believed by other physiologists.

Mechanism of ATP formation
Same as in Phosophosphorylation.

Balance sheet of energy in Respiration
(Account of ATP molecules synthesized). The following is a total of ATP molecules synthesized during complete breakdown of glucose to CO_2 and H_2O

Glycolysis

Reaction Sequence	ATP Used	ATP Synthesized
1. Glucose to Glucose 6 Phosphate	1	-
2. Fructose 6 Phosphate to Fructose - 1,6 - Diphosphat	1	-
3. 1.3 Diphosphoglyceric acid to 3 Phosphogyceric acid (1 x 2)	-	2
4. Phosphoenol pyruvic acid to pyruvic acid (1 x 2)	-	2
5. Dehydrogenation of Glyceraldehyde - 3 Phosphate to 1.3 Diphosphoglyceric acid (through ETS with participation of NAD) 3 x 2	-	6
Total	**2**	**10**

Net gain of ATP during Glycolysis = 8 molecules.
(Note : Reactions 1 and 2 are for entire glucose molecule, while, 3,4 and 5 are only for a triose, hence they are doubled)

Pyruvic acid Oxidation

Reaction Sequence	ATP Used	ATP Synthesized
Conversion of pyruvic acid to Acetyl CoA; Dehydrogenation through ETS with initial acceptor being NAD (3 x 2)		6

Krebs Cycle
(There is no ATP loss in Krebs Cycle)

	ATP Synthesized
1. Isocitric acid to oxalosuccinic acid dehydrogenation through ETS with NAD participation (3 x 2)	6
2. Conversion of Ketoglutaric acid to succinyl CoA, ETS with NAD as the initial acceptor (3 x 2)	6
3. Conversion of succinyl CoA to succinic acid; substrate level phosphorylation No ETS (1 x 2)	2
4. Conversion of succinic acid to Fumaric acid ETS beginning with FAD and not Nad. Hence only two sites of ATP synthesis (2 x 2)	4
5. Conversion of malic acid to Fumaric acid. ETSbeginning with NAD (3 x 2)	6
Total	**24**

Total ATP used during the entire aerobic respiration = 2

Total ATP produced during glycolysis, pyruvic acid oxidation and Krebs Cycle. 10 + 6 + 24=40

Net gain of ATP molecule for aerobic oxidation of one molecule of glucose 40-2=38

Difference between photophosphorylation and oxidative phosphorylation

No.	Character	Photo-Phosphorylation	Oxidative Phosphorylation
1.	Occurrence	Photosynthesis	Respiration
2.	Cell organelle involved	Chloroplast	Mitochondrion
3.	Occurrence within the organelle	Thylakoid	Inner membrane Memgranes of cristae
4.	Need for molecular oxygen	Not needed	Needed
5.	Source of energy	Light	Reduction-oxidation reactions
6.	Output of ATP molecules	Less	More
7.	Fate of ATP molecules	Used up for the Reduction of CO_2	Released into cytoplasm and available for all metabolic reactions.

Efficiency of Respiration

A molecule of glucose consists of a total amount of 6,86,000 (▲ F) calories of energy in the form of chemical bond. During the complete oxidation of a glucose molecule 6,74,000 (673.6 Kcal) calories of energy is released. Out of this, only a part is trapped in the energy rich phosphate bonds of ATP molecules, while the rest is released as heat. Each ATP molecule has about 7,600 calories of energy. The total amount of energy trapped during the breakdown of glucose depends on the number of ATP molecules formed.

In case of oxidative phosphorylation, the total number of ATP molecules produced will be 38. This means a total of 38 x 7600 x 100/6,86,000, or about 40% of the total energy will be negotiable, while the rest (60%) is wasted. In the alternate pathways of glucose breakdown (pentose phosphate shunt), a similar amount of energy will be made available in the form of molecules of ATP. This gives an efficiency of 40% for aerobic breakdown of glucose.

In anaerobic respiration, only two molecules of ATP are formed per glucose molecule. The efficiency will be

$$C_6H_{12}O_6 \text{———} 2C_3CH_4O_3 + 4H$$

(F = 52 Kcal)

= (2 x 7.6/52) x 100 = **29.2%**

The energy efficiency for alcoholic fermentation would be (here also only two ATP molecules are formed)

$$C_6H_{12}O_6 \text{———} 2C_2C_5OH + 2CO_2$$

(F = 52 Kcal)

= (2 x 7.6/52) x 100 = **29.2%**

OXIDATION OF INORGANIC COMPOUNDS

In chemolithotrophs, energy is obtained by oxidation of inorganic compounds. Chemolithotrophs are generally identified by their ability to grow in mineral media and to obtain their cellular carbon from carbon dioxide.

Chemolithotrophs may be classified into obligate chemolithotrophs and facultative chemolithotrophs. Obligate chemolithotrophs generally cannot utilise organic compounds as their carbon source. It was earlier believed that for these organisms organic compounds are toxic, but some of the recent studies have shown that organic compounds are not completely toxic eventhough they partially inhibit the growth of microorganisms. In facultative chemolithotrophs, the main source of cellular carbon is CO_2; they can use even organic compounds as the source of their CO_2 requirement.

Oxidation of inorganic compounds is the chief source of energy both for obligate and facultative chemolithotrophs. Inorganic substances like ammonia, nitrate, hydrogen, sulphur compounds, iron compounds etc. get oxidised and release to glucose. All chemolithotrophs can also be called autotrophs. In the biosphere, they are the chief non-photosynthetic primary producers. The following are some of the examples of oxidation of inorganic compounds.

1. Oxidation of nigtrogenous compounds

A. Ammonia Oxidation

This is seen in organisms such as *Nitrosomonas* and *Nitrosocystis*. Here, the electron donor is ammonium ion and the electron acceptor is oxygen. The oxidation of ammonium ion to nitrate provides the necessary energy for CO_2 fixation. During the process, water is a byproduct. The reaction takes place as follows :

$$NH_4 + 3/20_2 \text{ —— } No_2 - H_2O + 2H^-$$

Two steps are involved in the oxidationo of ammonia to nitrate. In the first step, ammonia is converted into hydroxylamine involving the expenditure of energy. Between hydroxylamine and nitrate, there is atleast one intermediate compound. The electrons liberated by hydroxylamine pass through a series of electron acceptors such as falvin, cytochromes (b,ca) and produce ATP. An ATP dependent reverse electron flow helps in the reduction of NAD to NADH2. The reduction of NADH is apparently at the cost of ATP. The nature of the intermediate compound between hydroxylamine and nitrite is still uncertain.

The conversion of ammonia to nitrite is called nitrification. This process also involves two steps. In the first step, ammonia is converted into nitrate by bacteria such as *Nitrosomonas* and *Nitrosocystis.* In the second step, the nitrite is oxidised to nitrate by nitrite oxidising bacteria like *Nitrobacter.* Nitrification is of great importance in the maintenance of nitrogen balance in nature.

The conversion of ammonia to nitrate and the release of ATP takes place as shown in the following equation:

NH_4 + ——— NH_2OH ——— Intermediate NO_2^-

|

e

Flavin ——— Cyt b ——— Cyt c ——— Cyt a ——— O_2

|

ATP

B. Nitrate Oxidation

As has been mentioned above the bacterium involved in nitrite oxidation is *Nitrobacter.* This bacterium utilises nitrite as the source of electrons and oxygen as the acceptor of electron. Oxidation of nitrite to nitrate is achieved at one step as follows :

$NO_2^- + 1/2\,O_2 \quad NO_3^-$

It has been found that Nitrobacter can obtain energy for its growth from the oxidation of formate. Enzyme such as nitrite oxidase and formate dehydrogenase are found in the same respiratory particle.

Oxidation of sulfur compounds

Many of the chemolithotrophs oxidise/reduce sulphur compounds to obtain their energy. These chemolithotrophs include sulphur bacteria both green and purple. The inorganic compounds that get oxidised are sulphide, thiosulphate, thiocyanate. and elemental sulphur. *Thiobacillus* can oxidase sulphur and sulphur compounds such as sulphides and sulphites.

Oxidation of sulphur (S^0)

Many of the species of *Thiobacillus,* such as T.*Thiooxidans,* can convert both sulphur and sulphite to sulphite to sulphate. *T.Thiooxidans. T.Concretivorous* and *Thioparus* can aerobically oxidise elemental sulphur to yield sulphate. In the oxidation of sulphur to sulphate, sulphate is the intermediate compound. Experiments conducted with cell-free extracts of *T.Thiooxidans* shows that the elemental sulphur first reacts with oxygen to produce sulphite. This reaction is cartalysed by the enzyme oxygenase and is dependent on glutathione. The reaction is as follows:

$$SO_8 + 12O_2 + 8H_2O \text{ ——————— } 8H_2SO_3$$

The sulphite that is produced as a result of the above reaction noenzymatically reacts with elemental sulphur to yield thiosulphate as per the following reaction:

$$H_2SO_3 + SO \text{ ——————— } H_2S_2O_3$$

Oxidation of Sulphide

Some Thiobacilli are known to use sulphide. During the oxidation of sulphide to sulphate 8 electrons are released. Green and purple sulphur bacteria can anaerobically oxidise reduced sulphur compounds like H_2S. Both sulphide and elemental sulphur are metabolised due to the interaction of the enzyme bound thiol or dithiol. Sulphite and elemental sulphur react with a sulphydryl group to yield an intermediate RSSH.

This intermediate is oxidised to sulphite by sulphide oxidise. There is no generation of ATP here as the reaction is not linked to cytochrome electron transport. At the next step sulphite is oxidised to sulphate by the enzyme sulphite oxidase. As this reaction is cytochrome linked it can generate ATP through ADP.

Oxidation of sulphite to sulphate

Many of the thiobacilli are known to bring about oxidation of sulphite to sulphate. This reaction takes place through an AMP-dependent APS - (adenosine phosphorulphonate) reductase pathway.

APS is a high energy compound and it is formed due to the interaction of sulphite and AMP (Adenosine Monophosphate). This reaction is catalysed by the enzyme APS reductase.

$$2SO_3 + 2\,AMP \xrightarrow{\text{APS reductase}} 2\,APS + 4e^-$$

At the same time inorganic phosphate is esterified to ADP, and the reaction is catalysed by the enzyme ADP sulphurlyase.

$$2APS + 2Pi \xrightarrow{\text{ADP sulphurlyase}} 2ADP + SO_4$$

ATP is now generated from ADP by the action of the enzyme adenylate kinase.

$$2\,ADP \longrightarrow AMP + ATP$$

The sum of the above reactions can be studied as follows :

$$2SO_3 + AMP + 2Pi \longrightarrow 2SO_4 + ATP + 4e^-$$

Iron oxidation

Some chemolithotrophs such as *Thiobacillus, Ferroxidans* can oxidase ferrous to ferric iron as follows :

$$4\,Fe_2^+ + 4H^+ + O_2 \longrightarrow 4\,Fe_3^+ + 2H_2O$$

As a result of the oxidation, electrons that are released will flow from ferrous compound to cyt c and cyt a, and ultimately react with moleculare oxygen. A molecule of ATP is formed between cyt a and O_2.

Oxidation of hydrogen

Hydrogen is produced during the anaerobic degradation of organic substances in soil. Many chemolithotrophic bacteria have the ability to utilise this hydrogen. In order to oxidis hydrogen, bacteria should possess enzymes such as hydrogenase. It has been seen that hydrogen oxidising bacteria have evolved independently and the enzyme hydrogenase occurs in at least three different forms –

1. Cannot reduce NAD and is inhibited by CO. This is the commenest one.
2. It is called the nitrogenase type and it reacts with ATP and Fd resulting in

the release of molecular hydrogen.

3. It is called the *Hydrogenomonas type,* it reduces NAD and is inhibited by CO.

Oxidation of hydrogen is of the following type:

1. Oxidation reduction
2. Nitrate reduction
3. Sulphate reduction
4. Methanogenesis

1. Oxidation reduction

This is seen in the bacterium *Hydrogenomonas.* Oxidative phosphorylation linked to the reduction of oxygen by molecular hydrogen provides energy necessary for growth.

$$H_2 + 1/2\ O_2 \text{ ——————— } H_2O$$

The energy released is used for the reduction of carbon dioxide to cellular carbon. Oxidation of hydrogen provides energy through which cellular synthetic activities take place. Hydrogenase enzyme can split hydrogen as follows :

$$H_2 \text{ ——————— } 2H + + 2e^-$$

The electrons released have a low electrifcal potential, hence the carriers also should have a low redox potential.

In *Hydrogenomonas,* the electron transfer chain takes place as follows :

H_2 —— Intermediate carrier —— Flavoprotein —— Cytochromes —— O_2

|

NAD

2. Nitrate reduction

This has been seen in bacteria commonly called denitrifying bacteria (Micrococcus denitrificans). These bacteria use nitrate as the terminal acceptor of electrons instead of oxygen. Nitrate is reduced with molecular hydrogen. The bacteria possess type 1 hydrogenase enzyme. As NAD cannot be reduced directly, an ATP dependent reduction system is used. The reduction of nitrate is believed to be a process occurring in several steps as follows:

$$NO_2 \text{ —— } NO_2 \text{ —— } NO \text{ —— } NO_2 \text{ —— } N$$

3. Sulphate reduction

Two types of sulphate reduction have been noticed in bacteria. These are *Assimilatory sulphate reduction,* and *Dissimilatory sulphate reduction.*

Assimilatory sulphate reduction

In this process, sulphate is reduced to sulphide which is used for the synthesis of sulphur containing amino acids like cysteine and methionine. This is called

Assimilatory sulphate reduction because the sulphate is assimilated into the amino acids. The following steps are involved in Assimilatory sulphate reduction.

1. Activation of sulphate by ATP. In this process, inorganic sulphate reacts with ATP and is catalysed by the enzyme ATP sulphurylase resulting in the formation of APS.
2. Phosphorelation of APS. APS is phosphorelated with the help of ATP to produce a compound called PAPS (3 phospho adenosine, 5 phospho sulphonate).
3. Reduction of sulphate. The sulphate in PAPS is reduced to sulphite by the action of the enzyme PAPS reductase. This reduction involves the transfer of electrons from NADH to PAPA.
4. Reduction of sulphite. NADPH reduces sulphite to sulphide under the influence of the enzyme sulphite reductase.

$$SO3 - 3NADPH + 3H^{+} \text{———} H_2S + N3NAD + + 3H_2O$$

The hydrogen sulphide released in the above reaction is immediately utilised for the formation of ammino acid cysteine. This reaction is catalysed by the enzyme sulphydrase.

Dissimilatory sulphate reduction

A few species of gram negative (*Desulfovibrio*), and gram positive (*Desulfotomaculum*) bacteria are capable of dissimilatory sulphate reduction. These bacteria are strictly anaerobic.

Dissimilatory sulphate reduction produces hydrogen sulphide which is found in the cytoplasm of bacteria such as *Desulphovibrio*. These bacteria can also use pyruvate, lactate etc. as substrates. *D.Desulfuricans* can also utilise molecular hydrogen for the reduction of sulphate as follows :

$$SO_4 - + 4H_2 \text{—} S - + 4H_2O$$

The first step in the reduction of the sulphate is the formation of APS. The reaction is similar to what is seen in the Assimilatory sulphate reduction. In the next step, APS is reduced to sulphite and AMP under the influence of the enzyme APS reductase. Further reduction of sulphide is brought about by the enzyme sulphite reductase. In the above process, Cyt c3 is an electron carrier.

4. Methanogenesis

In certain bacteria, oxidation of hydrogen takes place and carbon dioxide acts as a terminal electron acceptor. Reduction of carbon dioxide here results in the production of methane. Bacteria which reduce carbon dioxide to methane are